The Legacy of the Mastodon

The
Legacy
of the
Mastodon

*The Golden Age of Fossils in
America*

Keith Thomson

Yale University Press   New Haven & London

Set in Bulmer Roman by Binghamton Valley Composition.
Printed in the United States of America by Sheridan Books.

Library of Congress Cataloging-in-Publication Data

Thomson, Keith Stewart.
  The legacy of the mastodon : the golden age of fossils in America / Keith Thomson.
    p.   cm.
  Includes bibliographical references and index.
  ISBN 978-0-300-11704-2 (alk. paper)

1. Mastodons—North America.   2. Mammoths—North America.   3. Mammals, Fossil—North America.   4. Paleontology—North America.   5. Paleontology—United States—History.   6. Paleontology—United States—History—18th century.   7. Paleontology—United States—History—19th century.   I. Title.
  QE882.P8.T46   2008
  560.973'09033—dc22   2007037329

A catalogue record for this book is available from the British Library.

The paper in this book meets the guidelines for permanence and durability of the Committee on Production Guidelines for Book Longevity of the Council on Library Resources.

10   9   8   7   6   5   4   3   2   1

Contents

List of Maps and Tables   ix
Preface   xi
Acknowledgments   xv

PART ONE: The Jeffersonians

ONE
Fossil Hunters on the Frontier   3

TWO
Big Bone Lick   10

THREE
Franklin, Jefferson, and the Incognitum   24

FOUR
Jefferson's "Great-Claw" and a World About to Change   34

FIVE
The First American Dinosaurs: An Eighteenth-Century
Mystery Story   41

SIX
Fossils and Show Business: Mr. Peale's Mastodon   46

PART TWO: Fossils and Geology

SEVEN
Fossils and Extinction: Dangerous Ideas   57

EIGHT
Mary Anning's World   72

NINE
An American Natural Science 86

TEN
An American Geology 98

ELEVEN
Bad Lands: No Time for Ideas 105

TWELVE
Dr. Leidy's Dinosaur 122

THIRTEEN
Ferdinand Vandiveer Hayden 126

PART THREE: Giant Saurians and Horned Mammals

FOURTEEN
Kansas and a New Regime 147

FIFTEEN
Entry of the Gladiators 155

SIXTEEN
Riding the Rails 168

SEVENTEEN
The First Yale College Expedition 177

EIGHTEEN
The Competition Begins 191

NINETEEN
Buffalo Land: Who Was Professor Paleozoic? 200

TWENTY
1872: The Year of Conflict 211

TWENTY-ONE
The Case of the Great Horned Mammals 229

TWENTY-TWO
Going Separate Ways 242

TWENTY-THREE
Two into Four Won't Go 254

TWENTY-FOUR
To the Black Hills 263

TWENTY-FIVE
To the Judith River 271

PART FOUR: Toward the Twentieth Century

TWENTY-SIX
The Rise of Dinosaurs 279

TWENTY-SEVEN
The Good, the Bad, and the Ugly 295

TWENTY-EIGHT
Going Public 310

TWENTY-NINE
1890: The End of the Beginning 317

Appendixes
The Geological Column 335
Leidy on Evolution 338
Cope on Evolution 341

Notes 345
Index 371

Maps and Tables

The Oregon Trail, showing the Missouri River
and its tributaries   8

The geological time scale   69

The Late Cretaceous seaway of North America   95

The Bad Lands   118

Early Tertiary fossil basins of Wyoming, Utah, and Colorado   138

Wyoming trails leading to Fort Bridger   142

The Fort Bridger region of Wyoming   214

Preface

This book is intended in large part as a tribute to Alfred Sherwood Romer (1894–1973), my professor at Harvard and the greatest student and teacher of fossil vertebrates of the twentieth century. Romer taught me not only to be a paleontologist but also to love the history of this science. One of the delights of collecting Early Permian fossils with him in the hardscrabble country of north-central Texas was his daily recounting of stories of the early collectors there, such as Jacob Boll, a Swiss immigrant who collected for Louis Agassiz, the first director of the Museum of Comparative Zoology, in the early 1870s.

Perhaps Romer's best story about himself (and about bone hunting) concerns the day in the 1950s when he was unearthing the bones of a fossil reptile far out in the dry north Texas cattle country. Along came a cowboy riding the line, checking for downed sections of barbed wire fence. He was the authentic item: lariat on the saddle, fence tools in a saddlebag, a rifle in a scabbard plus a six-shooter on his hip.

COWBOY: What 'yer doing?
    Romer replied that he was collecting fossils.
COWBOY: What's that fer?
ROMER: Well, these rocks are full of the remains of creatures that lived here (hesitating over the subject of a biblical age for the earth) many, many years ago.
COWBOY: Yup, why do you do that?
ROMER: Well, I am a professor from a college back east and I take them back there and study them.

COWBOY: What fer?

ROMER (getting desperate): So that I can see how these animals lived a long time ago.

COWBOY: Why do you do that?

Then inspiration struck. Romer said: "The government pays me to do it."

At that the cowboy's face lost its frown of suspicious concentration and he nudged his horse into a walk. "Yessir. Y'all take care. Bye now."

Romer was not only the greatest paleontologist of the twentieth century: he was also a disarmingly modest man whose friendliness, generosity, and fondness for a good story made him beloved around the whole world. People are supposed to grow to resemble their pets; Romer had a very large nose that made him, in later years, bear a startling similarity in profile to one of his favorite Permian reptiles.

He spoke with such a strong New York accent that, when I first went to study with him, I could not understand his lectures, punctuated as they were with the Latin names of fossil creatures that I knew only from books. Most of them I had never before heard pronounced but, judging from the principles of Greek and Latin, they probably should not have been pronounced like *that*. The janitors in the Museum of Comparative Zoology at Harvard all called him "Al." The most familiar I ever became was the hopelessly contrived "ASR," but I could never have addressed a letter to him, or referred to him in conversation, as Al.

For as long as I knew Romer, two small photographs hung in his office at the museum. One, labeled *Romer,* shows him in the black suit he typically wore around the university. With typical economy and lack of pretense, he avoided issues of fashion by always wearing a version of the same outfit: black suit, white shirt, black tie, black socks and shoes. That way, he assured me, he was ready for any event. For the special occasion of this photograph he looks completely comfortable with the addition of academic cap and gown. Also typical for Romer, however, is the fact that the setting is not the leafy spaces of Harvard Yard, but the back steps of the museum, where every afternoon at four o'clock a brains trust of faculty (Romer, George Gaylord Simpson, Bryan Patterson, and Ernest Williams) would gather for coffee and a cigarette with the technicians and graduate students.

The second photograph, titled *Roamer,* shows him with a big grin sitting on the running board of an ancient field vehicle. He has on a disreputable khaki shirt, grubby pants, and field boots; he has just taken off a sweaty bandana and laid it on his knee. On his head is a filthy old straw hat. The picture was probably taken in the early 1950s, earlier than the academic one. Not only do these two images show a contrast between two sides of a man, they show a deep paradox in the field of study to which he devoted his life. On one hand, the study of fossil vertebrates is serious, rigorous science, conducted in the laboratories of the finest universities and museums in the world. On the other hand, vertebrate paleontology is adventure, exploration, and discovery, accomplished at the expense of fingernails and clothing, and experienced with a dash (not too much) of danger. Romer was perfectly at ease in the comfort of the Harvard Faculty Club or around a campfire deep in the Argentinean wilderness. However, he could not live with only one side of this duality—the professor or the cowboy, the scientist or the romantic. He had to be both.

This dichotomy was not typical of Romer alone; it is really the story of this whole subject. No matter the level of abstraction of the evolutionary theories they support or generate, the study of fossil vertebrates is dominated by the collecting of the fossils themselves. While an art historian does not have to have acquired a serious reputation as a painter or sculptor, most vertebrate paleontologists still earn their spurs in the field; explorations and discoveries are as much a driving part of their credentials as the theoretical papers, replete with mathematical formulas, they publish in the best journals. And they still head out west every year (or north or south or east) to live the life of a cowboy or a gold prospector in some remote region, searching for tiny mammal teeth, ancient fishes, or every kind of fossil reptile—all the elusive clues to the history of life on earth and of the earth itself.

Today, vertebrate paleontology, like all of science, is truly international in every respect, but collecting the fossils remains the most glamorous part of the whole subject. And that usually means travel to remote places, following the Willie Sutton principle (when asked why he robbed banks, Sutton replied, "Because that's where the money is"). Paleontologists everywhere share the same wanderlusts, and so when you ask a paleontologist why he heads off to the Great Plains and purple-headed mountains every summer, the answer is, "That is where the fossils are." But this is only part of the answer. In the United States, the rest has to do with intangibles: participation in a long-standing tradition and (less overtly admitted) the American sense of nation, of westward opportunity, of limitless possibility, a oneness with the glorious days of nineteenth-century western exploration and the establishment of the United States as one nation from Atlantic to Pacific.

Acknowledgments

I owe debts of gratitude to very many people who assisted, sometimes unwittingly, as I wrote this book. First among them is my friend and former colleague Professor Jim Kennedy at Oxford. Kennedy is someone whose knowledge of paleontology is so deeply detailed, his interest in science so wonderfully broad and authoritative, and his insistence on perfection so complete as to be positively irritating at times. Whenever I had a question (and there were many), he answered it. I look forward to helping him, at least in some small way, as he finishes his book on William Buckland.

I owe a similar debt to the staffs of two Philadelphia institutions, the American Philosophical Society and the Ewell Stewart Library of the Academy of Natural Sciences, the extraordinary archival resources of which were made available to me by a wonderful group of people who patiently answered my every query. Martin Levitt, Roy Goodman, Charles Greifenstein, Victoria-Ann Lutz, Earl Spamer, and Joseph-James Ahern at the Philosophical Society and Robert Mc-Cracken Peck, Eileen Mathias, and Mary-Gen Davies at the academy made this book possible. I am particularly grateful to Eileen Mathias, archivist at the academy, for her help at every stage. Ted Daeschler, curator of vertebrate zoology at the academy and custodian of its historically important fossil collections, was also always available with help and advice.

I am deeply appreciative of the public access to the historic photographs of William Henry Jackson and others, made available by the photographic libraries of the United States Geological Survey, the

National Parks Service, and the Kansas Geological Survey. Barbara Narendra, archivist at the Peabody Museum of Natural History, Yale University, helped find copies of key illustrations, as did Lisa Keys at the Kansas State Historical Society and Sarah Ligochi at the Wyoming State Archives. I am grateful to all their institutions for permission to publish photographs from their collections.

I am grateful to the following for permission to publish quotations from material from their archival collections: Yale University Library, for the Papers of Othniel Charles Marsh correspondence housed there, and the American Museum of Natural History, for its collection of the letters and notebooks of Edward Drinker Cope. Other Cope letters were consulted by permission of the American Philosophical Society and the Academy of Natural Sciences. The academy archives (which contain some copies of letters belonging to the College of Physicians, Philadelphia) are also the source for quotations from the correspondence of Joseph Leidy and F. V. Hayden.

Jefferson Looney and Andrew O'Shaughnessy at the International Center for Jefferson Studies, at Monticello, provided useful information and assistance. Glenn Matlack (Ohio University) provided unique information about his distinguished forebear Timothy Matlack. From his intimate knowledge of Leidy's work, Dennis Murphy pointed me to Leidy's statements on spontaneous generation and evolution. Ruth Lauritzen (Sweetwater County Museum, Green River, Wyoming) and Linda Newman Beyers (Fort Bridger State Historic Site, Wyoming) were also helpful with my enquiries. Thanks are also due to Kevin Walsh at Oxford for allowing me to reproduce one of his photographs of the White River Bad Lands, to Brent Breithaupt (University of Wyoming), who escorted me to key paleontological sites in Wyoming, and to Katherine Woltz (University of Virginia and the International Center for Jefferson Studies at Monticello) for her political and historical insights, especially into the painting *The Exhumation of the Mastodon* by Charles Willson Peale. Don Cresswell of the Philadelphia Print Shop was helpful with access to old maps. I am very grateful to Jim Kennedy, Kevin Padian, Katherine Woltz, Ted Daeschler, Anthony Fiorillo, Jessica Thomson Fiorillo, Elizabeth Thomson, and Linda

Price Thomson, who patiently read all or part of the manuscript. Linda Price Thomson also drew the original maps.

Literary agents Felicity Bryan and George Lucas, and editor Jean Thomson Black (no relation) at Yale University Press, provided invaluable support as ever.

PART ONE

The Jeffersonians

Fossil Hunters on the Frontier

What man in the world, I would ask, ever ascended to the pinnacle of one of
Missouri's green-carpeted bluffs, and giddily gazed over the interminable and
boundless ocean of grass-covered hills and valleys which lie beneath him, where
the gloom of silence is complete—where not even the voice of the sparrow or
cricket is heard—without feeling a sweet melancholy come over him, which
seemed to drown his sense of everything beneath and on a level with him?

GEORGE CATLIN, 1844

From the time of the early Spanish explorers onward, travelers in
America have responded in various ways to the "ocean of grass" that
covers the great prairie lands west of the Mississippi and east of the
Rocky Mountains. The modern traveler looks down from an airplane
and sees a checkerboard of farms and settlements. The early transcontinental migrants in their canvas-topped Conestoga wagons ("prairie
schooners") saw a seemingly endless and possibly dangerous obsta-
cle. When scientists first explored westward, they "saw" beneath the
grassy seas and found a huge geological puzzle and, more figuratively,
an opportunity.

With the Louisiana Purchase from France in 1803 the United States
doubled its territory by adding lands that extended from the Mississippi River to the Rocky Mountains; before midcentury the nation's
borders reached literally "from sea to shining sea."[1] This new western
half of the country was a cornucopia of wildlife migrating across seemingly limitless grazing land, magnificent stands of timber, fabulous silver and gold fields, rich arable lands, abundant water in some places,
severe deserts in others. At the end of the almost endless plains was a

3

mountain barrier, the grandeur of which made the Alps seem puny. Beyond the mountains was promised an Eden against the ocean. It all seemed a place so vast that there, surely, the presence and handiwork of man would always be insignificant, but first the steamboat and then the railroads reduced immense distances to manageable short hops among the new towns and cities. The Indians discovered that the white settlers could not be trusted in the way that the early traders and trappers could, and all too soon this promised land could be seen as a paradise lost. Hundreds and thousands of settlers learned the hard way that the promise "rain follows the plough" was a land agent's cruel hoax.[2]

Whatever it meant to fur trappers, gold miners, Indian traders, fortune hunters, or farmers and settlers, the West was also a scientific treasure house. Among its most exciting secrets were ancient fossils—the remains of hitherto unsuspected kinds of animals like birds with teeth, the diminutive ancestors of horses and camels, strange cattlelike creatures with claws on their feet, and over a hundred different kinds of dinosaurs. Between 1739 and 1890 a small group of scientists discovered and described thousands of previously unknown kinds of fossil animals from the American West, and a great number also from the eastern states. Along the way, they helped decipher and describe the geological structure and history of an entire continent. They took a little-known science, championed in the new nation by Thomas Jefferson and in Europe by Baron Georges Cuvier, and, especially when they discovered dinosaurs, transformed it into part of the American experience.

The doctrine of Manifest Destiny (technically referring to the inclusion of former Spanish possessions into the United States, but used here more figuratively) was Manifest Opportunity for science, and if it was accomplished only through extremely hard work, against heavy odds, there was more than a dash of glamour thrown in, fed by images of the West expressed in popular novels, many of which were written as quite blatant propaganda for the land companies seeking to attract settlers from the East. Many different kinds of people were involved in the scientific opening of the West. This book deals with a group of fewer than a dozen men who monopolized the study of fossil vertebrates, whether strange new mammals from the Dakota Bad Lands, flying reptiles and gigantic fishes from the Kansas chalk, dinosaurs from the Jurassic and Cretaceous cliffs of Wyoming and Montana, or ten-thousand-year-old elephants from Kentucky.

These findings depended on exploration and discovery on a greater scale than anything attempted in Europe, carried on during times of adversity and adventure when, for these men, prospecting for fossils meant carrying a pick in one hand and a rifle in the other, tackling hostile country and equally hostile Indians. And also keeping a careful eye on one another, because the rivalries among these explorers were intense. Theirs is a story of high adventure, and sometimes a far-from-noble ambition, all in the cause of serious science.

The question most often asked of any fossil collector is: how do you know where to go? What the paleontologist emphatically does not do is wander off into some strange wilderness without any prior clue as to where and why he is going. In fact, the specialist student of fossils is almost always dependent on someone else to have made the first discovery. In the vast reaches of this new land the first signs of the rich fossil beds lying out beyond the Missouri River came in the form of isolated specimens picked up by frontiersmen, fur traders, government surveyors, army personnel, and mining men. The much-honored Lewis and Clark expedition of 1804–6 collected very few fossils (or other geological specimens). They were not often in the right places and scarcely had the capacity to drag hundreds of pounds of rocks around with them for two years. They did, however, bring back a few small mineral specimens and at least one piece of a fossil fish, found in a bank of the Missouri River.

The first consistent discoveries of fossils in the West were a byproduct of the fur trade and were collected by people linked to the series of trading forts that sprang up along the frontier, sustained by the activities of the fabled trappers known as mountain men. When the first steamboat ascended the Missouri River as far as the Yellowstone River in 1831, a new era began, and by the 1840s enough people had penetrated into the West for hints of amazing fossil beds to find their way back east. By 1853 Dr. Joseph Leidy of Philadelphia had enough material at hand to write the first treatise on western fossil reptiles and mammals. And that, in turn, stimulated further efforts.

Although significant discoveries were being made, there still remained the material problem of getting out to the sites and bringing the specimens back. Collecting was greatly stimulated in 1853, when Congress authorized the surveying of routes for east–west transcontinental railroads. The surveyors consequently set out along the thirty-second, thirty-fifth, and forty-seventh parallels, as well as another charting a course between the thirty-eighth and forty-first. Congress authorized further surveys in 1856 for exploration of the Upper Missouri and Yellowstone Rivers and the route for a wagon road from Fort Riley to Bridger's Pass. Eventually five transcontinental railroads were built: the Northern Pacific, roughly following the forty-fifth parallel, the Union Pacific along the forty-second, the Missouri Pacific, Denver, and Rio Grande (Western Pacific) following the thirty-seventh parallel, the Atchison, Topeka, and Santa Fe on the thirty-fifth, and the Southern Pacific along the thirty-second parallel.

With a significant number of trained surveyors and geologists being employed on the surveys, the flow of specimens back east increased. Then, when the West was effectively opened to easy travel from the settled states via the Kansas Pacific and Union Pacific railroad lines, specialist fossil collectors seized the opportunity. Now, after only a week's travel, they could meet up with their army escorts and outfit their expeditions with wagons and horses at places with famous names like Fort Laramie, Fort Pierre, Fort Wallace, and Fort Bridger. But a new difficulty had to be faced. The Indian tribes, both indigenous and those forced out from the eastern states, began to contest with one another and to resist the invasions of farmers and gold miners who, with the connivance or encouragement of the U.S. government, were dispossessing them of their lands.

By the end of the Civil War, given access by the railroads and protected by the army, America's first professional paleontologists—intensely ambitious men whose behavior was at best idiosyncratic and at worst simply reprehensible—were intensely active in the western fossil fields. One example of the keenness and even bravery (or was it foolhardiness?) of these men who ventured into the wilder lands of the West was that in 1876, just a few weeks after the defeat of Custer's troops by Sitting Bull at the Battle of the Little Bighorn, Edward Drinker Cope from Philadelphia was out collecting on the Judith River in Montana, sure that "since every able-bodied Sioux would be with Sitting Bull . . . there would be no danger to us."

But this is not simply a story about fossils and the men who collected them. The story of the discovery and exploitation of the western fossil fields is in every sense tied directly to, and contingent upon, the greater history of the opening and population of the American West. Its context is therefore nothing less than the emergence of the new nation and a new cultural tradition in the era after the world of Jefferson and Franklin gave way to a newly populous and restless America. It involves the early history of exploration of the West, the role of the state and national geological surveys and economic aspects of geology, the opening of the West to waves of emigration and development, and the role of the new railroads. It ends with the announcement of the closing of the American frontier in 1890, with bone hunting fully launched into its modern mode of vertebrate paleontology.

The study of fossil vertebrates became the narrative of a uniquely American science through the intertwining of three threads: a straightforward practical empiricism appropriate to a new and rapidly expanding country; the adventure of exploration and discovery and attitudes toward the land and nature growing out of the Romantic Movement; and changing ideas about science itself, both nationally and internationally, as a new professional *natural science* based in geology emerged out of an older *natural history* tradition associated with medicine. In parallel with all this—as both a cause and a result—grew a changing relationship between knowledge of fossils and popular religious beliefs, accelerated by Darwinism.

On one hand, as their field journals, diaries, and letters show, the story played out as a dialogue between the people and the land they explored. On the other it is a story of the relationships between individual personalities and the science they developed. A third theme is more

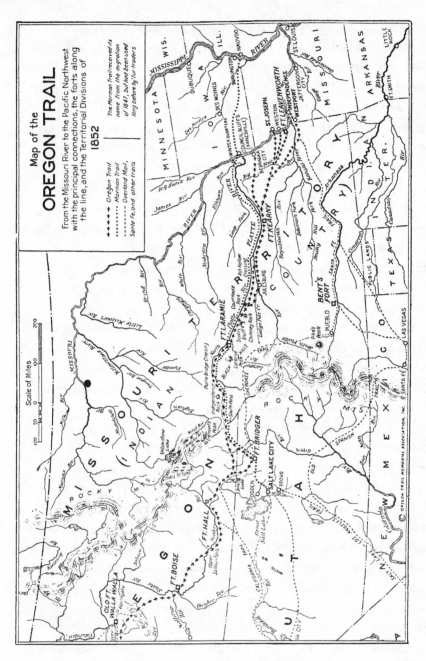

The Oregon Trail (and others), featuring the principal tributaries of the Missouri River, forts, and migration routes before 1860 (from Robert Ellison, *Fort Bridger, Wyoming: A Brief History*, 1931)

literary, concerning the very language that scientists used to describe their finds and the geological environments of the West, together with the ways in which books, newspapers, and museums promoted and encouraged the new science of paleontology, all leading eventually to the cult of the dinosaur.

Big Bone Lick

Great numbers of very large pronged teeth of some vast animals are [found] which have no resemblance to the molares, or grinding teeth, of any great animal yet known.

WILLIAM HUNTER, 1769

The American mastodon was a relative of modern elephants, with enormous curved tusks, and has been described as a fossil that helped shape America's sense of nationhood. That would be a unique role for any animal, living or fossil, in any culture, and the idea may be a little overstated. It would certainly have seemed so to Mary Draper Ingles.

In 1755, twenty-three-year-old Mary Ingles, with her husband William and sons Thomas, age five, and George, two, were homesteading on a stretch of western Virginia land called Draper's Meadows, high in the Appalachian Mountains near present-day Blacksburg. The Draper and Ingles families had settled there seven years earlier. It was a place where buffalo and deer were plentiful, the soil in the river bottom was productive, and the forests provided an unending supply of fuel. The settlement stood on a sharp horseshoe bend of the New River, just before it cut north through the folded mountain ridges to fall, via a series of gorges, into the Kanawha and thence into the great Ohio River at what is now Point Pleasant. Thus Mary Ingles's story was played out beyond the eastern continental divide—on the northwest side of the Appalachians, in the watershed of the Ohio and the Mississippi. It was, in 1755, in every sense the contemporary American frontier. Young Thomas is said to be the first white child born west of the mountains.

This was a time when Buffalo was the biggest western city, and Pittsburgh was still in French hands and called Fort Duquesne. The Appalachian frontier was not a place for the faint-hearted, nor one where political philosophy was nearly as important as sheer survival. But in fact it was near a major geopolitical epicenter—the earthquake being the French-Indian wars, a series of conflicts between England and France that lasted from 1689 to 1763. The last phase of the conflict began in 1754, at a time when the French controlled the St. Lawrence River valley, parts of the Great Lakes, and the whole eastern Mississippi Valley together with their southern lands around the mouth of the Mississippi. The British possessions in America were all on the Atlantic seaboard, east of the Appalachians. A huge disputed area lay between the two; French trappers and traders freely crisscrossed modern Ohio, and—as with the Draper and Ingles families—there was growing pressure from the east in the form of British-American settlers. The local Indians played both sides to their advantage, mostly tending to favor the French.

The Treaties of Utrecht (1713) and Aix-la-Chapelle (1748) had produced temporary stalemates, but along the Allegheny River in 1753 the French built a series of forts, which, in the following years, forces under Lieutenant General George Washington tried to capture. In 1754 Washington was defeated at Fort Necessity, and in 1755, the year of Mary Ingles's tribulations, he and General George Braddock were defeated at Fort Duquesne (a particularly painful loss, as Washington had tried to establish a fort there in 1753 but had been driven off then, too). Eventually, between 1758 and 1760, British-American troops took Louisburg and Forts Duquesne, Frontenac, Niagara, and Ticonderoga; General James Wolfe took Quebec; and Lord Jeffrey Amherst took Montreal. In 1763, through the Treaty of Paris, Britain gained all of the French territory east of the Mississippi, including Ohio and parts of what is now Canada. The result was a massive addition to the territory of the future United States. On the mainland, France retained New Orleans and the vast country west of the Mississippi.

For frontier settlers like the Ingles family and their neighbors, the immediate daily battle had little to do with events in London or Paris.

Instead they were working out a very personal destiny—as settlers in contested lands and among still uncivilized peoples. But they were nonetheless a small cog in a mighty machine of westward expansion, part of the wave of settlers that had started to cross the Appalachians. In one sense it was people like the Ingles and Draper families, who wanted to settle on the land rather than simply trap and trade, who in a sense forced the war. And in a very real respect it was they, as much as Washington and Braddock's soldiers, who were on the front line. No one worked in the fields without a gun ready to hand, because bands of roaming French and Indians, and even some renegade English "adventurers," were a constant threat.

On Sunday morning, July 8, 1755, a raiding party of Shawnee Indians descended on the tiny settlement at Draper's Meadows, killing four, including Mary's mother and her brother's infant son. William Ingles, who was out in the fields harvesting wheat, rushed to their aid but was attacked by two Indians who chased him into the woods, where he escaped. When it was safe to return to the cabin, he discovered that in addition to the killings, the Indians had ransacked the place for food and useful tools and had taken five of the settlers alive. His wife Mary (in some accounts said to be eight or nine months pregnant), her sister Bettie Draper, with her arm broken by a musket shot, their small sons Thomas and George, and a man named Henry Lenard had all been dragged off into the deep forest.[1] The raid happened the day before General Braddock's defeat.

For a month the party of captives traveled through the virgin forest, until they reached a Shawnee village on the banks of the Ohio where there was a salt spring, as well as some other English prisoners. The women were put to work boiling down salt, a valuable trade commodity. Then further tragedy struck. Both of the Ingles children were taken away to other, more remote villages, with the older boy, Thomas, ending up near Detroit. Bettie Draper was dragged off by the chief of yet another camp. Finally, two white traders, Frenchmen, arrived and took Mary Ingles and an older captive, referred to as "the Old Dutch Woman" (probably German) even farther into the wilderness. They trekked another 150 miles west along the Ohio River until, near modern-day Cincinnati and a few miles south of the river on the present Kentucky side, they reached a large encampment of Indians and renegade whites. Here the slavery of the women, naked and brutalized, continued. And here fossil vertebrates enter the story.

Once again the women were put to work boiling down salt. This swampy area was another place with sulfurous saline springs; it had for millennia been a place where animals came for salt. Buffalo were still common, and in the past many had been trapped in the swamps; their bones still lay all around. Mixed with the buffalo remains were other bones also—huge bones, far larger than those from a buffalo. Most of these giant bones were buried deep in the ground, but the milling buffalo had exposed a good number.

The spot was called the Licking Place, or Big Bone Lick, and those huge bones were fossils, ten thousand years or more old (as we now know), the remains of a huge species of animal unlike anything alive today. There were not only giant limb bones, but also tusks that would have been instantly recognizable to anyone who had ever seen an elephant, and baffling to someone who had not. And scattered in with the bones and tusks were enormous complex teeth with roots half a foot or more long and lumpy protuberances on their surfaces.

The Shawnee were very familiar with these giant bones and had various myths about them, calling their owner *le père des Boeufs* (father of cattle). In their legend, "when the Great Manitou descended to earth to see if the creatures he had created were happy, the bison replied that he would be happy in the prairies where the grass grew as high as his belly, except that he had constantly to be watchful lest the *père des Boeufs* should come down from the mountains in a fury and devour him and his like."[2]

Thomas Jefferson, in his *Notes on the State of Virginia,* reported the Delaware Indian story that "in ancient times a herd of these tremendous animals came to the Big-bone licks, and began a universal destruction of the bear, deer, elks, buffaloes, and other animals which had been created to the use of the Indians. . . . [T]he Great Man above, looking down and seeing this, was so enraged that he seized his lightning, descended on the earth, seated himself on a neighboring mountain, on a rock of which his seat and the prints of his feet are still to be seen, and hurled his bolts among them till the whole were slaughtered, except the big bull male, who . . . wounded in the side . . . bounded over the Ohio, over the Wabash, the Illinois, and finally over the great lakes, where he is living at this day." In another Shawnee legend, these animals had once lived with a race of equally huge human giants, but God destroyed those too.

How much attention Mary Ingles, in all her tribulations, would have paid to the paleontological oddities beneath her feet, we cannot tell. Simple survival was her first priority, and for weeks on end she endured the filth of the Indian camp and the abuse of her captors while constantly plotting to escape with the old Dutch woman. By the time the leaves began to turn, it was obvious that they had to act; their chances of surviving a winter under those conditions would be minimal. Mary Ingles was a strong, determined woman, and eventually the Indians came to trust her to gather nuts, berries, and wild grapes in the surrounding forest. As soon as the Indians relaxed their guard enough, the two women slipped out of camp and flitted through the forest as fast and silently as they could. Their plan was to backtrack Mary's earlier route, following the Ohio River east to the Kanawha, and then to head south through the forest, following the dangerous gorges of the New River. Through all her appalling forced march westward, Mary Ingles had kept her wits about her and noted landmarks along the way. At every moment they had to be afraid both of the Indians who might be pursuing from behind and those into whose clutches they might stumble.

It was early October and they had no weapons or tools except for a hatchet; they were naked save for a couple of blankets. The food they could gather was meager, but at least there was plenty of water. Mary Ingles soldiered onward, knowing that, as long as they survived, the New River would eventually lead them home. It soon became clear that the Dutch woman had gone mad. At one point they found a stray horse in a deserted cornfield, but then they lost it. The Dutch woman fell behind. It was November and cold, but amazingly, after some forty days and 450 miles, an emaciated Mary Ingles staggered out of the snowy woods into the fields of her neighbor Adam Harmon. She had found her way back to Draper's Meadows. Days later the Dutch woman wandered in on the horse.

Six months after Mary Ingles's epic escape, and in a distinctly more civilized place, the French Royal Academy of Sciences in Paris published a description of a huge tooth that had come from the place where she had been enslaved—the Big Bone Lick on the Ohio. The author was a French scholar and mineralogist named Jean-Etienne Guettard. He had first read his paper to the academy as early as 1752, and the specimen itself had been collected as far back as 1739, by a French officer, Lieutenant Charles Le Moyne, second baron de Longueuil.

Longueuil had been based at Montreal and was dispatched south with a party of men to meet up in present-day Tennessee with a second French force moving north from New Orleans. Their mission was to subdue the Chickasaw Indians of the Lower Mississippi Valley. As his party descended the Ohio River, Longueuil came across the Big Bone Lick site, which seems then to have been uninhabited (unless he simply went to a different part of the Licks from where Mary Ingles was later enslaved). He collected a tusk, a thigh bone (femur), and several of the huge teeth. When he reached New Orleans, Longueuil arranged to travel back to Paris and took with him the fossils that he had carefully preserved since leaving the Ohio. He presented them to the French royal collection.

Jean-Etienne Guettard concluded that the remains belonged to a kind of elephant, but he did not know what it could be. What he did know was that the *père des Boeufs* from America was not the only fossil elephant known to science. For more than a century, travelers' tales from Siberia had reported the existence of the remains of giant elephants preserved in the permafrost. These reports were confirmed when fossil ivory tusks were brought back and skeletons were collected. This animal was the Siberian mammoth. And indeed, some mammoths were preserved in the ice with their red woolly fur and meat intact.[3]

Those Siberian elephant-relatives were already a great puzzle. Modern elephants live in warm climates, but the Siberian mammoths were found in a very cold region where it seemed impossible that they could have lived and thrived. Over the next century, however, other kinds of normally warm-climate mammals, such as the rhinoceros, also turned up in deposits from the Arctic. What seemed more certain, however, was that no Siberian mammoths still lived. Now there were additional questions to add to the problems they presented. Not only had a second kind of fossil elephant been found, also living in a cold-temperate rather than a tropical climate, there was the issue of whether the European mammoth and the new American beast were one and the same species. And was the animal represented by the bones at Big Bone Lick still living somewhere in the North American wilderness?

In 1764 the French physician and naturalist Louis-Jean-Marie Daubenton restudied the bones that Longueuil had collected from Big Bone Lick, comparing them with those of a modern-day elephant and a fossil mammoth from Siberia. He used side-by-side comparisons of their leg bones and concluded that the three were so similar to each

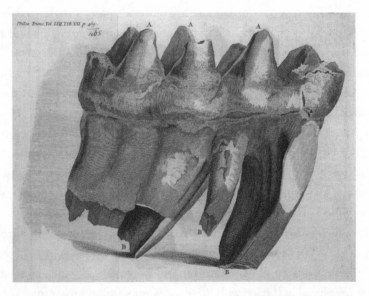

Mastodon molar tooth (from Peter Collinson, "An Account of Some Very Large Fossil Teeth, Found in North America," *Philosophical Transactions of the Royal Society,* vol. 57 [1768])

other that they must come from the same species. But since the teeth from Big Bone Lick were very different from those of either a mammoth or a living elephant, he concluded that they indicated the presence of a second kind of large mammal. His conclusion was that the big grinders (molar teeth) from the Ohio probably came from a hippopotamus.

For the next hundred years or so, the teeth of what came to be called the "American incognitum" (American unknown) continued to be a puzzle. It seemed unlikely that they came from a hippopotamus; no remains of a hippo had ever been found in the Americas. The incognitum should have been some sort of elephant, but the teeth of the living elephants and the mammoth both have many parallel cross ridges for milling plant food, whereas the Big Bone Lick teeth had lumpy surfaces as if for crushing rather than grinding. To some authorities, therefore, they almost looked as though the animal was a carnivore. Herbivore or carnivore, one species or two, the bones from Big Bone Lick turned out also to be much more than just a scientific puzzle; its resolution turned out to play a role in the development of the myths of American nationhood.

As numbers of fossil remains from Big Bone Lick continued to be picked up by travelers and would-be settlers along the Ohio River, many found their way to European collections, especially in Britain. Scholars were fascinated by them. The New World was a place of tremendous interest anyway, but these huge fossil teeth seemed to be evidence of yet a further world, an ancient world that preceded our own and was populated by different creatures, unless of course the giant creatures to which those teeth belonged were actually still alive. Either way, extinct or alive, it was a scientific sensation.

As the fossils became better recognized, it turned out that travelers had collected such specimens at the Big Bone Licks quite frequently since the 1740s. Evidently specimens were to be found over quite a wide geographical area, rather than a single small site. One early collector was a man called Christopher Gist, who is most famous for having been a scout and aide to Major George Washington on a journey through the western wilderness in 1753 to treaty (fruitlessly) with the French command near

Lake Erie. Before the hostilities of the French and Indian Wars began, Gist had been employed as an agent for the Ohio Company (a consortium of English and Virginian interests), charged with surveying the land as far west as the "falls of the Ohio," and advising on the state of the Indian tribes. On his first visit to the Ohio Valley, he reported: "This Ohio country is fine, rich, level land, well-timbered (and full of) meadows abounding with turkeys, deer, elk, and most sorts of game, particularly buffaloes. In short, it wants nothing but cultivation to make it a most delightful country." Such country was ripe for what we today call development: the Ohio Company wanted to buy the land from the British and sell it in parcels to settlers like the Draper and Ingles families. That would first mean dealing with the Indians; if they would not agree to treaties in which they gave up their lands, they would have to be forced to move. The French wanted the same sorts of treaties. All sides knew that the Indians would resist.

In 1751 Gist made his journey west by going directly overland, crossing the mountains to the Miami River, where he succeeded in making a temporary alliance with the Miami Indians, to the exclusion of French emissaries on the same mission. He returned to the East Coast via the Ohio River. Traveling upriver on horseback, he came to the Big Bone Lick site, or at least came near enough to it to have been given two teeth by a local trader named Robert Smith. Gist gave one of these teeth to the Ohio Company. In all probability that specimen found its way to London.

Perhaps the most significant early collections of fossils from Big Bone Lick were made by yet another person involved with the press to open up lands west of the Alleghenies to British-American settlers. George Croghan was an enterprising (and not always overly scrupulous) Irish trader, explorer, and land speculator who had traveled all over the western lands (Ohio and Illinois) since 1746. Because of his familiarity with the country and ability to negotiate with the Indian people, Croghan was often very useful to the British and Pennsylvanian governments as they tried to make treaties with the Indians. Benjamin Franklin, for example, dealt with Croghan in 1756 while serving on a commission attempting to get a fort built at Carlisle in southwestern

Pennsylvania.[4] In 1756 Croghan was formally employed by Sir William Johnson, the king's superintendent of Indian affairs, to find ways of opening up the Illinois country.

With the conclusion of the French and Indian War in 1763, the long-sought settlement of the western lands of Ohio and Illinois seemed possible, as long as some kind of accommodation could be achieved with the Indians. As Croghan had married a high-ranking Mohawk woman ("head of the Turtle Clan"), he was ideally suited to treat with the Iroquois, or Six Nations (a league consisting of the Mohawks, Oneidas, Cayugas, Senecas, Onondagas, and Tuscaroras). His travels to lay the groundwork for treaties with Indian nations took him as far west as Detroit, and he wrote back to Franklin and others, time after time, about the importance of making a fair settlement with the Indians lest a terrible new set of hostilities break out. In a somber foreshadowing of discussions that would take place a hundred years later, and concerning territory a thousand miles to the west, Croghan was heavily involved in negotiations for the establishment of a boundary zone between the British settlers and the Indian lands, "fixed as far back as the Ohio," that would create "a sufficient extent of Land for Colonization, and put an End to dangerous Disputes, respecting our Frontier People's hunting, on their Ground."[5] Meanwhile the number of hostile incidents grew and the Shawnee, Delaware, and Six Nations were steadily pressured to flee westward.

Croghan was at the Big Bone Lick site as early as 1762, traveling with a Philadelphia trader named Joseph Greenwood, a man who had explored the Ohio country extensively for Thomas Penn (colonial proprietor of Pennsylvania and son of the founder William Penn) ten years before and had created a map from his travels. He sent news about the Big Bone Lick to the London wool merchant Peter Collinson, a keen student and collector of natural history who had long since appreciated the range of natural wonders that could be found in the Americas. Collinson's many business interests in the New World kept him fully informed on political matters in the colonies.

Collinson was also the principal patron of the Philadelphia botanist John Bartram. As a leading light in early American science, Bartram was a friend of many of the great men like Benjamin Franklin, who, legend has it, conducted his kite experiments in the meadow next to Bartram's house.[6] Bartram, one of the greatest botanists and explorers of the age, traveled quite fearlessly on horseback across the Alleghenies in periods when the Indian tribes were relatively quiescent. He was responsible for getting two to three hundred native American plants into cultivation, and introducing fifty or more to Europe, through Collinson.[7]

Croghan and Greenwood's explorations fired Collinson's interest in the creature from Ohio. He wrote to Bartram on June 11, 1762, asking, among other things, for "some more particular observations of the Great Buffalo. Their bones or skeletons are now standing in a licking-place, not far from the Ohio, of which I have two of their teeth. One GREENWOOD, an Indian trader, and my friend GEORGE CROGHAN, both saw them, and gave me relation of them; but they omitted to take notice of what hoofs they had, and what horn. These two material articles known, would help to determine their genus or species. Prithee, inquire after them, for they are wonderful beyond description, if what is related of them may be depended upon. I heartily wish to be informed of them, and the place they were found in."[8]

On July 25 the impatient Collinson wrote again: "I forget if I ever mentioned two monstrous teeth I had sent me by the Governor of Virginia. . . . One other has Dr Fothergill, and T. Penn another. One Greenwood, well known to B. Franklin, an Indian trader Knocked some of the teeth out of their jaws; and George Croghan has been at the licking-place, near the Ohio, where the skeletons of six monstrous animals were standing, . . . the Indian tradition is, that the monstrous Buffaloes (so called by the Indians) were all struck dead with lightning at this licking-place. But is it likely to think all the race were here collected, and extinguished at one stroke?"[9]

Bartram had never ventured as far as the Big Bone Lick himself, but during 1761 and 1762, as part of a long collecting foray westward, he was based in Fort Pitt (as Fort Duquesne had become). Colonel Boquet at Fort Pitt had told Bartram of receiving from traders an elephant's tooth, weighing six and three-quarter pounds, and a large piece of tusk. Bartram was also shown specimens that had been brought in by Indians. When Collinson sent his inquiry, Bartram wrote to his friend James Wright, a Quaker naturalist in Susquehanna, a small town in western Pennsylvania, inquiring what he knew about the place where the bones came from.

Wright reported that he had talked with "two Sencible Shawanese Indians who described the place where such bones could be found." They had said that "there appear to be the remains of 5 Entire Sceletons, with their heads All pointing towards Each other." The animals had been huge, the size of a house. The Shawnee also recounted what seems to be one of the earliest versions of the Indian legends about the great beast to which the bones belonged. No one had ever seen the creatures alive, but in their tradition there had been a race of men in older times who were equally giant in stature and who had hunted them. "They had seen Marks in rocks, which tradition said, were made by these Great and Strong Men, when they sate down. . . . [W]hen there were no more of these strong Men alive, God Killed these Mighty Creatures."[10]

Bartram wrote rather acidly to Collinson: "As for these monstrous skeletons on the Ohio, I have wrote thee largely, just before I set out for Caroline, and since my return. But by thy letter thee seems to think the skeletons standing in the posture the beasts stood in when alive, which is impossible. The ligaments would rot, and the bones fall out of joint, and tumble confusedly on the ground. But it's a great pity, and shame to the learned curiosos, that have great estates, that they don't send some person that will take pains to measure every bone exactly, before they are broken and carried away, which they will be soon, by ignorant, careless people, for gain."[11]

In 1765 George Croghan made another exploring and trading expedition from Fort Pitt, down the Ohio and thence west. On May 31 he came to the Big Bone Lick, but he did not mention any Indian camp in his diary. "We passed through a fine timbered clear wood . . . into a game road which buffaloes have beaten, spacious enough for two wagons to go abreast, and leading straight into the Lick. It appears that there are vast quantities of these bones lying five or six feet under ground, which we discovered in the bank, at the edge of the Lick. We found here two tusks above six feet long; we carried one, with some other bones, to our boats."[12]

Fully aware of the importance of the bones, Croghan collected them carefully and headed farther down the Ohio. However, he fared little better than Mary Ingles. A week later a "group of Kickapoos and Musquattimes" attacked his party, killing five of them and taking the rest captive, including Croghan himself. "I got the stroke of a hatchet on the head . . . but my skull being pretty thick, the hatchet would not enter." Eventually, Croghan was released by the rival "Pondiacs," but he had lost everything, including the fossils. Not one to be deterred easily, a year later he was back at Big Bone Lick, this time with an army escort. Also traveling with Croghan on this trip were Ensign Thomas Hutchins (who later became the maker of important early maps of the American midwest) and a trader from Philadelphia named George Morgan. Morgan, at least, traveled in some style, taking with him "Gun, Pistols, Sword, Spy Glass, Speaking Trumpet, Pipes, Tea Chest, Compass, Pen & Ink & Chest of Drawers."[13]

Like Longueuil before him, Croghan's route this time eventually took him all the way down to New Orleans, from where he shipped some of his fossils to London. On February 7, 1767, he sent Lord Shelburne "two of the largest tusks, or teeth, one whole and entire, above six feet long, the thickness of common elephants teeth of that length [and] Several very large forked or pronged teeth; a jaw-bone, with two of them in it." Shelburne was minister for the colonies in the British government; his influence became crucial in the issuing of land grants for the settlement of the West.

Croghan sent a bigger parcel of specimens to Benjamin Franklin, who was then in London as representative of the Pennsylvania Assembly. As later reported by Collinson to the Royal Society, Franklin received "four great tusks, of different sizes; One Broken in halves, near six feet long; one much decayed, the center looks like chalk or lime. A part was cut off from one of these teeth, which has all the appearance of fine white ivory. [Also] a joint of the vertebrae; Three of the large pronged teeth; one has four rows of fangs."[14]

Around this time another set of bones from Big Bone Lick appeared in the collection of the Tower of London. It is unclear who collected these or how they came to be in London, unless they were obtained from Croghan and Greenwood in 1762. In his list of specimens sent to Shelburne and Franklin, Collinson noted the existence of yet more fossils: "Captain Owry, an Officer who served in the country during the last war, now living at Hammersmith, hath a small tusk, as if of a calf elephant, the surface of a fine shining chestnut colour, and a recent look; and a great pronged tooth, larger than any of the above, which were brought from the same licking place." But still no one could be sure what the Ohio animal was, or whether there were two different kinds.

THREE

Franklin, Jefferson, and the Incognitum

In a letter to Croghan, thanking him for the mastodon specimens, Franklin observed:

They are extremely curious on many accounts; no living ele-phants having been seen in any part of America by any of the Europeans settled there, nor remembered in any tradition of the Indians. It is also puzzling to conceive what should have brought so many of them to die on the same spot; and that no such remains should be found in any other part of the conti-nent, except in that very distant country Peru, from whence some grinders of the same kind formerly brought, are now in the museum of the Royal Society. The tusks agree with those of the African and Asiatic elephant, in being nearly of the same form and texture; and some of them, notwithstanding the length of time they must have lain, being still good ivory. But the grinders differ, being full of knobs, like the grinders of a car-nivorous animal; when those of the elephant, who eats only veg-etables, are almost smooth. But then we know of no other animal with tusks like an elephant to whom such grinders might belong. It is remarkable, that elephants now inhabit naturally only countries where there is no winter, and yet these remains are found in a winter country; and it is no uncommon thing to find elephant's tusks in Siberia, in great quantities, when their rivers overflow, and wash away the earth, though Siberia is still more a wintry country than that on the Ohio; which looks as if the earth had anciently been in another position, and the climates differently placed from what they are at present.[1]

One can scarcely imagine any American more likely to have interesting insights about the remains of the giant animal from the Ohio than the polymath Franklin. The breadth of Franklin's scientific interests was already legendary. With his accomplishments as one of the leading inventors, experimenters, observers, and theoreticians of the Enlightenment, not only was Franklin bound to be keenly interested in the incognitum, he would inevitably have something interesting and new to contribute. From the very first, Franklin made the eminently sensible assumption that all the bones had come from a single species, rather than two. In his letter to Croghan, although the bones had been in his possession for only a few weeks, he had seized upon all the questions that would dominate scientific discussion for the next fifty years. Was this animal now extinct? What was a kind of elephant doing in America? What did the occurrence of an elephant tell us about ancient climates and the history of the earth? What kind of elephant was it? How was it related to the European mammoth and was it an herbivore or a carnivore?

Franklin was also mistaken in two respects. As noted already, several American Indian tribes had a traditional knowledge of these creatures, although of course they did not know to identify or describe them as "elephants." As for his statement that "no such remains" had been found elsewhere on the continent, there were at least two prior examples. In 1743 the pioneering naturalist and artist Mark Catesby had noted that "at a place in Carolina called Stono [presumably the present Stono River] was dug out of the earth three or four Teeth of a large animal, which, by the concurring opinion of all the Negroes, native Africans, . . . [were] the Grinders of an elephant."[2]

This animal was probably not a mastodon but a mammoth (or the Africans would not have so readily identified the teeth). And, as far back as 1705, bones similar to those subsequently found at Big Bone Lick had been discovered in New York State, on the banks of the Hudson River. News about them spread and, on learning about them from Governor Joseph Dudley of Massachusetts, Rev. Cotton Mather (always referred to as the eminent divine—"divine" being used as a noun) wrote describing them to Dr. John Woodward and Richard Waller at the Royal Society in London. They in turn reported to the Royal Society, in what may be the first published account of a fossil vertebrate from North America, Mather's account of

> a Tooth brought from the Place where it was found to New York, 1705, being a very large Grinder, weighing 4 pounds and three quarters, with a Bone, suppos'd to be a thigh-bone, 17 foot long. He also mentions another Tooth, broad and flat like a fore-tooth, four Fingers broad. . . . He then gives a description of one, which he resembles to the Eye Tooth of a Man; he says it has four Prongs, or Roots, flat, and something worn on the top it was six inches high lacking one eighth, as stood upright on its Root, and almost thirteen inches in circumference; it weigh'd two pounds four ounces Troy weight. There was another near a pound heavier found under the Bank of Hudson's River, about fifty leagues from the Sea, a great way below the surface of the earth, where the Ground is of a different Colour and Substance from the other Ground, for seventy five Foot long, which they suppose to be from the rotting of the Body, to which these Bones and teeth did, as he supposes, once belong.[3]

Woodward also noted: "It were to be wished the writer has given an exact figure [drawing] of the Teeth and Bones." But the original discoverers of the remains had made a very modern observation in noting the shadow of decayed soft tissues from the carcass around the bones.

Evidently, if Franklin was aware of Cotton Mather's report, he failed to make the connection to the Big Bone Lick remains. Franklin was in any case more concerned with asking informed questions about the incognitum than making descriptions of its anatomy.

Franklin did, however, mention a key anatomical detail—the knobbiness of the teeth—that finally led to the naming of the animal more than thirty years later. Describing the American incognitum in 1799, the French zoologist Baron Cuvier coined the term "mastodonte." This name referred to the resemblance of the knobs on the teeth to small breasts, as in the dugs of a pregnant bitch, and from this came the name by which we now know the creature: mastodon. (The Latin root of mast-o-don, or breast-tooth, will therefore be familiar from such terms as mastectomy and mastitis.) The formal name of the mastodon is *Mammut americanum,* this name having been bestowed by the German scientist Johann Friedrich Blumenbach, also in 1799, when it was still thought to be a kind of European mammoth.

With the arrival of Croghan's specimens in London the focus of discussion concerning the American mastodon had definitely shifted from Paris to that city. Very shortly after they arrived, Bartram's patron Peter Collinson delivered two short papers (published the following year) about the bones collected by Croghan, at meetings of the Royal Society. He compared the tusks with those of elephants and noted the obvious similarities but, like Daubenton and contrary to Franklin's view, thought the teeth told a different story. He reverted to the view that two different animals were represented. "As the biting or grinding teeth, found with the others, have no affinity with the molars of the elephant, I must conclude that they, with the [tusks], belong to another species of elephant, not yet known; or else that they are the remains of some animal that hath the [tusks] of the elephant, with large grinders peculiar to that species, being different in size and shape from any animal yet known."[4]

Like Franklin, Collinson puzzled over the existence of an elephant (the owner of the tusks) and the other elephant-like creature (the owner of the teeth) in North America. "As no living elephants have ever been seen or heard of in all America, since the Europeans have known that country, nor any creatures like them; and there being no probability of their having been brought from Africa, or Asia; and as it is impossible that elephants could inhabit the country where these bones and teeth are now found, by virtue of the severity of the winters, it seems incomprehensible how they came there." The only possibility, which Collinson found unconvincing, was that they had been carried north by Noah's Flood: "By the violent action of the winds and waves, at the time of the deluge, these great floating bodies, the carcasses of drownd elephants, were driven to the Northward, and, at the subsiding of the waters, deposited where they are now found. But what system, or hypothesis, can, with any degree of probability, account for these remains of elephants being found in America, where those creatures are not known ever to have existed, is submitted to this learned Society."

Collinson was not persuaded, however, that the mastodon was a carnivore: "the heavy unwieldy animals, such as elephants, and the rhinoceros, &c. are not carnivorous, being unable, from want of ability and swiftness, to pursue their prey, so are wholly confined to vegetable food; and for the same reason, this great creature, to which the teeth belong, wherever it exists, is probably supported by browsing on trees and shrubs and other vegetable foods."

Within months, the great English anatomist William Hunter presented to the Royal Society a different account based on Croghan's material from Big Bone Lick. His methods were sound. He had had expert ivory dealers examine the tusks, and they assured him that they were made of true ivory. He also compared the teeth and lower jaws of the elephant and the mastodon in the same way that Daubenton had compared limb bones. Whereas Daubenton saw all the similarities in the limb skeletons, by focusing on the teeth Hunter saw all the differences. He concluded that what he called the "American incognitum" was "an animal different from the Elephant," and that the tusks and teeth had come from a single species. Confusingly, however, he followed that by expressing the opinion that it was "probably the same as the Mammouth of Siberia." By this he must have been assuming that the evident differences in the teeth were simply the result of variation within a species.[5] Hunter, however, came down on the side of the theory that whatever the creature had been, it was "some carnivorous animal, larger than an ordinary elephant." He concluded by remarking that "if this animal was indeed carnivorous, which I believe cannot be doubted, as men we cannot but thank Heaven that its whole generation is extinct."

Hunter did not believe the theory that the remains of the mammoth and the mastodon had been transported to northern regions by the Flood. Rather, he adopted the view of a small but growing number of geologists about the ever-changing history of the earth. The existence of the mastodon, for Hunter, "seemed to concur with many other phenomena, in proving, that in former times some astonishing change must have happened to this terraqueous globe; that the highest mountains, in most countries now known, must have lain for many ages in the bottom of the sea; and that this earth must have been so changed with respect to climates, that countries, which are now intensely cold, must have been formerly inhabited by animals which are now confined to the warm climates."

The studies of Collinson and Hunter evidently made Benjamin Franklin rethink the problem of the mastodon, and he was now inclined to change his mind about the animal being a carnivore. In 1768 he sent one of the teeth from Big Bone Lick to his friend the Abbe Jean-Baptiste Chappe d'Auteroche in Paris and asked his opinion. Chappe was an astronomer and an expert on Siberia and therefore had first-hand knowledge of the European mammoth. Franklin wrote: "Some of Our naturalists here . . . contend that these are not the Grinders of Elephants but of some carnivorous Animal unknown, because such Knobs or Prominences on the Face of the Tooth are not to be found on those of Elephants, and only, as they say, on those of carnivorous Animals. But it appears to me that Animals capable of carrying such large heavy Tusks, must themselves be large Creatures, too bulky to have the Activity necessary for pursuing and taking Prey, and therefore I am inclin'd to think those Knobs only a small Variety. Animals of the same kind and Name often differing more materially, and that those Knobs might be useful to grind the small Branches of Trees, as to chaw Flesh. However, I should be glad to have your opinion, and to know from you whether any of the kind have been found in Siberia."[6]

Franklin's argument had two strands: he explained the shape of the "Knobs" as functioning for grinding plant material—as Collinson had suggested. And he followed Hunter in the notion that the differences between the teeth of the mastodon and the mammoth were due to individual variation—then still very much an unknown subject. That is to say, they were merely localized variants of the same animal, adapted for different diets. Chappe's reply to Franklin has been not been preserved. But the "carnivore" idea would not die.

If Franklin was one American intellectual heavyweight of the eighteenth century, the other was obviously Thomas Jefferson. Like Franklin, he was both a practical man (surveyor, architect, inventor, farmer, and statesman) and passionately interested in philosophy. Throughout his life he had a keen interest in minerals and fossils. He eventually kept an array of specimens from Big Bone Lick in the entrance hall of Monticello.

In 1781, Jefferson set out to amass examples of the richness—and (where possible) the sheer brute size—of American wildlife for his *Notes on the State of Virginia,* a book intended (as we shall see more fully in the following chapter) to counter French claims that animal life in America was inferior to that of Europe.[7]

If anything would show that American fauna included beasts larger and more ferocious than anything in Europe, the mastodon would do it. Obviously it would have helped Jefferson's cause if the mastodon had been a carnivore. He did not directly claim this, but instead offered the slightly circumspect statement that the "tradition of the Indians" was "that he was carnivorous."

In his discussion of the mastodon Jefferson allowed himself some wonderfully sarcastic comments about scholars like Daubenton and Collinson who thought that the grinders belonged to a different species from the tusks, and he dismissed the notion that the teeth from Big Bone Lick belonged to a hippopotamus. "It will not be said that the hippopotamus and elephant came always to the same spot, the former to deposit his grinders, the latter his tusks and skeleton. For what became of the parts not deposited there? We must agree then, that these remains belong to each other." And since the hippopotamus does not have tusks, then obviously the creature was not a hippopotamus.

Jefferson's second conclusion was: "That this is not an elephant, I think ascertained by proofs equally decisive." Rather, the mastodon was genuinely something very new and different in science. Always looking for examples of the superiority of the American fauna compared with that of Europe, Jefferson wrote: "The skeleton of the mammoth (for so the incognitum has been called) bespeaks an animal of five or six times the cubic volume of an elephant, as Mons. de Buffon has admitted, . . . [with] grinders five times as large . . . square . . . the grinding surface studded with four or five rows of blunt points."[8]

But Jefferson followed a very conventional argument concerning the presence of the mastodon on the Ohio: "from the known temperature and constitution of an elephant, he could never have existed" at the latitude where these fossils had been found. Therefore, Jefferson argued, anyone who thought that the elephant and the mastodon were the same must believe either that an elephant could survive the cold winters of North America, or (with Hunter) that the earth had once been much warmer than it is now, with places like Ohio having been tropical in climate. Jefferson was not ready to accept either of those conclusions.

Jefferson arrived at the opinion that the mastodon was an animal "resembling the elephant in his tusks and general anatomy, while his nature was in other respects extremely different." Nature seemed to have drawn "a belt of separation between these two tremendous animals" and around the globe had assigned the elephant to the south of a zone between about thirty and thirty-six degrees north latitude, and the incognitum to the north. However, the theory that the biblical Flood had deposited mastodon bones on the Ohio and the mammoth in Siberia was far too convenient (and pious) to die easily. John Whitehurst, in *An Enquiry into the Original State and Features of the Earth* (1792), wrote: "The great assemblage of bones discovered upon the banks of the river Ohio, have been described, with much reason, to the effects of a deluge of water gradually rising, from which the animals fled for safety into a small spot of ground; but the water increasing upon them, reduced the larger animals to the necessity of trampling down the smaller for their own preservation. But after every possible effort to preserve their own lives, the largest and the most powerful of them perished with the smaller and weaker animals: for the heap of bones is thus circumstanced, the bottom thereof is composed of the smaller, and the upper part of the larger and most powerful animals."

Although the fossils from Big Bone Lick represented a being that was unknown anywhere else in the world (unless it was merely a variant of the Siberian mammoth), still no American scholar had made a formal study of the beast. In 1786, however, an anonymous article in *Columbian Magazine* described "a thigh bone, part of a jaw, with the grinders, and part of a tusk" that had been collected in Ohio by a Major Craig, and put on exhibition in the library of the American Philosophical Society in Philadelphia. The anonymous author was a fervent biblical creationist (to use a modern term), and he assumed that Noah's Flood had brought all the remains together at Big Bone Lick. But he took exception to Hunter's observation about the incognitum being extinct: "This . . . I apprehend, conveys an idea injurious to the Deity; who, at the creation, wanted neither foresight to discover how detrimental so powerful an enemy must prove to the human, as well as animal race, or benevolence to prevent the evil."[9]

Soon thereafter the Philadelphia judge and naturalist George Turner wrote a rather sparse and highly conjectural account that did little to move things forward. His whole essay was centered around Hunter's insistence that the mastodon was "some huge carnivorous animal." He therefore distinguished the mastodon as a separate species from the herbivorous Siberian mammoth. Furthermore, "I have often expressed a belief, that whenever the entire skeleton should be found, it would appear to have been armed with claws." Turner had no problem with the concept of the mastodon being extinct: "In my mind it is highly probable, that both species of incognita in question, have long since perished. . . . The benevolent persuasion, that no link in the chain of creation will ever be suffered to perish, has induced certain authors of distinguished merit [this was a thinly disguised jibe at Thomas Jefferson and the British zoologist Thomas Pennant] to provide a residence for our Mammoth in the remote regions of the north. Some of the North American Indians also believe in the now existence of this animal."[10]

Turner had evidently visited the Big Bone Lick site, where he found a "Stratum of bones of the buffalo and other smaller animals. . . . But, judge of my surprise, when attentively examining them, I discovered, that almost every bone of any length had received a fracture, occasioned, most likely, by the teeth of the Mammoth, while in the act of feeding on his prey." This allowed Turner to indulge in flights of fancy no one else had risked. "May it not be inferred, that as the largest and swiftest quadrupeds were appointed for his food, he necessarily was endowed with great strength and activity?—that, as the immense volume of the creature would unfit him for coursing after his prey through thickets and woods, Nature had furnished him with the power of taking it by a mighty leap?—That this power of springing to a great distance was requisite to the more effectual concealment of his bulky volume while lying in wait for prey? The Author of existence is wise and just in all his works. He never confers an appetite without the power to gratify it."

To explain that the mastodon was nonetheless apparently extinct, Turner produced a new and interesting argument: "With the agility and ferocity of the tiger; with a body of unequalled magnitude and strength, it is possible the Mammoth may have been at once the terror of the forest and of man! And may not the human race have made the extirpation of this terrific disturber a common cause?"

Two years later, Georges Cuvier in Paris, beyond argument the greatest zoologist and paleontologist of his age, brought the discussion back to the realm of observable facts and rational conclusions, settling the matter of the identity of the mastodon once and for all. To us it seems commonsensical. There were three kinds of elephants, he concluded: the living elephants (the African and Asian species); the so-called Siberian mammoth, which actually had a range that included North America (just to confuse things, there were true mammoth remains at Big Bone Lick in addition to the mastodon); and finally there was the American mastodon, found in the northern parts of North America but nowhere else. They were all "elephants" and all were herbivorous, the mastodon feeding on much rougher plant material than the others. The mammoth and the mastodon had been physiologically adapted to live in cold climates, and they were both extinct. These conclusions were all very simple and straightforward, but not necessarily easy, however, for many people to accept.[11] More dangerously, they signified that the earth was not a fixed set of structures with a single unvarying population of organisms, but that all was subject to change.

Jefferson's "Great-Claw" and a World About to Change

Before the discovery of the giant, apparently extinct, mastodon from Big Bone Lick established America as a place of importance in the study of fossils, all serious intellectual work concerning fossil vertebrates had been conducted by European scholars. Then in 1797, a second set of ancient oversized bones, together with the energy of Thomas Jefferson, completed the launching of American paleontology. The American frontier had shifted well westward from its position in Mary Ingles's day. The whole of Virginia, including much of what later became West Virginia, was settled. At a meeting of the American Philosophical Society in Philadelphia—the U.S. equivalent of London's Royal Society—Jefferson announced the discovery of some large fossil bones in a cave being mined for saltpeter in "western Virginia" (Greenbrier County, now in West Virginia). His account of them was published by the society in 1799. He could not be sure just to what species, or even to what group, the animal belonged. "These bones only enable us to class the animal with the unguiculated quadrupeds" (which simply means animals with claws), he reported, and then continued: "and of these the lion being nearer to him in size, we will compare him with that animal."[1]

The bones, lying today in a drawer at the Academy of Natural Sciences in Philadelphia, are, at first, unprepossessing—merely some limb and toe bones—there is no skull or teeth. But they are indeed very large and, from the great curved talonlike last segments of the toes, it is clear that the animal evidently had possessed massive claws. And Jefferson was right: not only had the whole animal, whatever it was, been huge: it was unlike anything known to be living in North America.

Jefferson never came out and said that his great-claw was a lion, but it is clear that he really wanted it to be just that: a huge American kind of lion.[2] "We may safely say that he was more than three times as large as the lion: that he stood pre-eminently at the head of the column of clawed animals as the mammoth stood at that of the elephant, rhinoceros, and hippopotamus, and that he may have been as formidable an antagonist to the mammoth as the lion to the elephant." To establish the superior size of the animal, Jefferson presented tables of comparative measurements, made possible by the relatively recent detailed descriptions of the lion by French zoologists.

Jefferson measured the largest claw at 7.5 inches—huge compared with that of a lion at 1.41 inches—but he lacked the technical knowledge to make a full anatomical description and in any case was more interested in what the fossils might mean. In particular: "A difficult question now presents itself. What is become of the great-claw? Some light may be thrown on this by asking another question. Do the wild animals of the first magnitude in any instance fix their dwellings in a thickly inhabited region? Such, I mean, as the elephant, the lion, the tyger? As far as my reading and recollections serve me, I think they do not: but I hazard the opinion doubtingly, because it is not the result of full enquiry."

With this bit of rhetoric, Jefferson was not leading into a philosophical discussion of extinction; instead he was heading toward one of his favorite subjects, arguing that the great-claw, like the mastodon, might still live in, or have recently migrated from, the eastern forests. It could probably be found still in the great continental hinterland—"Our entire ignorance of the immense country to the West and North-West, and of their contents, does not authorize us to say what it does not contain." Jefferson saw the trans-Mississippi country, although still belonging to the French, as logically being part of one great American nation.

In the same volume of the *Transactions of the American Philosophical Society* for 1799 that Jefferson's paper was printed, the great physician

and anatomist Caspar Wistar, his colleague and successor as president
of the society, wrote a scholarly analysis and description of the bones of
the great-claw, among other things gently correcting Jefferson's measurements.[3] (Jefferson had used "a slip of paper" to take off the dimensions; Wistar used proper surgeon's calipers.) Wistar was a distinguished
physician and could describe the remains in formal anatomical terms
and, even more importantly, could write about them as part of a living
animal. His study has rightly been called the beginning of technical paleontology in America.[4]

In a brilliant piece of scientific deduction, Wistar read the bones
for what they told him about the behavior of the beast. He used a
forensic approach to describing the bones that allowed him to make
reconstructions of the life of the animal that extended far beyond the
layman's vocabulary of "ferocious" and "formidable." This approach
quickly became one of the main elements of all investigative paleontology and was something that could be done much more effectively
and extensively with a vertebrate, with its complex skull, teeth, and
limb bones, than something as apparently simple and featureless as a
fossil shell.

"From the shortness of the metacarpal [wrist] bones, and the form
and arrangement of the other bones of the paw, and also from the form
of the solitary metatarsal [foot] bone, it seems probable that the animal
did not walk on its toes, *it is also evident that the phalanx was not re-
tracted.*" Wistar's italics emphasize that this was not a lion, which, like a
domestic cat (but unlike, say, a dog), could withdraw its claws. He continued: "The particular form of [phalanx] No.2, and its connection
with the metatarsal bone, and with No.3, must have produced a peculiar
species of flexion in the toes, which, combined with the greater flexion
of the last phalanx upon the second, must have enabled the animal to
turn the claws under the sole of his feet: from this view of the subject
there seems to have been some analogy between the feet of the animal
and those of the bradypus—having no specimens of that animal I describe these conclusions from the descriptions of the feet given by M.
Daubenton of the great skeleton found lately at Paraguay."[5] The great-

claw, as Jefferson also really knew, was without doubt a giant sloth. In fact, Jefferson had seen a copy of the *Monthly Magazine* with a copy of a reconstruction of the skeleton that Wistar referred to. Cuvier had named the South American animal *Megatherium*.[6]

A great part of the importance of the great-claw from Virginia and the mastodon from Big Bone Lick is how Jefferson and Wistar highlighted broad "philosophical" issues that were to remain central themes in the history of American paleontology for the next hundred years or more. The first of these was Jefferson's almost visceral need to describe American fossil creatures in terms of their ferocity and power (a conceit copied, and indeed exaggerated, by George Turner). Against the evidence, Jefferson had wanted the mastodon to be a mighty living predator of the distant eastern forests. Similarly his great-claw might have been—ought to have been—a similarly impressive, ferocious carnivore. These two incognita could then be projected as powerful symbols of the new nation and described in terms matching the belligerence and strength that allowed America to become independent from its European forefathers.

In both cases, the partisans were to be disappointed. Most contemporary intellectuals followed Franklin and Cuvier in recognizing that the mastodon was an herbivore, large and probably slow moving. It certainly had great tusks, but an image of it as a rampaging predator of the forests was difficult to sustain. And Wistar showed that the great-claw was a sloth. Its scientific name became *Megalonyx*. Although a provoked sloth uses its claws to great effect on any would-be predator, it mostly uses them for grasping onto trees. The creature is, as its name suggests, the very paradigm of passivity. The great-claw certainly was no lion. Nonetheless, a hundred years later, and even today, we see the same issues of size, strength, and aggressiveness arise again with respect to the great American dinosaur discoveries of the West, as paleontologists vied with one another to find bigger and more ferocious species, giving them names like *Brontosaurus* (thunder lizard) and, of course, the most evocative name in all nature: *Tyrannosaurus rex* (tyrant lizard king).

A related pair of themes running through Jefferson's and Wistar's papers on the great-claw concerns the relationship between American and European science. Both men acknowledged the lack of collections of scientific materials and technical libraries in America, and resented their dependence on the published work of Europeans. One can only too readily imagine Jefferson chafing under such a restriction. It was particularly galling because of a parallel issue: the nagging insistence by French scholars on the inferiority of American fauna compared with that of Europe.

Georges-Louis Leclerc de Buffon (later Count Buffon), the great French mathematician and naturalist zoologist, disparaged the fauna and flora of the Americas in *Histoire Naturelle,* his encyclopedic ten-volume, and generally brilliant, review of world natural history. The New World's living things were in every way inferior to those of the Old World, he thought—figuratively and in many cases literally degenerate because of living in a cold climate and a landscape of wastelands. "In America," Buffon wrote in his fifth volume, published in 1766, "animated nature is weaker, less active, and more circumscribed in the variety of her productions; for we perceive, from the enumeration of the American animals, that the numbers of species is not only fewer, but that, in general, all the animals are much smaller than those of the Old Continent."

Buffon especially highlighted the failure of European breeds of domestic livestock to flourish in the New World. Another incontrovertible fact was that in every cultural sense the indigenous peoples of North America seemed so far behind the peoples of the Old World. Buffon turned this into a sneer: "In the savage, the organs of generation are small and feeble. He has no hair, no beard, no ardour for the female. Though nimbler than the European, because more accustomed to running, his strength is not so great. His sensations are less acute; and yet he is more timid and cowardly."

Buffon was the leading natural philosopher and author of his day, but he was poorly informed about American nature. He had never visited the

New World, north or south, and relied to an unfortunate extent on the dubious testimony of others, many of them equally ignorant and poorly traveled. Another problem was that he tried to generalize from information gathered in different parts of the New World, from Canada to Central America, to the Andes and the far south. For him, "America" was one place, just as today some people tend to see "Africa" as one country.

As famous as Buffon's calumnies was Jefferson's response, which he composed as a reply to the secretary to the French delegation in Philadelphia, the Marquis de Barbe-Marbois, who had written to Jefferson asking for, among other things, an account of North American nature.[7] The result of his query was Jefferson's *Notes on the State of Virginia*, based on data that he had been compiling for some years. Here, in his only book, the scholarly and meticulous Jefferson comes through—with page after page of tables comparing the sizes and weights of European and American animals.[8]

Most of Buffon's charges were easily answered. For example, where in Europe was a deer the size of the American moose? Jefferson arranged to have one sent to Paris. But it has to be admitted also that Buffon was right in some respects. European animals and plants transplanted to the Americas fared poorly—as might indeed have been expected. One area, however, where Buffon ironically ceded pride of place to the Americas was in reptiles and amphibians, especially poisonous ones, and the diversity of insects and spiders.

Jefferson had happily included the American mastodon in *Notes on the State of Virginia*, and by then mastodon fossils had been found far and wide in the thirteen original states. His great-claw, something unknown in Europe, allowed him a second opportunity for a diplomatic "I told you so," tempered with the magnanimity of the victor. In his essay from 1799, he asked: "Are we from all these to draw a conclusion, the reverse of that of M. de Buffon. That nature, has formed the larger animals of America, like its lakes, its rivers, and mountains, on a greater and prouder scale than in the other hemisphere? Not at all, we are to conclude that she has formed some things large and some things small, on both sides of the earth for reasons which she has not enabled us to penetrate; and that we ought not to shut our eyes upon one half of her facts, and build systems on the other half." This was a neat put-down and was not weakened in any way by the fact that Jefferson had to add a last-minute appendix to his paper in which he noted that the giant sloth *Megatherium*, recently discovered in Paraguay and described by none other than a French scientist, indicated that his beloved great-claw was also a sloth, not a lion.

Despite Jefferson's efforts, the French calumnies continued to find a ready audience. The sting in Buffon's literary tail was the more venomous because, for the next fifty years, author after author picked up his characterizations without question—indeed some did so with zest, even though, in later volumes in the series, such as the *Époques de la Nature* (1778), Buffon largely recanted his claims about American nature.[9] William Robertson, for example, in his popular *History of America*, first written in Scotland in 1777 and going into many nineteenth-century editions, repeated all of Buffon's claims, as well as those of other continental authors such as Cornelius de Pauw, in English for a wide audience. As late as 1807 the Philadelphia naturalist Benjamin Smith Barton felt it necessary to launch an attack on Robertson, describing him as someone who "with stronger and better lights to guide him, has deformed his History of America with the most palpable falsehoods and errors, concerning the physical condition of the continent, and of its inhabitants." But Buffon's claims were still in the twelfth edition of Robertson's work, published in 1812.[10]

Meanwhile Jefferson's interest in the mastodon continued. He sent William Clark, after his return from the great expedition with Meriwether Lewis, to make a concerted dig at the Big Bone Lick site. Clark came back with a huge collection of some three hundred bones of the mastodon, bison, musk ox, and deer. These were spread out on the floor in the (empty) East Room of the White House, and Wistar was summoned from Philadelphia to view them and select specimens for the Philosophical Society's collections. Jefferson gave a third to the society, kept a third for himself, and sent a third to Paris.[11] With typical graciousness, when sorting through the bones, Wistar did not take any mastodon head bones because the Philosophical Society already had some and he knew that the French did not. After Jefferson's death, his collection was dispersed; the Philosophical Society specimens are now held at the Academy of Natural Sciences in Philadelphia.

FIVE

The First American Dinosaurs
An Eighteenth-Century Mystery Story

It turns out that the mastodon and great-claw were not the only giant vertebrate fossils to be discovered in eighteenth-century America. At a meeting of the American Philosophical Society in Philadelphia on October 5, 1787 ("15 members present, Franklin presiding"), Casper Wistar and a prominent Philadelphia patriot named Timothy Matlack (one of the "fighting Quakers" and later clerk of the State Senate) made a presentation about a "large thigh bone found near Woodbury Creek in Glocester County, N.J." The minutes of the October meeting record that the authors and "Dr Rodgers" (presumably Dr. John R. B. Rodgers, later a physician at New York Hospital) were directed to "search for the missing part of the skeleton," although no record exists of such a further search.

No manuscript copy of the presentation exists from which one can learn details of the bone or the exact location of the discovery, but luckily the journal *American Museum* published notes on a number of Philosophical Society meetings, including the following report: "Paper from Timothy Matlack, esq. and Dr Wistar, of Phila; giving an account and description of part of a thigh-bone, of some unknown species of animal, of enormous size; lately found near Woodbury-Creek in Gloucester County, New Jersey. By a comparison of measures, it appears, that the animal to which this bone belongs, must have exceeded in size the largest of those whose bones have been found on the Ohio, of which we have any account, in the proportion of about ten to seven; and must have been nearly double the ordinary size of the elephant."

This little bit of extra information gives us something to go on. A typical mastodon femur from the Ohio country measures some thirty-six to forty inches. A bone bigger than thirty-six inches by a ratio of 7:10 would be about forty-eight inches long. At this size, the Matlack-Wistar thigh bone was truly enormous, and could only have been from a dinosaur. Forty-eight inches is, in fact, the typical length of the femur in the dinosaur *Hadrosaurus,* which was discovered seventy years later at nearby Haddonfield, New Jersey, and described by the great Philadelphia anatomist and paleontologist Joseph Leidy.[1]

The Matlack-Wistar specimen was presumably owned by Timothy Matlack: Wistar would not have needed a co-presenter if it had been his. Matlack had been born in Woodbury, New Jersey, to a large family living in Burlington, Camden, and Gloucester Counties. His ancestor William Matlack had arrived in the area as an indentured servant from Europe around 1677 and eventually became a successful farmer and landowner. An map of Gloucester County from 1849 shows a "J. Matlack" residence very close to Woodbury Creek, about three-quarters of a mile west of the main intersection in the center of Woodbury. The Matlack family owned approximately fifteen hundred acres along the "Woodbury-Moorestown Road" well into the nineteenth century.[2] It seems likely, therefore, that the bone was collected on the property of the Matlack family. But what happened to it?

Between February 21 and 26, 1797, Charles Willson Peale, artist, naturalist, and museum pioneer, made a trip to New Jersey with his friend the French natural historian Ambrose-Marie-François-Joseph Palisot de Beauvois, to collect rattlesnakes from the vicinity of Bridgeton (also Bridge town, Bridgetown), New Jersey, staying in the home of a Dr. Elmer, who was their guide to the rattlesnake dens. They spent the first night of the journey just across the Delaware River from Philadelphia in Camden, New Jersey, so as to make an early start the next day.

Peale wrote in his diary: "Rose before it was light and we soon began our journey—we breakfasted at [left blank] were [*sic*] I enquired after a large Bone which had been dug up, in digging a Ditch on a farm about 1 1/2 mile from Town—this Bone had long been promised to me for my Museum, some persons at this the Tavern here promised me to make enquiry for it, and I requested, that it might be brought to Town, and I would take it to Philada. on my Return in the stage on Monday next. The bone is supposed to be like those found at the salt lick on the Ohio—it was about 4 feet long, but by misfortune was broken by the Negro in digging it up. However I hope to find the relicks worth having, as they may serve with parts of a large Animal, which have been found in different parts of America to prove, that this enormous non descript animal has been in other parts, besides the Ohio."[3]

There is no diary record for Peale's return journey from Bridgeton, so we do not know whether he in fact collected the bone. We cannot be positive where it was found, because Peale omitted to write down the name of the town. However, internal evidence shows that the day referred to was February 22, and Peale wrote his diary entry on Sunday the 26th, presumably intending to fill in the missing name of the town later. Woodbury is some eight miles from Camden, where he had spent the night, along the main route from Camden to Bridgeton, so the assumption seems reasonable that they stopped over at Woodbury.

The obvious conclusion is that Peale went to collect the Matlack-Wistar specimen. But in that case it is interesting that Peale (who knew Matlack) had to send someone to "make an enquiry" for the bone(s) he was seeking—suggesting that he did not know the owner (although it could mean that he knew the owner was away). It is also interesting that Peale referred to "relicks," plural, although it is possible that he was referring to the multiple parts of a single broken bone. Peale said that the bone had "long been promised him," and ten years had elapsed since the first find at Woodbury. At four feet it would have been of similar length to the Matlack-Wistar thigh bone of *Hadrosaurus*. Possibly, then, this was the same bone. At least we can be reasonably sure that it had been collected from the same place. Unfortunately, if Peale did get it for his museum (which is likely if it had been "promised"), it has probably long since been lost in the various fires and other misfortunes that overtook Peale's collections following their purchase after 1850 by P. T. Barnum and Moses Kimball.

The Matlack-Wistar specimen was the first dinosaur bone discovered in America, and one of only a handful of eighteenth-century discoveries of dinosaurs anywhere in the world. There may even have been a second such discovery if the Peale specimen was different.

Sixty years later, in 1858, William Parker Foulke unearthed the complete skeleton of a dinosaur from a marl pit in Haddonfield, New Jersey; subsequently the great Philadelphia anatomist and paleontologist Joseph Leidy described the animal and named it *Hadrosaurus*. At a meeting of the Academy of Natural Sciences of Philadelphia on December 14, 1858, Foulke formally donated his specimens to the academy, and Leidy mentioned that other dinosaur bones had been found in New Jersey even earlier. "Occasionally uncharacteristic fragments of huge bones have been found in the green sand of New Jersey (of which we have several in the collection of the Academy), which I suspect to belong to *Hadrosaurus*. One of these specimens, exposed to the view of the members, indicates a much larger individual than the one whose remains have been presented this evening." The bone Leidy exhibited "to the view of the members" that evening seems to be the same one that he listed in another account of *Hadrosaurus,* published nine years later: "Another metatarsal bone, with its extremities mutilated, from Peale's Museum, formerly existing here in Philadelphia, and probably obtained from the Green-sand of New Jersey, was presented by Dr. P. B. Goddard. It appears to have had the same form as [another specimen from New Jersey], but was somewhat larger." Paul Beck Goddard was a physician, teacher, and chemist; as a pioneer in photography, he invented an improved daguerreotype process using bromine; in medicine he promoted the use of the microscope.

The bone in question still exists in the collection of the academy (catalogue number ANSP 15717) and has the number 473 stenciled on it, presumably representing its position in either the Peale Museum or Goddard's collection. That Leidy was correct in stating that ANSP 15717 was originally part of Peale's Museum can be relied upon, as Leidy was Goddard's student and later his assistant and friend. It was

Goddard who introduced Leidy to microscopy, and Leidy's name was once linked romantically with Goddard's sister.[4] Leidy would have had the information firsthand. The Peale Museum existed from 1786 until 1850, when its contents were sold off. Unfortunately, the date of Goddard's donation of the specimen to the academy is unknown.

At twelve and three-quarter inches ANSP 15717 is far too small to be the Matlack-Wistar specimen. It is, any case, a fourth metatarsal bone (a bone of the hind foot), and without question Dr. Wistar would not have mistaken a metatarsal, however large, for a femur. The reddish clay matrix adhering to the bone, however, has been identified as being from the Woodbury Formation of New Jersey. ANSP 15717 may therefore be yet another eighteenth-century dinosaur discovery in America.

It would help greatly if any dinosaur bone were accounted for in the available records of the Peale Museum, but the *Scientific and Descriptive Catalogue of Peale's Museum* (1796) by Peale and de Beauvois does not mention any unusually large bone specifically, and every fossil bone then in the collection was identified as being from the mastodon. There exists also an accession book for the years 1805 to 1837 that mentions several fossil items from New Jersey.[5] Two bones collected near Swedesboro and accessed on June 20, 1817, could have been from the Woodbury Formation. Intriguingly, a later note states that these bones had been "omitted sometime since." That might represent the Goddard acquisition, but such a suggestion is truly straining at gnats and swallowing camels. If ANSP 15717 is not one of the specimens mentioned in Peale's acquisition book, then it must have been added to the collection either before 1805 or after 1837.

ANSP 15717 may well be the oldest American dinosaur bone still surviving in a collection. The only possible rivals for this honor would be a specimen of *Hadrosaurus* in the New Jersey State Museum, and material of a small dinosaur (*Anchisaurus*) collected in Connecticut in 1818, now at Yale University.[6] Meanwhile, somewhere in an attic or at the bottom of a garden, the Matlack-Wistar (and, if distinct, the Woodbury 1797) specimen may still exist, waiting to be rediscovered.

SIX

Fossils and Show Business
Mr. Peale's Mastodon

The vast Mammot, is perhaps yet stalking through the western wilderness; but if he is no more, let us carefully gather his remains, and even try to find a whole skeleton of this giant, to whom the elephant was but a calf.

REV. NICHOLAS COLLIN, 1793

For all the excitement that the mastodon and great-claw engendered on both sides of the Atlantic, it seemed that the only way to resolve some of the uncertainties over their identity, their size, their behavior, and possibly even their philosophical meaning would be, as Reverend Collin, rector of the Lutheran Churches in Pennsylvania, observed, to find better material. In 1799 Jefferson organized a committee charged with the task of investigating the remains of "ancient Fortifications, Tumuli, and other Indian works of art" and finding "one or more entire skeletons of the Mammoth, so called, and of such other unknown animals as either have been, or hereafter may be discovered in America. . . . [T]he committee suggest to Gentlemen who may be in the way of inquiries of that kind, that the Great Bone Lick on the Ohio, and other places where there may be mineral salt, as the most eligible spots."[1]

At the time he and Wistar wrote about the Virginia "great-claw," Jefferson was, among other things, president of the trustees of the first natural history museum in North America. The Philadelphia Museum was the creation of one of the more remarkable and flamboyant men of that highly distinctive age: Charles Willson Peale. Peale was an artist, known to us now both for the number and brilliance of his portraits of

46

his contemporaries (a wonderful record of the age) and for the number and brilliance of his children (subtly named Rembrandt, Titian, Rubens, and so on). He was also a gifted naturalist. Peale was curator of the Philosophical Society's collections, and as a staunch political supporter of Jefferson he was naturally one of the members of Jefferson's committee on the mammoth.

In 1783, Peale was asked to make a drawing of a mastodon tooth from Big Bone Lick. This tooth, along with other remains, had been brought back from Ohio by Dr. George Morgan, when he accompanied George Croghan on his second, and successful, collecting trip in 1766. When he got the specimens back to Philadelphia, Morgan gave them to his brother Dr. John Morgan, who was a keen collector of curiosities and artifacts. While they were in John Morgan's possession they were seen by the physician to the Hessian troops attached to Washington, Dr. Christian Frederick Michaelis, whose father (a distinguished scholar) had asked him to bring back some mastodon bones. Michaelis tried to pursue the discovery of bones in New York State on the farm of Rev. Robert Annan, in Orange County, during the war. Failing that, Michaelis asked Peale to draw the tooth from Morgan's collection.

It was while Peale had the Morgan specimens at his studio that they were seen by his brother-in-law and friend Colonel Nathaniel Ramsay, who told Peale that he was missing a great chance in not showing the collection to the public. Which Peale did, at his brand-new Philadelphia Museum, housed first in the Philosophical Society's rooms.[2]

In 1784, after long periods of blandishment by Michaelis (who died before he could see the result of his labors), Morgan finally agreed to sell his collection, which was purchased for the University of Groningen in the Netherlands. It was these specimens that were seen and described by Cuvier, among others.[3] By this time, Peale had acquired some other oddments of mastodon remains from the Big Bone Lick and elsewhere. As a showman as well as a scientist, he well understood the importance of finding, as Collin had said, a complete skeleton of the "vast Mammot." When he got word of the discovery of the skeleton of a mastodon on the farm of John Masten, near Newburgh in Orange County, New York, he rushed off to see for himself.

Occasional mastodon bones had turned up at various sites in Orange County since well before the Revolutionary War, and bones been found in a clay pit on Masten's farm since 1799. When he arrived, Peale learned that already many had been lost to souvenir hunters. In his typical decisive, if not overbearing, way Peale saw the opportunities and, with the help of Jefferson and five hundred dollars from the American Philosophical Society, took over the whole enterprise. Within weeks a major excavation was under way. Famously it involved an elaborate bucket-wheel apparatus for draining water out of the pit, as recorded later in Peale's painting of the scene.[4]

The pit on Masten's farm turned out not to be very productive, although a tusk nearly eleven feet long was recovered. Interested locals pointed Peale to other sites in the region, including the farm of Peter Millspaw, a few miles away, where a remarkably complete skeleton was found. Altogether they collected the remains of three individual mastodons. The bones were shipped to Philadelphia, where Peale and his son Rembrandt assembled the best of them, with a little fudging by adding elements carved from wood, into a fully mounted skeleton. It was America's first reconstruction of a fossil vertebrate.

The great incognitum now had a new role to play: that of a public spectacle in Philosophical Hall. Peale's mastodon skeleton was revealed on Christmas Eve, 1801. In a broadside for the exhibit, Peale announced: "Ninety years have elapsed since the first remains of this Animal were found in this country. . . . Numerous have been the attempts of scientific characters of all nations to procure a satisfactory collection of bones; at length the subscriber has accomplished this *great object,* and now announces that he is in possession of a SKELETON of this ANTIQUE WONDER of North America. . . . [N]o other vestige remains of these animals; nothing but a confused tradition among the natives of our country, which states their existence, *ten thousand Moons ago;* but, whatever might have been the appearance of this ENORMOUS QUADRUPED when clothed with flesh, his massy bones can alone lead us to imagine; already convinced that he was the LARGEST of *Terrest[r]ial Beings!*"[5] Rembrandt Peale gave his best account of the Indian legends concerning the mastodon (first related by Jefferson) in his pamphlet *Account of the Skeleton of the Mammoth: A Non-descript Carnivorous Animal of Immense Size, Found in America,* published in London in 1802 for the skeleton's European tour, and dedicated to Sir Joseph Banks.

Peale gave responsibility for promoting the mastodon to Rembrandt, who proved his own skill at showmanship by holding a dinner for thirteen people under the ribs of the reconstructed skeleton.[6] Later in 1802 Rembrandt Peale took it on a tour to Europe, where it was intended to be both a moneymaker and a final vindication of Jefferson's rebuttal to Buffon over the supposed inferiority of American wildlife. However, it was necessary that it be even more than that. Peale, a great salesman as well as showman, was soon promoting the mastodon as a symbol of the strength of the new nation, a strength growing in part from its ancient history—normally the sort of claim that Europeans made for themselves. In Peale's new view of the mastodon, America had something to silence the Buffons of this world forever.[7] There was also a new adjective: "mammoth" entered the English language as a rival to "gigantic."

Peale's American chauvinism would of course have been helped if the mastodon had actually been a carnivore rather than a gentle herbivore. By this time even Jefferson himself had concluded that it was an herbivore—or rather, an "arbivore," browsing on trees. Yet, nothing daunted, in advance of the European tour Charles Willson Peale wrote to Sir Joseph Banks at the Royal Society in London, "my Sons . . . carry with them the Skeleton of the Mammoth, so called, but what may properly be named the *Carnivorous Elephant of the North*."[8] And Rembrandt Peale argued: "Was it an elephant, it would be an astonishing monument of some mighty revolution in our globe . . . but it is not an Elephant, having held among Carnivorous, the same rank the Elephant holds among Grammivorous [grass-eating] animals."[9]

In reviving the herbivore-carnivore argument, Rembrandt focused discussion on the orientation of the tusks. After his return from Europe, he reconstructed the tusks as curving menacingly downward like weapons. He published a pamphlet titled *An Historical Disquisition on the Mammoth, or Great Incognitum, an Extinct, Immense, Carnivorous Animal, Whose Fossil Remains Have Been Found in North America,* in

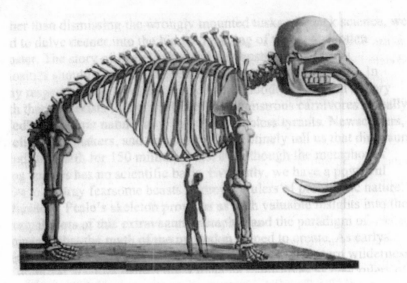

Rembrandt Peale's reconstruction of the mastodon with reversed tusks
(from Edouard de Montulé, *Travels in America*, 1821)

which he defended the downward curve of the tusks and conjectured
that the mastodon had used them for "rooting up shellfish." This was
never very convincing, and to many eyes (including his father's), upward curving tusks were more terrifying and lethal looking than the reverse. The tusks were finally restored to their correct upward orientation
in a redescription of the material by Rembrandt Peale's son-in-law, the
naturalist Dr. John Godman, in 1826. Invoking the authority of Cuvier,
he wrote that nothing could "justify us in placing these tusks otherwise
than in the elephant, unless we find a skull which has them actually im-
planted in a different manner."[10]

In turn-of-the-century America, just as today, science, politics, and religion readily became entwined, and controversially so. Jefferson's fascination with natural history was inseparable from his views of the
democratic value of science and contrasted with the more elitist views of
his Federalist opponents, recently defeated in the election of 1800. His
outlook could also be made to seem faintly ridiculous, as one polemicist

demonstrated at the time: "Nowadays, a man need only discover a mammoth pit to be celebrated as a great scholar or philosopher."[11]

Party politics had come to America with a vengeance, and Jefferson was lampooned for his apparent preoccupation with old bones, which his antagonists portrayed as symbolic of him having got everything wrong—from his philosophy and his francophilia to wasting money on the Lewis and Clark expedition and the Louisiana Purchase, to his views on the role of industry and capital in the American experiment. John Adams complained: "The country is explored too quickly and planted too thinly. . . . [S]peculations about mammoths are all pitiful bagatelles when the morals and liberties of the nation are at risk, as I believe them to be at this moment."

In a letter to the mystic philosopher Francis Adrian van der Kemp, Adams wrote: "I can afford you no ideas on the subject of the mammoth because I have none. The Spirit of Political Party has seized upon the Bones of this huge Animal, because the head of a Party has written something about them, and has made them a subject of more conversation and Investigation than they merit. The Species may yet exist in America and in other quarters of the globe. They may be carnivorous, or they may subsist on the Branches of the trunks of Trees: but as I see no means of determining these questions, I feel little interest in them, till a living Individual of the kind be found."[12]

In *Notes on the State of Virginia,* Jefferson had used the mastodon to demonstrate that the fauna of American was not as weak and depauperate as Buffon had claimed. Two decades later, the eagerness of the Peale family to promote the mastodon as an aggressive symbol of American power may have had something to do with defending Jefferson against the snide attacks of his enemies.

Many such attacks were published in *The Port Folio,* edited in Philadelphia by Joseph Dennie. Satiric verse was the weapon of choice, as in an anonymously published poem from 1802 by Samuel Ewing that went after both Jefferson and Peale. The immediate object of scorn was the dinner that Rembrandt Peale had held under the mastodon skeleton, when political songs had been sung. There is no need to rehearse the whole poem; the following extract gives the flavor of the piece.

Thou know'st, sweet Orpheus! that this Mr. Peale
has sent his Raphael and his Rembrandt round
wherever toe-nails of a flea are found
to sense, *without reward,* the common weal!
. . . Yet when they only Skeletons could find
They bought the bones, but left the life behind.[13]

Perhaps an even more vicious attack on Jefferson came from William
Cullen Bryant, the child-prodigy son of a Massachusetts doctor and later
the author of *Thanatopsis,* in a work titled *Embargo,* part of which reads:

Go, wretch, resign thy presidential chair,
Disclose thy secret measures, foul or fair,
Go search with curious eyes for horned frogs,
Mid the wild wastes of Louisiana bogs;
Or where the Ohio rolls his turbid stream
Dig for huge bones, thy glory and thy theme.

This would be bad enough had it not continued:

Go scan, Philosopher, thy (*****) charms
And sink supinely in her sable arms;
But quit to abler hands the helm of state,
Nor image ruin on thy country's fate.[14]

The missing word is presumably "Sally's," a reference to Sally Hemings, who even then was being widely gossiped about as Jefferson's
slave mistress (she was, perhaps not coincidentally, the half-sister of
Jefferson's late wife, the two women sharing the same father).

In this connection, discussion has been focused on interpretations
of Charles Willson Peale's famous painting *Exhumation of the Mastodon.*
On the surface this seems to be a simple depiction of the scene where
the marl pit was being excavated. Peale's pose is that of the classical
Apollo Belvedere, a symbol of victory—but which victory? The storm
clouds, the central role given to the elaborate human-powered wheel

pump (democratic labor) being used to drain the water from the flooded pit, the curious collection of onlookers, some of whom are Peale's own family (some deceased by that date) while others are dressed as English country squires (Federalist elite), together with an absence of the mastodon bones themselves (used to attack Jefferson), have all suggested that a number of narrative and political themes underlay the work.[15]

Peale's backing of Jefferson cost him financially. He had requested federal support for his museum. The Federalists blocked it, but he did manage to persuade the Pennsylvania legislature to allow him to move his growing enterprise, including nearly a thousand mounted mammal and bird specimens, into the tower and top floor of the former State House. Mastodon Hall remained at the Philosophical Society's rooms.

Peale's specimens later had a checkered history. The mastodon exhibit remained in place at Philosophical Hall, but when the once-great museum started to fall into bankruptcy, the first mastodon specimen was sold, around 1848, to some German speculators. They tried to sell it to the king of France (Louis-Philippe) for the Jardin des Plantes in Paris, the country's most important botanical garden, but that fell through when the king abdicated. Next it was offered unsuccessfully both to the British Museum and the College of Surgeons in London. After being exhibited in London but remaining unsold, it was bought by Dr. J. J. Kaup for the Geology and Mineralogy Museum in Darmstadt, where the skeleton now resides. The second specimen ended up at the American Museum of Natural History in New York.[16]

Meanwhile, the naturalist Benjamin Smith Barton had very early on reported that the mastodon was a vegetarian. The Right Reverend Bishop James Madison of Virginia (a cousin of the later president) had written to him in 1805 about a specimen from his state, and Barton passed on the information to Cuvier: "What renders this discovery unique among others, is that in the midst of the bones was found a half triturated mass of small branches, of gramina, and of leaves, among which it was believed that a species of reed still common in Virginia could be recognised, and that the whole seemed to be enveloped in a sort of sac, which was considered as the stomach of the animal, is that there was no doubt but that these were the very substances upon which the animal had fed."[17]

The surge of interest in the mastodon caused by the Peale specimens also produced the remarkable observation that some soft tissues of the mastodon had been preserved. Benjamin Smith Barton reported: "As late as 1762, which was, in all probability, several centuries after the extinction of the species in America, the proboscis (trompe) of one of the animals was preserved; for the Indians, in their account of the discovery, said, that the head of one of the Mammoths was furnished 'with a long nose, and the mouth on the underside.' "[18] A combination of increased public awareness of fossils and continued development of the land led to a growing stream of mastodon discoveries after 1820, principally in New York State, although specimens also turned up all through the eastern United States, as far south as the Carolinas. At present, mastodon remains have been found at more than six hundred locations in eastern North America.

If there was any doubt left among scholars or the public concerning the carnivorous or herbivorous habits of the mastodon, they were silenced finally by another discovery in Newburgh, New York, in 1845. At long last, the first complete mastodon fossil was found, just as Nicholas Collin had hoped for more than fifty years earlier. In the words of a geology textbook published a decade and a half after the find, the specimen "occupied a standing position, with the head raised and turned to one side, and the tusks were thrown upwards—the position natural to a quadruped when sinking in the mire. In the place where the stomach lay, and partially inclosed by the ribs, there were found about seven bushels of vegetable matter—i.e. bruised and chopped twigs and leaves."[19]

So the mastodon really was an herbivore. But it would still have been impressive enough as a peaceful giant, always ready to defend itself against attackers with its huge tusks: not a bad image for any nation.

PART TWO

Fossils and Geology

SEVEN

Fossils and Extinction
Dangerous Ideas

We have so little of final causes that no certain conclusion can be drawn from the Wisdom of the Creator against the Extinction of a Species. There may have been reasons for their existence at one time, which may not remain at another.

JOHN ADAMS TO F. A. VAN DER KEMP, 1802

The English physician and collector Dr. John Hunter had written of the mastodon: "We cannot but thank Heaven that its whole generation is extinct." But how could anyone be sure that creatures like the European mammoth or the American mastodon and great-claw really were extinct? And if they were, what would be the consequences for our ideas about the history of the earth and life upon it?

Today the concept of an animal or plant species being extinct is commonplace. We are comfortable with the fact that the fossil record documents the ancient existence of millions of extinct species that lived and died during the two and half billion years (give or take a few million) that life has existed on earth. But in Jefferson's time the "record" of fossils was neither full enough nor continuous enough to present anything like a documentary history of changing life on earth. The conventional view was that the earth had been formed in its present state, just as the book of Genesis said, and only some six thousand years ago. As for extinction, although there seemed to be empirical evidence that the earth had once been populated by creatures that no longer existed, there was no coherent theory about how or why they could have disappeared. Extinction could also be thought contrary to biblical teaching.

The discoveries of the two American incognita, the mastodon and the great-claw, therefore focused attention on some dangerous ideas and put people like Thomas Jefferson into a tricky logical box. Although it was politically important that the American incognita be defined as new and different from their European counterparts, that would mean flirting with the issues of extinction and an ancient earth unlike our own. For centuries it had been philosophically inconceivable and theologically unacceptable that any species could be considered extinct. Buffon, Jefferson's literary sparring partner, compromised over this issue by concluding that the American mastodon was the only terrestrial animal that had truly become extinct.

But that was an evasion of the main problem. If God had made all the animals and plants of the earth during the few days of Creation, as described in the first chapter of the book of Genesis, how could there be *any* extinct creatures of a different sort from those we know today, buried in the rocks? If extinction were real and widespread, did that imply that God had made mistakes or that the Bible was fallible? Extinction—if a true phenomenon—opened up the possibility that there had been previous, perhaps multiple, events of creation not mentioned in the Bible and a longer history for the earth than six thousand years.

Dr. Robert Plot, the first curator of the Ashmolean Museum in Oxford (and the earliest describer of a dinosaur bone), wrote in 1677: "I shall leave it to the Reader to judge whether it be likely that Providence, which took so much care to secure the Works of Creation in Noah's Flood, should either then, or since, have been so unmindful . . . as to suffer any one Species to be lost."[1] A hundred years later, Jefferson framed the issue the same way in his *Notes on the State of Virginia*, using, as was a common practice of the day, the term "nature" as a neutral proxy: "Such is the economy of nature, that no instance can be produced, of her having permitted any one race of her animals to become extinct; of her having formed any link in her great work so weak as to be broken."

Some modern authors give the great French geologist and paleontologist Georges Cuvier the credit for having been the first to make extinction a credible concept. For *Recherches sur les Ossemens Fossiles de Quadrupèdes* [Researches on the Fossil Bones of Quadrupeds], his great compilation of contemporary knowledge of fossil bones first published in 1812, Cuvier wrote an introduction, or "Preliminary Discourse," in which he pointed out that any animal as large as a mammoth, mastodon, or giant sloth would surely have been seen by now if it were still alive.[2] In fact, Cuvier had simply made the question one of greater immediacy. Extinction was an old problem. When the Oxford scholar Edward Lhwyd discovered trilobites in 1698, it was clear that none were still living, and neither were any of the commonest kinds of fossil shells that could be found in the ground all over Europe. These included the hundreds, if not thousands, of species of ammonites (from what we now call the Mesozoic era), popular with many collectors for their "Cabinets of Curiosities." Cuvier seized on the fact that while these might have been easy to explain away as "formed stones" (some kind of sport of nature), creatures like the mastodon and mammoth were obviously more modern, more real, and self-evidently closely allied to well-known living animals. And equally obviously extinct. And problematic.

In his essay from 1797 on the great-claw, Jefferson's argument against the idea of the giant sloth's extinction was, typically, empirical and lawyerly: "The bones exist, therefore the animal has existed. The movements of nature are in a never-ending circle. The animal species which has once been part of a train of motion, is still probably moving in this train. For if one link in nature's chain might be lost, another one and another might be lost, till this whole system of things should evanish by piece-meal; a conclusion not warranted by the local disappearance of one or two species of animals, and opposed by the hundreds and thousands of instances of the renovating power constantly exercised by nature for the reproduction of all her subjects, animal, vegetable, and mineral. If this animal has once existed, it is possible that he still exists."

Jefferson was a deist, and nature was a crucial part of his personal philosophy: nature was the embodiment of all that was good, therefore it could never be wrong.[3] For him and for many others religiously and philosophically opposed to the concept of extinction of any creature, a possible solution to the absence of living mastodons and great-claws in the known world lay in the fact that so much territory, especially in North America, was still unexplored. If the mastodon did still live, the only real possibility was that it had migrated into the unexplored West. Part of Jefferson's sense of national identity was his conviction that the great sweep of the West and Northwest making up the Louisiana Purchase was another form of incognitum—a *terra incognita*—that would turn out to hold all manner of scientific wonders.

Therefore, when he dispatched the Lewis and Clark expedition of 1804-6 to survey the newly acquired territory and to look for a link between the mighty Mississippi and the Pacific Ocean, thereby making a huge political statement in opening up America as one great continent, Jefferson charged the explorers to search for animals like the mastodon—creatures "the remains and accounting of any which may be deemed rare or extinct." In the event, while the expedition failed to yield the mastodon or the great-claw, the new lands of the West eventually did produce a vast array of extraordinary fossil creatures— including dinosaurs—that made American scientists the leading paleontologists in the world by 1900. Ironically, it turned out that the dinosaurs were all extinct too.

The question of whether the mastodon (or any other creature) was truly extinct, when put in the larger context of European and American scholarship in the Age of Enlightenment, gives us a glimpse into a world of science (or natural philosophy, as it was still called) very much on the cusp of a major revolution. Around 1800, ideas that easily traced their origins back to before the seventeenth century were being tested and thrown aside in favor of seemingly radical ones, particularly with reference to the nature and history of the earth itself. This often led to curious inconsistencies. It was common, for example, to follow philosophers like Descartes who showed that the earth had a common origin with the sun as a fiery molten mass, or to understand with Buffon, Cuvier, and others that the earth's surface and all the living creatures on it have gone through massive revolutionary changes—and yet at the same time to believe that the earth was created as recently as 4004 B.C. The result was that, for quite a long time, balancing respect for religious authority with rational inquiry meant that it was necessary to behave like the White Queen in *Through the Looking Glass:* "Why, sometimes I've believed as many as six impossible things before breakfast." The reason for all the vacillation is obvious. It is a problem that we still engage every day—the absence of definitive proofs.

A classic case of this failure of proof concerned the nature of fossils themselves. With very few exceptions, everyone today recognizes that fossils are the remains of once-living organisms preserved in, and more or less completely transformed into, rock. The processes by which the carcass of an animal or the remains of plants become incorporated into new sediments and undergo a variety of slow chemical transformations can be studied with great precision. We do not need to rehearse the details of those processes here, but it is interesting to note the variety of the resulting fossils. Within the vertebrates alone, some bone may become changed to opal and some remain almost unaltered, even retaining traces of colors and residues of organic materials such as amino acids. In the very youngest fossils (less than a hundred thousand years old) one can occasionally find remnants of DNA. Sometimes only a trace of the fossil is preserved, as a footprint or a burrow. Sometimes a fossil was formed deep in the rock and then dissolved away, leaving only a perfect natural mold for us to discover.[4]

Knowing all this, it is no problem for us to accept that fossils were once real, living things, however strange they might have been and however different from those that still exist. But that has not always been the case. In the seventeenth and even the eighteenth centuries, because fossils seemed to offer awkward evidence about the history of life on earth—facts that did not fit into a literal interpretation of the words of the first chapter in Genesis—many arguments were raised about what fossils actually represented. Many thought that they were artifacts of the rocks and merely mimics of real organisms. One of the key arguments against fossils being the remains of real animals and plants, or even that they were simply the remains of creatures drowned in Noah's flood, was that, while most known fossils are of seashells of various sorts, they are found high up hills and on the sides of mountains, often thousands of feet above sea level. If there had been a conclusive proof of mechanisms by which mountains (and the remains of sea creatures) might be raised up out of the sea bed, then the nature of fossils would have been accepted more readily. Equally, if there had been independent proof of the organic nature of fossils, then the raising up of mountains would have been more believable.

If the philosophical problem in natural science was to explain the stupendous diversity of life on earth (and in fossils), and the religious problem (one of many) was to account for or explain away extinction, the geological problem was to explain the extraordinary complexity of structure in the earth's crust as exposed in quarries and, above all, in mountains like the European Alps, the Appalachians, and the Rocky Mountains. If God had made the earth in one week in the year 4004 B.C., as Bishop James Ussher, primate of All-Ireland, had calculated, why had he made it so complicated?[5]

Why had he fashioned the earth into contorted shapes and structures that, to a rational eye, suggested the action of great forces, forces that could throw rocks into folds and raise up whole mountains, creating, as Rev. Thomas Burnet put it in 1681, "a broken and confus'd heap of bodies, plac'd with no order to one another . . . a World lying in its own rubbish"?[6] If, on the other hand, all this had happened naturally, what force could have been (could still be) so powerful?

Dr. John Whitehurst's *Theory of the Earth*, published in 1795, was one of the last expressions of a long-standing "physico-theological" theory that at Creation the surface of the earth had been flat and whole, rather like an egg, and that all the elements of geological structure—folding, faulting, mountain building, deposition of sediments in deltas—together, not incidentally, with the formation of fossils—were the result of vast upheavals produced at the Great Flood of Noah when the fountains of the deep had opened (meaning that the earth's crust fractured). All this devastation, moreover, had a reason: it was (as God told Noah) a punishment for man's wickedness. Such a view had a long history and a broad following even though it flew in the face of the biblical statement that mountains had existed before the Flood (Genesis 7:19–20). The various versions of such theories suffered from the drawback that they were caught between a view of the earth that was stable, fixed in its present "confus'd condition" (either at Creation or at the Flood), and the opposing evidence of the continuing operation of geological processes—if nothing else, then the earth is being changed by earthquakes, volcanoes, and the inexorable drip, drip, drip of erosion.

In 1785, the Scottish polymath genius James Hutton (agriculturist, manufacturer, geologist, philosopher) pointed to the way out of the dilemma. In a short essay, Hutton published the first version of a radical new theory about the history of earth and the processes that continually shape it.[7] He took two concepts that had long been established in natural philosophy and put them together. The first was the obvious one that the earth is constantly being eroded by the action of water, frost, and winds. Millimeter by millimeter, the mountains and plains are constantly being denuded and reduced to dust. The sediment so formed is carried by rivers to the seas or to inland lake basins and there builds up as layers of new sediment. Over time that sediment is converted to rock. Philosophers of science had long since pondered the consequences of this universal erosion. The conventional view (eminently compatible with the Judeo-Christian tradition of a beginning and a final Judgment Day) was that eventually the earth would be made flat and then rupture and come to a fiery end. As Thomas Burnet wrote in his *Sacred Theory of the Earth,* the earth eventually would be reduced to "the uninhabitable form in I do not say ten thousand years, though I believe it would but take twenty, if you please, take an hundred thousand, take a million."

Hutton borrowed his second idea from work of Robert Hooke (published a hundred years earlier), and proposed that instead of progressing inexorably to an end the earth might be in a rough sort of balance: the erosion of older rocks being matched by uplift of the new sediments and the products of other forces such as volcanoes depositing ash and lava. In the process, old sedimentary rocks containing fossils, once laid down in shallow seas, were raised up in mountains. In that case, the earth was continually recycling and, instead of being mortal and heading for a final end, was potentially immortal. In Hutton's view, all this recycling of the earth was designed by the deity to create and sustain an environment fit for life, especially for humans. But Hutton, like Hooke and many others before him, had difficulty in finding the actual causal processes underlying mountain building. He concluded that the agents were volcanoes, earthquakes, and the action of the earth's inner heat. Not until the 1950s did we have evidence for the processes of plate tectonics, responsible for continental drift and the folding and faulting of massive portions of the earth's surface. But this theory turns out not to be far different from an idea of Benjamin Franklin, who had come amazingly close in 1793 with the surmise "that the internal part [of the earth] might be a fluid . . . [and the solid crust] might swim in or upon that fluid. Thus the surface of the earth would be a shell, capable of being broken and disordered by any evident movements of the fluid on which it rested."[8]

Convinced that the earth had not been shaped by the operation of miraculous or extraordinary cataclysmic events (the theory known as catastrophism), Hutton tried to determine its age from measurements of the rates of the natural processes of erosion and sedimentation as he was able to observe them at the time. He concluded, however, that such calculations were not possible, memorably finding in the geological record "no vestige of a beginning, no prospect of an end." By this lyrical pair of phrases, Hutton did not mean that the earth had had no beginning or would have no end (a view that Aristotle among others espoused) but simply that it was then beyond man's capacity to calculate them. But by this time Hutton had set in place the final necessary elements for a new theory of the earth, and the conclusion that the world is extremely old—dated in terms of hundreds of millions instead of thousands of years—and populated by a constantly changing set of animals and plants, with humans arriving on the scene only, figuratively speaking, at the last minute.

After Hutton's death, his ideas were taken over by his colleague John Playfair. Then, in the 1820s and 1830s, Charles Lyell, another

Scot, developed them into what is essentially our modern view of the material history of the earth (even though he too did not know the mechanical causes). Geologists then had an intellectual framework on which to build, layer by layer, fossil by fossil, a general understanding of the history of the earth. To do that, they also needed the practical tools that had been provided by the work of a seventeenth-century Catholic saint and an eighteenth-century English canal surveyor.

Even the most casual inspection of the rocks making up a cliff or hillside shows that they exist in layers. It seems implicit to us that the layers of different rock types within the earth's surface represent a time sequence. Unless something extraordinary has happened, the deeper rocks are older; closer to the surface the rocks are younger. And that applies also to the fossils they contain. But that insight was quite late in coming. We owe its origin to an extraordinary Dane named Niels Stenson (known by the Latin name Steno), who was the first to set out the simple rules for unraveling the history of the rocks. Steno started out as an anatomist, first in Leiden around 1660 and then in the court of Ferdinand II de Medici in Florence. Grand Duke Ferdinand had been the last of Galileo's patrons, and Steno published his revolutionary ideas there only thirty-five years after Galileo's trial.[9]

Steno's first rule (like so many really new ideas) was laughably simple when you thought about it his way: if one layer of rock lies on top of another, then the lower one was deposited first and played no part in the formation of the upper one. His second rule was no less obvious: all beds of rock were originally laid down horizontally. The third rule was that beds of rock are laterally continuous until they are replaced by another bed of the same age. These ideas were revolutionary. In practical terms they meant that layered rocks represented a time sequence, and that the rocks exposed at any one small locality were not a separate set of structures resulting from unique processes, but were in principle related to other exposures.

With his simple rules in mind, Steno studied the landscapes of Tuscany as a single unit and developed a theory that the earth had gone through a number of stages (one of which was caused by the Flood), essentially presaging Hutton's view of the recycling earth. While generally conforming to a rather liberally interpreted Creation sequence, his ideas therefore created a view of a changing, rather than static, earth. The world might have been created by God initially but it has been changed by natural processes operating over time.

Steno, having made the great intellectual breakthrough in geology, proceeded to give up science completely, converted from Lutheranism to Catholicism, and became a priest. He died in poverty back in northern Europe ministering to the oppressed Catholic minorities of Protestant Germany. Beatified by the church in 1988, he became the closest thing to a patron saint for geology.

The person who turned the work of Hutton and Steno into practical geology was the simply named William Smith (1769–1839). Smith worked as a surveyor for mining and canal companies. His great insight, borne out of years of practical experience, was the realization that different strata were characterized by different arrays of fossils. Each layer had its own fingerprint of fossils and, as a result, could be traced for miles across the countryside even though it outcropped only here and there in road cuts and quarries. Because most of the strata were also tilted, a transect along the earth's surface showed successive strata neatly arranged in order; this order could be inferred even when only one layer was revealed at a particular spot. For the first time it was possible to make a detailed map of the surface geology of a country the size of England, and also to record a great deal of the underlying geology.[10]

This was not just a theoretical exercise; in fact, it depended on no theories except the simple rules that Steno had outlined. Smith was concerned with practical matters like being able to situate canals in the best place. He also could use his map to predict where coal, limestone, and other minerals would be found. And, equally usefully, where they would not.

With this array of new ideas, scientists in the first decades of the nineteenth century had a powerful set of tools with which to work. They also needed facts, and as they dug into the earth they found ever more wonderful, ever more challenging kinds of fossils representing creatures that no one could have imagined ever existed. And so they also needed new theories.

Enlightenment scholars, and indeed philosophers since at least the time of Aristotle, sought the underlying meaning behind the evident patterns of similarity and difference among living organisms. One result of their inquiries was the formulation of the hierarchical classifications of like with like that we use today: sparrows, crows, woodpeckers, and geese form separate groups within a larger group, "birds," and birds are part of the vertebrates, which are part of the animals, and so on. Jefferson was much taken with the idea that all life, so classified, forms a "Great Chain of Being." This was a concept that went back to Plato and Aristotle. In this chain, everything in creation can be assigned a position relative to an ideal hierarchical pattern extending from nothingness at the base to God at the top. Man is next to God and the angels, the apes next to man, and so on. The lowliest forms of life, at the base, were just above minerals. In this hierarchy, each kind was more complex and more perfect than, and in some way contingent upon, the one beneath. The chain was static, the whole having been created by God, and it represents the perfect symmetry of his creation. Any living organism can be placed appropriately in the chain; potentially, any new discovery would readily fit into its ordained spot among the others.

The recognition of a vast world of fossils first supported and then challenged this view, as did the burgeoning scholarship and the first-hand knowledge of the living world produced by the explorations of the globe from the sixteenth century onward. Soon there were too many kinds of organisms; at the least, instead of one chain, there must be many, and the Chain of Being became more like a Tree of Life. Any fossil discovery should fit neatly into the Chain of Being, which would have no gaps. In the eighteenth century, however, while the growing fossil record closed up many gaps among groups, it opened up new ones and disclosed the existence of entirely new (extinct) groups. In the nineteenth century, extinction became a critical issue because, if real, it would show that the chain, or chains, could be broken. Many of the kinds of fossils that we will meet in the following chapters—giant reptiles like mosasaurs, pterosaurs, or ichthyosaurs, and invertebrates like trilobites, or graptolites—were dead ends. Perhaps most lines were. And many familiar living groups had extinct members, among the most dramatic of which were the mammoths and mastodons, unmistakably species of elephants but no longer living.

Eventually, the concepts of continuity and gradation that lie at the heart of the Chain of Being made it logical for philosophers to ask whether organisms were not also related in the genetic sense, through a process of transmutation, or evolution: the Chain of Being became a Chain of Becoming.

The final pieces of the puzzle concerning the history of life on earth follow rather logically from the facts of extinction and an ancient, rather than young, earth. If animals and plants have constantly become extinct in past ages, then new species must have steadily arisen to replace them. The simplest solution to this was, of course, to posit that, instead of a single creation event, God had serially created wave after wave of new species, allowing all in turn to become extinct and be replaced yet again. Perhaps, as many people like Buffon thought, there had been a whole series of floodlike catastrophes, and that would have been the origin of the many layers of fossil beds in the earth. In the Age of Enlightenment, however, other causes of the changing history of life on earth were sought. And the first evidence that the processes and causes might be gradual rather than episodic, and both material and lawful rather than supranatural, was that the fossil-bearing strata of the earth were simply too numerous.

The fossils also seemed at first to indicate that the earth was cooling. During the Coal Measures, a major division of the Upper Carboniferous period some 300 million years ago, most of the northern hemisphere had been tropical and humid. Very recently in Europe, lions and rhinoceroses had roamed the land. Mammoths and mastodons had lived in places where the temperature is now too cool for any elephant relative to survive. Steadily, however, other evidence appeared to show that the earth had gone through many climatic changes; there had been massive glaciations in the Permian period, for example.

The geological time scale. Ages in millions of years before the present mark the boundary between intervals. The earliest dinosaurs appeared during the Triassic period, well over 200 million years ago; the extinction of the dinosaurs took place some 65 million years ago, at the boundary between the Cretaceous and Tertiary periods.

Eon	Era	Period	Epoch	Millions of years before the present
		Quaternary	Holocene	
				0.01
			Pleistocene	
				1.8
			Pliocene	
				5.3
	Cenozoic		Miocene	
				23.8
		Tertiary	Oligocene	
				33.7
			Eocene	
				54.8
			Paleocene	
				65
Phanerozoic		Cretaceous		
				142
	Mesozoic	Jurassic		
				206
		Triassic		
				248
		Permian		
				290
		Carboniferous		
		Pennsylvanian		323
		Mississippian		
				354
	Paleozoic	Devonian		
				417
		Silurian		
				443
		Ordovician		
				495
		Cambrian		
				545
Precambrian				
Proterozoic				2500
Archean				

More evident from the fossils was the impression that the change in species during the earth's history had been progressive. First there were simple organisms, then more complex ones. While the details took some working out, in the early decades of the nineteenth century the fossil record started to show that, for example, among vertebrates, fishes came first (in the oldest and deepest rocks), followed by four-legged land animals, with birds and mammals coming last. Man either came last of all or was a special case, and there were as yet no human fossils known. However, there was plenty of room at first for argument about this apparent progression, starting with the question of what "more" or "less" complex or advanced might mean. For example, dinosaurs are extinct, but were they in some cosmic sense less advanced than mammals? (After all, we now know that they survived in various guises for some 150 million years.)

As late as 1851, it was possible for Charles Lyell, the greatest geologist of the age, to question the reality of a progression of complexity in life. Concerning the giant extinct "sauroid fishes" of the Devonian and Carboniferous periods, he wrote: "Although true fish, and not intermediate between fish and reptiles, they seem undoubtedly to have been more highly organised than any living fish, reminding us of the skeletons of true saurians by the close sutures of their cranial bones, their large conical teeth, striated longitudinally, and the articulations of the spinous processes with the vertebrae." Furthermore, even if a doctrine of successive development had been paleontologically true, "the creation of man would rather seem to have been the beginning of some more and different order of things. . . . By the creation of a species, I simply mean the beginning of a new series of organic phenomena, such as we usually understand by the term 'species.' Whether such commencements are brought about by the direct intervention of the First Cause, or by some unknown Second Cause or Law appointed by the Author of Nature, is a point upon which I will not venture a conjecture."[11]

But this was all a rearguard action. The concept of organic evolution (transmutation of species) was well and truly in the air, and among liberal thinkers the whiff of revolution that it promised—change and the capacity for individuals to rise above any "predetermined" station in life—was irresistible. Charles Darwin's own grandfather Erasmus Darwin, philosopher, physician, poet, had published various versions of a theory of transmutation around the turn of the century. His ideas were taken up and developed by Jean-Baptiste de Lamarck, Buffon's successor at the Jardin des Plantes in Paris. Fifty years later Charles Darwin articulated the theory of evolution by means of natural selection. In Darwin's theory, all organic change is driven by the occurrence of natural, inherited variation (in every biological feature), tested in the crucible of the environment. The simple fact that all organisms tend to produce far more offspring in their lifetime than are necessary to maintain the population means that not only does each individual have to contend with the vicissitudes of the environment, but also to compete against all others in the "struggle for existence." Only those that are best fitted for particular conditions survive to produce the following generation.

A logical consequence of the theory of evolution is that all organisms, living and fossil, are related to one another in a series of branching family trees. Furthermore, because the earth is constantly changing, extinction is also explained: it is the eventual fate of all species as they are replaced by others. Not only species but whole lineages (ichthyosaurs or trilobites, for example) may reach a dead end, while others persist in greatly changed form. This theory superseded all previous scientific ideas about life on earth, and it forms the central integrating principle of modern biology. It also provided the intellectual foundation for biological paleontology. Fossils are the documentary evidence of all those family trees stretching back in time. They also present a crucial test of the theory. If they did not document the pattern of branching trees, or if they showed that change occurred in violent leaps, a different theory would be needed. With the advent of evolutionary theory, if the science of discovering fossils had ever seemed like glorified stamp collecting, now it was at the documentary forefront of science. But that was the 1850s, and getting ahead of our story.[12]

EIGHT

Mary Anning's World

In regard to quadrupeds . . . every thing is precise. The appearance of their bones in strata, and still more of their entire carcasses, clearly establishes that the bed in which they are found must have been previously laid dry, or at least that dry land must have existed in its immediate neighbourhood. Their disappearance as certainly announces that this stratum must have been inundated. . . . It is from them, therefore, that we learn with perfect certainty the important fact of the repeated irruptions of the sea upon the land . . . and, by careful investigation of them, we may hope to ascertain the number and epochs of those irruptions of the sea.

GEORGES CUVIER, "ESSAY ON THE THEORY OF THE EARTH," 1813

At the same time that the American mastodon was being puzzled over, two dramatic and scientifically important fossil reptiles were discovered in Europe. One was very big, one was small. Sometime between 1770 and 1780 "the aquatic Reptile, the *Mosasaurus,* or Lizard of the Meuse," was discovered in one of the underground galleries of limestone quarries in St. Peter's Mountain, Maastricht, the Netherlands. Quarry workmen exposed what seemed to be a large skull and then, luckily, stopped working and called for advice. Dr. J. L. Hoffmann, a surgeon in the town who had long collected fossils in the quarries, realized its importance and supervised while they took out a huge complete block containing the skull. The jaws, with some ferocious-looking teeth, measured no less than four feet long.

Hoffmann got the block home, wanting it for his own collection, but soon lost it in a legal battle with Canon Godin, who owned the land. The canon's triumph was short-lived, however, as Napoleon's troops besieged Maastricht in 1795. The canon hid the skull in his cellar, but to no avail.

Hoffmann had corresponded with Cuvier and others about the find and the French had issued instructions that the specimen must be seized; a bribe of some bottles of wine soon showed the French troops where it was. It was taken off to Paris as war booty, where it remains to this day.[1]

The Maastricht skull was by far the largest and most complete fossil of a reptile then known. It still dwarfs many dinosaurs. In Paris, the great skull and jaws, with their rows of crocodile-like teeth, were studied by Cuvier, who showed the animal to be a kind of marine lizard. He gave it the name *Mosasaurus* (after the River Meuse). It had lived during the Cretaceous period, some 85 million years earlier. Mosasaurs have not been particularly plentiful in subsequent fossil collections in Europe, but they turned out to be particularly important in early discoveries in the American West, where Kansas became the center of mosasaur finds.

The second discovery was perhaps even more spectacular. In Bavaria, the Solnhofen limestone quarries are famous both for the fine grain of their stones—yielding superb lithographic stones—and for the exquisite preservation of their fossils. Evidently the limestones were laid down in a shallow, rather poisonous marine lagoon into which animals were carried by wind or water. Among the most famous of all the fossils that have been found there was *Archaeopteryx*, intermediate in structure between a reptile and a bird. The first *Archaeopteryx*, however, was not discovered until 1861 (a feather was found in 1860). Sometime in the late eighteenth century, workmen at Solnhofen had found, on the exposed surface of a slab they had split open, the remains of a delicate little reptile. Once again it was Cuvier who described it, showing that it was something completely new: a reptile indeed, but a flying reptile with a wing span of some twenty inches. Its extraordinary wing was created by a web of skin extending from a single elongated finger to the side of the body. Cuvier called it a ptero-dactyle (wing-finger).

A third discovery was not a physical fossil, but a whole intellectual approach. The Solnhofen pterodactyl came from quarries that were packed with fossils. It was such places, where fossils were found in large numbers in a definable geological context, that, as Cuvier preached, proved to be the key to understanding the history of the earth. However, the fossils that were being found in greater and greater diversity

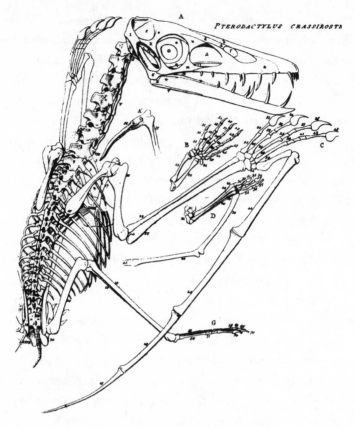

Cuvier's Pterodactyle skeleton from Solnhofen
(from William Buckland, *Bridgewater Treatises,* 2nd ed., vol. 2 [1837])

prompted scholars all over Europe to ask disquieting questions about the planet's age and development. As we have seen, the mastodon was one such dangerous fossil, and the pterodactyl was another.[2]

The key to understanding the earth was not only theoretical, however, but practical. As is usually the case in such matters, it turns out that the English geologist and surveyor William Smith was not the only one to learn to read the correlation of strata by their signature fossils. Another breakthrough happened with the study of beds in and around Paris (the Paris Basin). From early collections there, Georges Cuvier had thought that the deposits recorded two phases of geological history, with a major revolution separating earlier and modern faunas. Then, by careful surveying and collecting in widely dispersed quarries, Cuvier and a young geologist colleague named Alexandre Brongniart uncovered evidence of a succession of deposits, alternating between marine and freshwater origins, and containing a succession of faunas. Of the mammals they found, one was particularly numerous; Cuvier called it *Palaeotherium* (ancient animal) and thought it was something between a tapir and a camel. Like Smith, Cuvier and Brongniart used fossils (mostly the shells) to identify and map the successive strata. Over two decades of work they built up the first detailed picture of a segment of an ancient world, or rather a sequence of worlds, in change—a story told both in the rocks making up the earth's crust and in the animals that lived in those long-off times.[3] Paleontology had progressed from the stage of collection and description to one of systematic analysis. Cuvier's "Preliminary Discourse" (later revised and published as *Essay on the Theory of the Earth*) became one of the most influential books on geology and paleontology in the early decades of the nineteenth century, fully espousing the idea of extinction and the fact of an ancient, constantly changing earth in which fossil shells on mountainsides and elephants in Ohio were to be understood through rational, material explanations.

Back in England there was a similarly fortunate coincidence of the discovery of major new kinds of fossils, and the emergence of a group of young paleontologists keen to study them, untrammeled by the thought habits of an older generation. Most of the new breed of scholars had been university trained under Adam Sedgwick at Cambridge University or William Buckland at Oxford. All were heavily influenced by Cuvier. Significantly, except for Gideon Mantell, a Sussex doctor, they were all geologists by training and inclination. These men played a major role in the study of fossil vertebrates, including the first description of a dinosaur, although their contributions to vertebrate paleontology had a more humble beginning than either the grandeur of the Museum d'Histoire Naturelle in Paris or the richly paneled halls of the ancient English universities. It began out on the cliffs and beaches of England's south coast, and with a twelve-year-old girl named Mary Anning.

Mary Anning, by sheer economic necessity, was one of the world's first full-time professional fossil collectors. It was she who "sold sea shells by the seashore." And most of her "sea shells" were fossils. The Anning family lived in Lyme Regis, Dorset. The town, essentially a fishing village, had become a popular coastal resort at the turn of the century, and one of the attractions was the cliffs, from which waves and weather constantly eroded a variety of interesting fossils. The Jurassic-age Blue Lias at Lyme Regis consists of layers of shale and limestone marl originally laid down 195–200 million years ago in a shallow coastal sea. The fossils in the limestones are preserved uncrushed and were specially sought after. Ammonites (called "snake stones" from their resemblance to a coiled snake) were common, along with large isolated vertebrae and what looked like huge crocodile teeth.

As a young child Mary Anning collected fossils on the beach to sell to the visiting gentry, as did other Lyme residents. A good ammonite might fetch half a crown (about six pounds, or ten dollars, in today's

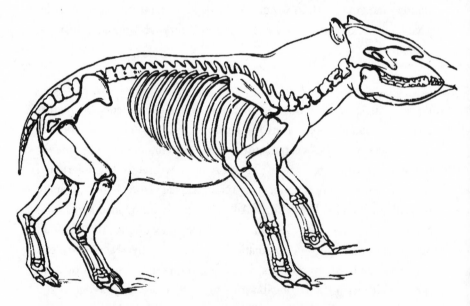

Cuvier's reconstruction of the skeleton and body outline of *Palaeotherium* from the Paris Basin (from William Buckland, *Bridgewater Treatises,* 2nd ed., vol. 2 [1837])

terms and therefore well worth the effort). Her father Richard was a carpenter but often out of work in the difficult economic times following the wars with France. He collected on the beach too, as did his wife, also named Mary (leading to some confusion as to which Mary collected a particular specimen). After Richard Anning died, his twelve-year-old daughter spent most of her time on the beach and in the lower cliffs searching for fossils. She was barely literate, although in her later years she taught herself a little French so as to be able to keep up with developments on the other side of the English Channel.

The cliffs at Lyme Regis are soft and new fossils are constantly being exposed by erosion from the weather, especially in winter storms. In 1811, Mary's brother discovered a set of bones that he carefully concealed from other collectors. Then, over a period of a year, he and Mary solved the question of what animal the strange "crocodile teeth" from Lyme Regis belonged to. They found that those teeth came from a kind of large reptile, up to fifteen feet or more long, with an elongated snout, remarkable paddlelike limbs, and a long tail. They had excavated what turned out to be a large, quite complete ichthyosaur: the Jurassic equivalent of a toothed whale. This was not the world's first ichthyosaur, but it was the most complete specimen and became the first to be described properly by scholars. Mary, having taken over the collecting operation, sold it for twenty-three pounds to a local landowner.

Where others continued to find the commonplace fossils at Lyme Regis, Mary Anning possessed a gift essential to any good field paleontologist—she had "the eye." Out in the field, fossils do not simply leap out of the rocks or lie there gleaming and pristine, waiting to be picked up. They have to be picked out in a background of a thousand confusing shapes, colors, and textures. A fossil collector with "the eye" will spot the potential in a slight curve to a layer of rock or a trifling discoloration. Where any other mortal would simply walk by, the person with the eye finds the treasures.

Obviously Mary Anning also had extraordinary local knowledge, having walked the foreshore and climbed (insofar as was safe) the cliffs for years. Her second great discovery was in some ways even more dramatic than the ichthyosaur. Over the following years she unearthed several examples of a different kind of marine reptile, one with a very long neck and tiny head. Much debate ensued over this creature, as there was nothing remotely resembling it among living reptiles. (Later zoologists would describe these creatures as resembling a snake threaded through a turtle.) Cuvier thought it most likely that a mistake had been made, and bones from different animals had been mixed. But Cuvier, for once, was wrong. Another entirely new kind of fossil vertebrate had been found. Mary Anning sold the first specimen, collected in 1823, to the parson-geologist Rev. William Conybeare, who named a whole new group for it: the Plesiosauria.

Within a decade the young woman had become famous for her discoveries and her little shop of curios. She was a familiar sight on the beach with her dog, her thick cloak, and her hammer. When visiting Lyme Regis, the great and the good would make it a point to meet her. In 1824, an English gentlewoman, Lady Sylvester, wrote (somewhat patronizingly but on the whole admiringly, given the times) in her diary: "[T]he extraordinary thing in this young woman is that she has made herself so thoroughly acquainted with the science that the moment she finds any bones she knows to what tribe they belong. . . . It is a wonderful instance of divine favour—that this poor, ignorant girl should be so blessed, for by reading and application she has arrived to that degree of knowledge as to be in the habit of writing and talking with professors and other clever men on the subject, and they all acknowledge that she understands more of the science than anyone else in this kingdom."

Later, Mary Anning discovered the first English flying reptiles (pterodactyls, like the ones that had been found in Germany). Next, she found a relative of the sharks that seemed to be a link to the skates and rays. Wealthy patrons vied to buy these new treasures, and paleontologists in turn competed for the right to study and describe them. The status of fossils had subtly changed; they now had serious monetary value. This might be direct value, as in the case of a purchase, or indirect. University scholars could not contend with the wealthy aristocracy who wished to add these new fossil discoveries to their cabinets. The opportunity to describe one of these creatures, however, might set up a young man's academic career permanently. And if not fortune—after all, Mary Anning always lived on the edge of poverty—then with fossils might come fame.

No one craved the fame and credentials accorded to leaders among the new breed of geologist-paleontologists more than the Sussex doctor Gideon Mantell. Unable, as a religious dissenter, to attend university in England (which required membership in the Church of England), he used his whole income—and eventually lost both his medical practice and his wife—in the search for newer and more exciting fossil vertebrates. Beginning largely outside the academic mainstream of the Geological Society of London and the Royal Society, he built a major reputation with the discovery of fossils in the Tilgate Forest region of Sussex, in southern England. He might easily have been the first person to describe dinosaur fossils in England. In his book on fossils of the English South Downs he reported having found "teeth, vertebrae, bones and other remains of an animal of the lizard tribe of enormous magnitude, . . . perhaps the most interesting fossils that have been discovered in the country of Sussex."[4]

These turned out to be the remains of a large meat-eating dinosaur, but Mantell did not know what they were. Then he found some different, more or less leaf-shaped teeth, the identity of which was also baffling. Cuvier included a reference to these teeth in his great compilation *Recherches sur les Ossemens Fossiles* [Researches on Fossil Bones], but even he was unsure what Mantell's creature was: he thought the teeth might have been from a fish, although he wrote, "it is not impossible that they also came from a saurian [lizard], but a saurian even more extraordinary than all that we now know." Unable to obtain the imprimatur of Cuvier, and perhaps due to the competitiveness of William Buckland, Mantell did not publish a formal paper on his discoveries until early 1825.[5]

Buckland, meanwhile, a reader in geology at the University of Oxford, had been working sporadically for more than a decade on some fossils from Oxfordshire that he, too, was unsure about identifying. As far back as 1667 Dr. Robert Plot had described a large partial femur (now lost or strayed) from the Jurassic period found in Stonesfield, Oxfordshire. Not knowing what it was, Plot had decided that it came from a biblical human giant.[6]

By 1824 Buckland had a whole new suite of material from Stonesfield in Oxfordshire, including a jaw with teeth and parts of the pelvis and limbs. Prompted by Cuvier, who wanted to include the Oxfordshire animal in the new edition of his *Ossemens Fossiles*, Buckland finally described it in 1824, concluding—startlingly enough—that it was a kind of giant forty-foot carnivorous reptile, now known as *Megalosaurus*, or "giant lizard." (Later, when the Oxford University Museum moved into its new building in 1860, the *Megalosaurus* bones were put on public display, along with ichthyosaurs from Lyme Regis.)

Mantell, seeing that his Tilgate Forest carnivore was the same animal, had been beaten to publication. However, he quickly described the creature with the leaflike teeth. It was a thirty-foot plant-eating reptile that he named *Iguanodon* (because of the resemblance of the teeth to those of a living iguana). Seven years later he described a second dinosaur: *Hylaeosaurus* ("forest lizard"), a somewhat smaller, spiky creature and also herbivorous. It took a brash and even more ambitious young man to realize that these giant creatures were not simply overgrown lizards but members of an entirely separate kind of reptile. The anatomist Richard Owen, who later became the first director of the Natural History Museum in London, concluded that there had existed a whole separate category of these creatures, not lizards at all and quite different from other kinds of reptiles. In 1842, he gave them the name Dinosauria (*deino,* for fearful or terrible, and *sauria,* or lizards).

All these extinct, extraordinary but exceptionally real Mesozoic reptiles finally dispelled any possible notion that they, or any other fossils, were simply "formed stones"—quirks of the rock mimicking living organisms. Despite, or perhaps because of, their strangeness they entered the popular imagination without difficulty.

The 1820s and 1830s marked a peak of popularity for the movement of Natural Theology, in which the wonders of nature were studied as prime examples of the bounty and wisdom of God. William Conybeare saw his plesiosaur as "striking proof of the infinite richness of creative design." Some people still attempted to equate the new giant fossil reptiles with great mythical beasts like the Behemoth of the Old Testament.

Ichthyosaurs, plesiosaurs, pterosaurs, mosasaurs, and dinosaurs were documented decades before Charles Darwin's *On the Origin of Species* and before any coherent theory or mechanism of evolution became widely accepted. Conybeare, who speculated that plesiosaurs were related in some way to crocodiles, dismissed as "monstrous" the ideas of those who "have most ridiculously imaged that the links (from species to species) . . . represent real transitions." Nonetheless, whatever theory one might have had about the actual age of the earth or the role of God in creating it, all these discoveries of ancient fossil creatures established beyond any doubt that the history of life on earth had been complex. Clearly, long ago the earth had been populated by creatures totally unknown today—and not merely by different versions of living creatures such as the mastodon (which was related to living elephants). In fact, at least two threads could be traced through the comings and goings of fossil organisms. Some of them were apparently related to modern forms, and through them the fossil record could be seen as a sequential story, a record of continuity from age to age. Others—many of them bizarre to modern eyes—had arisen, flourished, and then expired without leaving any later progeny. As every new excavation had the capacity to reveal yet another glimpse into these ancient worlds, the role of the paleontologist in hunting the bones of ancient animals had become tremendously exciting.

The one place where actual results in the field lagged behind the heightened interest in fossils, however, was the United States. This might seem odd, given the flying start that the mastodon and great-claw, and Jefferson and Wistar, had given to the new republic, but the reasons for the gap are not hard to find. As will be discussed in the following chapter, fossil bones were scarcely at the front of anyone's mind, except for Jefferson and a coterie of Philadelphia physicians. There were many fewer pure scholars in America than in Europe, and most Americans were concerned with more practical issues. And there is an even simpler issue: the availability of specimens. The settled Atlantic states and the mountains of the immediate hinterland lacked the sorts of exposures of older sedimentary rocks and fossil bones that were found in the Paris Basin or the cliffs of Lyme Regis. Americans were not digging deep into the ground for quarries, canals (except for the Erie Canal), or roads. Whatever secrets the earth held deep belowground were not yet revealed, while the surface geology of most of the coastal states was either a fairly modern alluvium (soils and silts) or bedrock granite from which glaciers had stripped away the bulk of potential fossil-bearing strata.

For the first few decades of the nineteenth century, mastodon remains continued to turn up in alluvial deposits across the mid-Atlantic region, from Orange County in New York to Virginia and the Carolinas. Only in a few pockets, in these early days, were other kinds of bones, of older ages, being found. This contrasts sharply with today, when we know of many important, much older, localities on the East Coast where fossil fishes and reptiles abound.[7] For example, abundant fossils of Late Triassic age (140 million years ago) are now found along a long swath of territory from Virginia to Connecticut, but mostly only in cliffs and quarries. Interestingly, one such set of localities exists in the bluffs along the Hudson River at Weehawken, New Jersey, directly opposite Manhattan and the site of the infamous duel between Aaron Burr and Alexander Hamilton in 1804. These remains are preserved in ancient lake beds, formed in a set of rift valleys similar to those of modern East Africa. Not only are the specimens well preserved, they tell us a lot about life 200 million years ago.

Between 1800 and 1850, while discoveries of the mastodon continued to proliferate, the example of the great incognitum and the great-claw did not spawn a whole new era of discovery of fossil mammals. Rather, the field went into something of a decline. For the first decades of the nineteenth century, the study of fossils remained something of a sideline of physicians rather than a serious subject for scientists. Physicians were well able to make detailed anatomical descriptions of fossils, as had Wistar, although few could match his level of perception and insight.

Positioned figuratively between a careful anatomist like Caspar Wistar (who died in 1818) and the professionalism of scientists of the second half of the century was a group of physician amateur naturalists, mostly based in Philadelphia (by far America's largest city), who described the remains of fossil vertebrates in the 1830s and 1840s. Unfortunately they happened to be extremely quarrelsome. In the cold light of hindsight, they seem to have treated their fossils more as magnificent curiosities to enhance their reputations than as discoveries of progressive science. It is only fair to note that many of their specimens were found in very recent deposits (at best tens of thousands of years old) and did little to illuminate the great debates of the day about the age and structure of the earth. There was as yet no view (especially no American view) of an overarching scientific context, no theory of earth history or of organic evolution, within which to place them.[8]

In 1824, the Philadelphia physician and keen naturalist Richard Harlan made a full scientific description of the only surviving fossil—a jawbone from Iowa—that Lewis and Clark had brought back from their expedition. He thought it belonged to a reptile, possibly related to ichthyosaurs, and named it *Saurocephalus*. Soon thereafter, a Philadelphia contemporary, Isaac Hays, found similar material in New Jersey, called it *Saurodon,* and argued that Harlan's description had been so inaccurate that this name should be used instead. Among other fossils, in 1834 Harlan described what he thought was another ichthyosaur (*Ichthyosaurus missouriensis*) for a fragment of a reptile snout from present-day South Dakota. In fact, it was not an ichthyosaur but a mosasaur (a larger second portion, apparently of the same specimen, was described a decade later). In the same year Harlan also described a giant fossil sent from Arkansas as a plesiosaur, which he named *Basilosaurus* (king reptile) *cetoides* (whalelike), although in fact it really was a whale, so Harlan's science clearly might be called into question. However, he was merely doing his best with the material and information at hand, and the errors that he and other naturalists made give us a view of the difficulties of working with the contemporary state of zoological knowledge.[9]

Harlan became an important figure in early American natural science. He was notorious, however, for his querulous, argumentative manner, and the small group of men devoted to the study of fossils gathered together in Philadelphia at that time ended up being divided rather than united around their subject. Harlan was cordially detested by John Godman, for example, who accused him of plagiarism. An almighty row developed over a mastodon-like creature from Orange County, New York, that Godman described. Because it had four tusks, he named it as a new kind of elephant, *Tetracaulodon*.[10] Others, including Harlan, concluded that it was really only another variety of the mastodon, the four tusks probably being a juvenile feature. Isaac Hays vigorously defended Godman, who, regrettably, died before his paper came out. The affair reached Europe, where discoveries in America were becoming more widely noticed; in England, Richard Owen came down on Godman's side.[11]

The debate further highlighted the fact that there was as yet no factual base or theoretical understanding of issues like variation within a species, or knowledge of developmental stages of the mastodon. Inevitably, any odd variant fossils that were found tended to be described as new species and genera. Eventually there was consensus that *Tetracaulodon* really was a mastodon.

Unfortunately, the period was marked also by the activities of a rather shameless showman, Dr. Albert Koch, a German immigrant living in St. Louis, who had explored for "animal organic remains in the far west of the United States"—by which he meant Missouri.[12] In Benton County he found a large collection of fossil bones and assembled them into the monster to which he gave the scientific name *Missourium theristodon* (sickle tooth), comparing it with the leviathan of the Bible. ("Can'st thou draw out leviathan. . . . Upon earth there is not his like, who is made without fear"; Job 41.) This "Great Missourium," or Missouri leviathan, was displayed around the country and in London despite fairly obviously being a fake, as Richard Owen reported to the Geological Society of London (in the same paper in which he gave his favorable judgment on *Tetracaulodon*).[13]

Koch nonetheless succeeded in selling the skeleton to the British Museum, where it was eventually reconstructed properly as a rather fine mastodon. Five years later Koch produced his second marvel, this time from the Eocene epoch, in Washington County, Alabama, unveiling it with the name *Hydrarchos harlani*. This new monster, over a hundred feet long, turned out to be a composite of several skeletons of the whale that Richard Harlan of Philadelphia had previously described as (the plesiosaur) *Basilosaurus cetoides.*

Meanwhile, an interesting new source of fossils had emerged in the valley of the Connecticut River in Massachusetts. As the river flows south from Vermont it cuts through the Triassic "New Red Sandstone" that early geologists confused with the Devonian "Old Red Sandstone" of the Catskills. As the dense, dark red rock along the river valley began to be exploited for use in building, where it became known as brownstone, slabs were unearthed that were covered with three-toed footprints. Some footprints were a foot or so long, with a clear imprint of a number of segments and claws, as if made by a giant turkey or ostrich. These slabs were first noticed, apparently, by a boy named Pliny Moody on his father's farm at South Hadley, Massachusetts, as far back as 1802. In 1836 a local doctor saw more footprints on "flagging stones" from a quarry near Montague, Massachusetts, and drew them to the attention of Dr. Edward Hitchcock, president of nearby Amherst College, who had recently completed the first geological survey of Massachusetts. Hitchcock promptly described them as the footprints of five different species of giant birds—a conclusion that seemed more reasonable when, four years later, Richard Owen described the Moa, a giant flightless fossil bird from New Zealand.[14]

Despite much skepticism about his "birds" on the part of contemporary scientists, and the charge that footprints were not even real fossils, Hitchcock threw himself into the discovery and description of more specimens and founded the new science of these tracks and other traces left by animals, which he called ichnology. His work culminated in 1856 with the publication of *Ichnology of New England,* in which he described and named dozens of "track-way" species, including mammals, lizards, "batrachians" (amphibians), and chelonians, as well as "birds."[15] Eventually it would be realized that these were tracks made by Triassic dinosaurs. Other trackways were found in Europe at about the same time, and new sources from other ages soon appeared in the United States, including a reptile track from near Pottsville, Pennsylvania.[16]

NINE

An American Natural Science

In Europe, naturalists form an extensive community, are governed by the pure
love of the science of nature. There is not a branch of natural knowledge that is
not under investigation there, by men eminent in science. . . . It is painful to perceive what conspicuous blanks are yet left for America to fill up, and especially in
these important branches, American geology and American organic remains.

G. W. FEATHERSTONHAUGH, *MONTHLY AMERICAN JOURNAL OF*
GEOLOGY AND NATURAL SCIENCE, 1831

With a name like Featherstonhaugh (pronounced *Fanshaw*), the editor
of the fledgling *Monthly American Journal* could only have been English. George Featherstonhaugh was a well-off Englishman who moved
to America, married into society, and set about establishing himself as a
geologist. He hoped that the journal would make his scientific reputation. In his disdain he seems to have missed the point that the intellectual and empirical traditions of his adopted country were different from
those in Europe.[1]

Across the Atlantic, before the establishment of a few professional
positions for geologists and paleontologists in universities and museums, there had already been a long tradition of amateur naturalists of
great seriousness. Often, like Gilbert White, the author of *Natural History and Antiquities of Selborn* (1788), they were country parsons. In the
American context, on the other hand, there was a preoccupation instead with what Jefferson called "useful knowledge." Patterns of leisure
and personal wealth were different. There was no great moneyed leisure
class in America, and little stomach to accept such a lifestyle even in
those who might afford it. In the new century, although American

86

scholars assiduously kept up a correspondence with their British and French counterparts, there was also an intense patriotism and, as in Jefferson's time, a deep ambivalence regarding Europe. Featherston-haugh was certainly correct that the United States needed to develop its own libraries and scientific collections, and to train its own savants. Jefferson had long before said the same thing, as had Franklin before him. A frankly nationalistic motive lay behind Noah Webster's first American dictionary, in 1828, for example. But Jefferson also had to ar-gue vociferously against tariffs that made European books too expen-sive in America.

Among those calling for a new investment in natural science for nineteenth-century America, Rev. Nicholas Collin urged the American Philosophical Society to encourage research on native plants: "very few of them are well known as to the extent and peculiarity of their quali-ties, and a very small number is adopted either by the apothecaries, or regular physicians." Also important were insects and spiders, "because some of these do us remarkable mischief," and snakes, for the same rea-son. Of the native mammals he warned (presciently) that "wanton de-struction of the buffaloes on the Western country . . . should be checked." And he argued against destruction of small birds deemed to be of no value because of their potential importance in the overall "oe-conomy of nature." Museums and botanical gardens should be founded. He even voiced the complaint that we do not understand "changes in the atmosphere," the "irregularity of our seasons (being) a great impediment in the business of social life."

Even more influential than Collin was Benjamin Smith Barton, America's leading botanist after John Bartram and his son William. Bar-ton was also a physician, anthropologist, and archaeologist and had a keen appreciation of geology and mineralogy. A member of the Ameri-can Philosophical Society—the nearest thing to a scientific Establish-ment that America then had—he decried the absence of serious work in natural science in the United States, and criticized the little that was being done as too arcane and impractical.

Barton was a great one for lecturing people, as in the talk he gave in 1807 to the fledgling Philadelphia Linnaean Society—a gathering of like-minded naturalists—which was distributed as a small pamphlet: *A Discourse on Some of the Principal Desiderata in Natural History and on the Best Means of Promoting the Study of this* SCIENCE *in the United States.*[2] Here, in what amounted to a review of the scientific "state of the nation," he set out an American scientific agenda. He would have none of the gentlemanly butterfly collecting and classifying on the old English naturalist model. He defined natural history, for Americans, as more: "Natural History . . . is . . . or it ought to be, necessarily a Science of Facts. But no science more than this calls for systems and arrangements of acts, and for reasonings concerning them. One of the higher claims of Natural History is, that it so easily admits, in many instances at least, of just and happy arrangements; and of beautiful and correct theories; of theories, too, which are permanent, and not those false, those evanescent creations of a day, by which Medicine (not to mention other science) has sometimes been injured, and often sullied, disfigured, or disgraced."

Barton was scornful of those who clung to the idea that species had not (and could not) become extinct. Taking direct aim at Jefferson, he wrote in a letter to the French natural philosopher Lacépède (Bernard-Etienne-Germain de la Ville-sur-Illon, comte de Lacépède):

The American species [the mastodon] is unquestionably lost; for nature, it would seem, is much less anxious to preserve the whole of created species than some illustrious naturalists have supposed. . . . I speak of these animals as extinct. In doing this, I adopt the language of the first naturalists of the age. No naturalist, no philosopher; no one totally acquainted with the history of nature's works and operations, will subscribe to the puerile opinion, that Nature does not permit any of her species of animals, as of vegetables, to perish.

We are already in possession of a sufficient number of facts to establish this point, that the continent of North America was previously inhabited by several species of animals, which are now entirely unknown to us, except by their bones, and which, there is every reason to believe, now no longer exist. . . . For what can be more interesting than histories of the species which formerly inhabited the Globe, and have now entirely disappeared.[3]

Barton continued his attack on Jefferson's science by stating: "There is, without doubt, a harmony in the works of nature:—a harmony beautiful and divine! There is a passage by gradual and intermixing characters from species to species, and from genus to genus. BUT THERE IS NO SUCH THING AS A CHAIN OF NATURE: an absolute dependence (on this earth) of one species upon another. Plato's chain of nature is a dream."[4]

While Peale's Museum found favor with Barton as being "very respectable, both for the number and value of the articles which it contains," he saw that the greatest need was in the sciences of the earth. "Of all the branches of natural history, none, I think, is so little cultivated in the United States as mineralogy. This is the more remarkable, not merely by reason of the great utility of this branch of the sciences, but because its sister science, I mean chemistry, is ardently cultivated in Philadelphia." Philadelphia was still the nation's second largest city and the center of a thriving industry of manufacturing chemistry. Barton knew that Americans, as they moved from a Jeffersonian model of a society of farmer-citizens to one of city-dwelling citizen-burghers, would need coal to replace the wood that was already in short supply around the big cities. They would need limestone for burning, ironstone for smelting, and alum, mercury, lead, precious metals, even whetstones. They would need every kind of manufactured chemicals. All that would depend on knowledge of geological resources.

True to his ideal of mixing practical and theoretical science, Barton was possibly the first to describe a remarkable fact of American geology. In 1785 he wrote: "The strata in the countries west of the great Alleghany-mountains, are, in general, horizontally disposed, while the strata, of the same materials, in the countries between the mountains and the Atlantic, are almost all disrupted and placed at an angle of about 45 degrees. The very different arrangement, then, of the strata of stony material, of coal, of iron-ore, etc, in the countries on both sides of the

Alleghany-Mountains, is one of those great features in our country, for which we have not yet been able to give a satisfactory theory. But I doubt if such a theory is beyond the reach or grasp of science. . . . We shall, at some future period, possess a correct theory of the earth. But such a theory is not to be attained, by the mere aid of genius or imagination, in a cabinet of little fragments of stones, of earth, and of metals."[5] This discovery about the geology of the West, made even before the country had been properly explored, became one of the defining features of American geological science. If rocks of the same apparent age, on either side of the Alleghenies, were arranged so differently, powerful forces and unusual geological conditions must have prevailed there.

Before 1800, America took its intellectual cues from Europe. All the reference books available to American scholars, and textbooks in everything from medicine to agriculture, for example, were European and largely British. Serious scholars naturally expected to study and train in Europe and also to publish their work there. No American university yet taught courses in science. Among the major figures in American natural science of the time who had trained abroad were Barton (Edinburgh and Göttingen), the mineralogist and manufacturing chemist Adam Seybert (Edinburgh, Göttingen), the chemist and geologist Samuel Latham Mitchill (Edinburgh), the physician and zoologist James De Kay (Edinburgh), the chemistry and geology professor Lardner Vanuxem (Paris), and the physician, naturalist, and pioneer ethnographer Samuel George Morton (Edinburgh). Their choice of universities fairly accurately reflected the relative merits of European universities at the time. Oxford and Cambridge are conspicuously absent from the list; not only did they not admit dissenters, until the 1830s they were bogged down in old ways (classics and preparation for the church) rather than facing the challenges of late Enlightenment natural philosophy and medicine. In fact, Morton studied medicine at Edinburgh only a few years before a young English naturalist named Charles Darwin did exactly the same. In those days Professor Thomas Charles Hope lectured there on chemistry, being famous for his showy demonstrations and promotion of James Hutton's theory of the earth. Robert Jameson (an advocate of Abraham Gottlob Werner's opposing views on the origin of basalts) covered all aspects of natural history and geology and was enormously influential as the translator and editor of Cuvier's *Essay on the Theory of the Earth.*

But the focus began to shift. Not only were an increasing number of scholars transferring to a scientific career after having studied law or theology at home (Benjamin Silliman, James Dwight Dana, and Edward Hitchcock, for example), the country was significantly enriched intellectually by men who migrated from Europe, bringing their skills with them and helping to train the new generation of Americans. Much of this brain drain from Europe resulted from an awareness of the opportunities presented by the New World. The political and religious reaction against free-thinking that followed the excesses of the French Revolution drove others across the ocean, with the prime example being the chemist and Protestant theologian Joseph Priestley, who was forced to flee Great Britain and came to the United States in 1794. By the first decade of the 1800s the French Revolution had produced a great backlash of fear of social experimentation in England, Germany, and France. The whole nation of the United States, by contrast, was an exercise in religious and political freedom, and conducted through a different kind of democracy. It was a magnet for independent scholars.

Among the immigrants from England was the botanist Thomas Nuttall. One of the great explorers of the lands beyond the Mississippi, Nuttall made a very early series of expeditions across the West as far as Missouri. In 1809 he traveled from the Great Lakes down to St. Louis and then up both the Missouri and Arkansas Rivers. Like Barton before him, he was fascinated by the "near approach which the calcareous and other strata west of the Alleghany mountains make to the horizontal line," even though in terms of lithography they "presented not a single dissimilar figure [to] the mountain limestone of Derbyshire." Even this extremely observant scientist and careful field collector found very few fossils, however.[6]

Like Lewis and Clark and so many other western explorers, Nuttall had to stick close to the rivers along which he found his routes through the interior. And many of the potential exposures of fossil-bearing rocks along the banks of the Missouri and its major tributaries were covered by very recent alluvial sediments. To probe fully into the structure of the vast expanse of "horizontal" secondary rocks (what we call the Mesozoic and Tertiary layers) of the trans-Allegheny West that so intrigued Barton and others, and to find their true relation to the geology of Europe, let alone to find fossils in any numbers, would require exploring deeper into the canyons and badlands of the hinterlands where streams running off the Rockies had carved deep into the surface of the earth. And until that happened discoveries of fossil vertebrates would be rare.

Another important immigrant was America's first great ornithologist, Alexander Wilson, a radical who came from Scotland in 1797 and became a devoted ally of Jefferson. He complained (with all the conviction of the poacher turned gamekeeper) of "being obliged to apply to Europe for an account and description of the productions of our own country" and produced the first treatise on American birds since Mark Catesby's *Natural History of Carolina, Florida, and the Bahama Islands* of 1731.[7] But perhaps the most influential scientific immigrant of this period was an extraordinary man named William Maclure.

By any modern standard, Maclure was a millionaire, having made a fortune in business before leaving what he referred to later as "old Europe, for some time past in her dotage."[8] He immigrated to the United States in 1803 at the age of forty. A man equally interested in ideas of social justice, liberty, and education and in the value of modern science, he immediately set about pursuing his passion for geology and produced the first real geological map of North America.

The first map showing the occurrence of fossil remains in America was made by the botanist John Bartram. Dated to the 1740s, it is a simple freehand sketch of the eastern seaboard and indicates various places where he found "sea shells in stone."[9] (Ben Franklin wrote on the back of the map that it was "very curious.") A cursory survey of American geology had been made by the French intellectual Count Constantin-François de Volney, who fled France in 1795. "Saddened by the past and anxious for the future, I set out for a land of freedom, to discover whether liberty, which was banished from Europe, had really found a place of refuge in any other part of the word. . . . I beheld nothing but a splendid prospect of future peace and happiness, flowing from the wide extent of improveable territory."[10] Unfortunately, anti-French hostility forced him to return to Europe three years later, following passage of the Alien and Sedition Acts and, as Volney wrote, "so violent an animosity against France, and a war seemed so inevitable, that I was obliged to withdraw from the scene."

William Maclure single-handedly surveyed the eastern United States. In the same painstaking way that William Smith had produced his famous geological map in Great Britain in 1801, Maclure made his survey by horse and on foot and is said to have crossed the Appalachians no fewer than fifty times.[11] Like Volney, Maclure produced what might be called a physical geography of the United States, with much emphasis on such useful matters as climate and soil types. In this respect, Maclure's map, together with five rather conjectural cross-sections through the Appalachians, and the accompanying text perfectly reflect its American context. The book is a classic of useful science: of the four chapters, two describe the surface geology of the land, one discusses the breakdown of rocks to form soils, and the last relates all this to the fertility of soils.

For the first time, however, Maclure identified and mapped rocks according not just to their mineral type but to their status in the formal sequence (the "geological column") currently being elaborated. He delineated four classes, marking them in different colors: Primitive, Transitional, and Secondary rocks, and Alluvial deposits (described further in Appendix A). In a later revision he added a category of "Old Red Sandstone" for what he thought was a single band of primarily red-colored rocks running from the Connecticut River valley to the Catskill Mountains. While this might seem rather sketchy and superficial compared with the rich detail of Smith's British map, it was a superb achievement given the vast area to be covered and the fact that there was no history of geological discovery to build upon.

Like Barton, Maclure noted that to the west of the Appalachians, in

the Mississippi and Missouri country, the Secondary layer was all horizontal, and "for the extent of the surface it covers and the uniformity of
its deposition, is equal in magnitude and importance, if not superior, to
any yet known. . . . We have indeed every reason to believe . . . that the
limit of this great basin to the west, is not far distant from the foot of the
Stony mountains. . . . The foundation of most of the level countries is
generally limestone, and the hills or ridges in some places consist of
sandstone."[12]

Subsequent generations of geologists and fossil collectors would
amplify Maclure's sketchy notes into a major geological paradigm: the
convergence of a modern sea of grass with ancient ocean sediments belowground, to tell the story of the geological history of the West, of its
fossils, and of the diversity of life that they represented. Throughout
the following century this set of images would dominate accounts of the
geologists and bone hunters who explored the "Missouri country."

Perhaps just as important as Maclure's science was his role in encouraging and developing science in others. From 1817 until his death in 1840
he was president of the new Academy of Natural Sciences of Philadelphia. The founding of the academy in 1812 marked a significant step toward meeting the goals set out by Benjamin Smith Barton. A group of
Philadelphians, several of them young, all of them social reformers and
enthusiastic, but by no means accomplished, naturalists founded the
academy as "a society devoted entirely to the advancement of useful
learning." John Speakman was an apothecary and Thomas Say his junior partner. Jacob Gilliams was a dentist who had once treated George
Washington, John Shinn was a manufacturing chemist, and Nicholas
Parmentier was a distiller from France; Gerard Troost, a former phar-
macist from Holland, had a factory for manufacturing alum (used in
medicine and dyeing). Dr. Camillus Mann was the sole physician. Significantly, there were no members from the American Philosophical Society's social elite. This was definitely a different generation.[13] They
were all avid collectors—minerals and shells being the favorites—and
soon founded a museum in which they merged their collections.

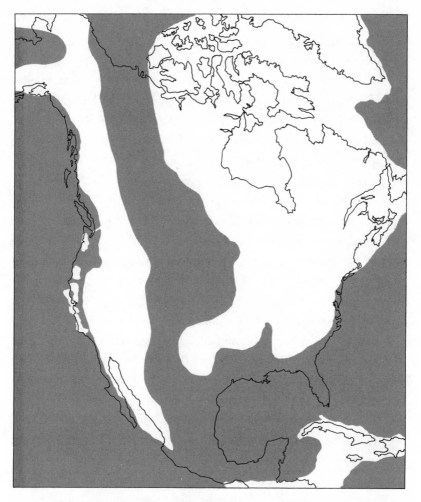

Maximum extent of the Late Cretaceous seaway (shaded) over North America, 75 million years before the present

Although its founders were all practical men rather than scholar-philosophers, the academy was to become one of the leading scientific institutions of nineteenth-century America.

In 1817, the academy established a journal of natural science (Maclure helped the young journal along by buying the academy its own printing press). The first volume of the *Journal of the Academy of Natural Sciences* mostly contained papers describing and classifying new species—of plants, crustaceans, fishes, and insects. The one biological essay was on the parasitic Hessian fly (one of the "injurious insects" that Collin had been concerned with). Eventually, the *Journal* and later the *Proceedings* became a major vehicle for publication of the discoveries of fossil vertebrates from the West. The library and collections of the academy quickly grew to fill a crucial gap in America's early resources in natural science. One of their important purchases was the mineral collection that Adam Seybert had brought from Europe, which made the academy the center for mineralogical study in America.[14]

Also in 1817, a group led by James Ellsworth De Kay founded the New York Lyceum of Natural Sciences (now the New York Academy of Sciences), with aims similar to those of the Philadelphia academy, although it was never as successful or influential. The following year, natural science was further encouraged when Benjamin Silliman at Yale began publication of the *American Journal of Science and the Arts*. Back in 1802 Yale University had taken the unprecedented step of appointing someone to teach sciences; the appointment of Silliman was engineered by Federalists to counter the Jeffersonian deist approach to science, which they saw as leading to heresy.[15]

Although he had trained as a lawyer and needed to travel to Philadelphia to learn some geology and mineralogy before he could teach the Yale students, Silliman soon became another of the dominant forces of early American geology. In introducing his new journal he saw natural history as comprising "three great departments of Mineralogy, Botany, and Zoology," which stood alongside "Chemistry and Natural Philosophy and their various branches: and Mathematics, pure and mixed." Science for Silliman, like Jefferson and Barton, was something inherently practical; of his new journal, he wrote, "while Science will be cherished for its own sake, and with all due respect for its own inherent dignity; it will also be employed as the handmaid to the Arts, its numerous applications to Agriculture . . . the Manufactures, both mechanical and chemical; and to Domestic Economy."

Silliman's Journal, as it was commonly known, published papers in paleontology right from the beginning, but its main subject was geology and mineralogy. In his inaugural issue, Silliman echoed Barton in noting that "Natural History has been most tardy in its Growth, and no branch of it was, till within a few years, involved in such darkness as Mineralogy, . . . notwithstanding the laudable efforts of a few gentlemen. . . . [O]nly fifteen years since, it was a matter of extreme difficulty to obtain, amongst ourselves, even the names of the most common stones, and minerals." For this he blamed the preponderance of teaching the classics in schools.[16]

TEN

An American Geology

"Mr Maclure has, with great ability, sketched the outline; but much labour is still needed in filling up the detail," wrote Silliman in the first issue of his new journal. But Maclure would not be the one to do it. Parker Cleaveland was professor of mathematics and natural philosophy at Bowdoin College (like Silliman, he trained first in law and theology). He produced the first textbook of American mineralogy in 1816, and for the first time there was a domestic rival to European textbooks, such as the enormously influential *Manual of Mineralogy* by Robert Jameson at Edinburgh.[1]

Meanwhile, Maclure had many interests other than science, and in 1826 he took a number of learned Philadelphians off to New Harmony, Indiana, to the utopian colony of scholars and educators founded there by another British emigrant, Robert Owen.

Within a couple of years the utopian experiment had failed, but the community of keen scholars continued to flourish there on more conventional terms. Being relatively isolated in Indiana, the community had to have its own library and scientific collections, including an excellent mineralogical collection with many reference specimens of European rocks. And it was from here that one of the first successes of American geology arose. Among those who went with Maclure was Thomas Say, a protégé of Maclure, now both a skilled entomologist and one of the leading experts in the identification of living and fossil shells. Say had taken it upon himself almost single-handedly to raise the level of American natural science to that of Europe.[2]

The first task in any geological surveying project is to identify the kinds of rocks through their detailed mineral makeup. Cleaveland had provided a mineralogical basis for identifying American rocks, at least in the Atlantic region. Next, as William Smith established, comes the work of stratigraphy—of comparing the signature of each layered rock type with others, regionally, nationally, or worldwide. And that depended on the fossils. As was often the case in the history of American geology and paleontology, deciphering the complexities of American strata and their relationships to those of Europe did not involve discoveries of strange and wonderful fossil vertebrates; rather, it was about finding the more lowly shells and other invertebrates. Say was the first American to extol the importance of William Smith's methodology—in an article on fossil shells in the first volume of Silliman's new journal.[3]

For a decade, however, nothing much happened on this front, until Lardner Vanuxem, a Philadelphian who had trained at the Ecole des Mines in Paris and had recently retired from teaching chemistry and mineralogy in South Carolina, began an intense study of the geology of New York, Ohio, Kentucky, and Virginia for the New York legislature (before the State Geological Survey had been organized). In 1828 he started to pull his thoughts together in a series of notes for a paper on the "Secondary, Tertiary, and Alluvial formations of the Atlantic Coast of the United States." When he departed for a long trip to Mexico, Samuel Morton edited Vanuxem's notes on the fossil invertebrates, principally an abundance of shells, to produce the first study using William Smith's methods to distinguish strata in the Atlantic region and to correlate them with European beds.[4]

As Vanuxem and Morton had noted, the surface deposits of the Atlantic states had "by most writers been referred to the *Alluvial* as constituting a single deposit; while by others they have been designated by the general name of *Tertiary*." Vanuxem and Morton instead demonstrated the existence of Secondary, Tertiary, and Alluvial formations and showed the relationships between the Secondary Cretaceous beds of America (principally the greensands and marls of New Jersey) and those of Europe. This was not only a milestone in the development of American geology and paleontology—it opened the way for future paleontologists exploring the vast array of Secondary beds in the West.

Soon, across the Union, state after state realized the importance of surveying its geological resources, both in terms of learning about soil types for agriculture and discovering commercially useful minerals—everything from building stone to coal, iron ore to limestone, and not forgetting gold and silver, of course. After some preliminary efforts in North and South Carolina (1823 and 1824), the first state geological survey was organized in Massachusetts in 1830, followed quickly by Tennessee and Maryland (1831), New Jersey, Connecticut, and Virginia (1835), and New York, Maine, Ohio, and Pennsylvania (1836).[5] At the same time, the federal government began to commission its own surveys. Its first venture was conducted by none other than George Featherstonhaugh, who styled himself "U.S. Geologist." He explored the mineral deposits of "the elevated country lying between the Missouri and Red Rivers, known under the designation of the Ozark Mountains," in 1834.[6]

Soon David Dale Owen of the New Harmony community (son of founder Robert Owen) was commissioned to make a geological survey of Indiana.[7] In 1839 the Treasury Department of the federal government recognized the need to survey the lands that it owned if they were to be sold or leased for mining, and Owen became the first director of a fledgling United States Geological Survey. Shortly thereafter he was commissioned to survey the mineral potential of Wisconsin, Minnesota, and Iowa.

For the purposes of our story, however, one of Owen's most important explorations was a repeat journey to Iowa and Wisconsin made between 1847 and 1850. As discussed in the following chapters, it was during this "Wisconsin Survey" that the full potential of the West for the discovery of fossil vertebrates was first realized.[8]

In the meantime, some of the most important geological discoveries had already been made in New York State by Samuel Latham Mitchill, who had started reviewing the geology of the region as early as 1798. The geology of New York turned out to be extremely complex and especially rich in rocks of what in Europe was being called the Transitional Series—except that they contained fossils. Suddenly the state became a hotbed of geology; one of its most colorful students was a man named Amos Eaton.

Eaton was neither a blue-blood Philadelphian nor someone who had fashionably studied science in Edinburgh or Paris. He was a homegrown and largely self-trained lawyer and naturalist from New York State who was guided into geology by Mitchill. His career now seems more film script than fact and a textbook example of what could be accomplished by intelligence and ambition in America. In his work, as in his personal life, Eaton constantly overreached himself, with the result that he put many wild theories and simple errors into print, and made many enemies. He had one prominent sponsor in the form of Stephen Van Rensselaer, however. When he was put in jail at one point for land fraud (probably a trumped-up charge), he even turned that to his advantage, coaching John Torrey, the son of the prison director, in natural history. Torrey later became one of America's greatest botanists. As a reward for exemplary behavior Eaton gained an early release from prison, but it came with banishment from the state of New York. So Eaton went off to Yale, at the age of thirty-one, to learn more geology. Van Rensselaer (by then the governor) later repealed his exile and put Eaton to work at the Rensselaer School, now Rensselaer Polytechnic Institute, in Troy, New York.

Apart from the fact that Eaton was one of the most colorful, if eccentric, often wrong, and sometimes downright obnoxious characters in American science, we can confidently look for good things in him because G. W. Featherstonhaugh hated him so, lambasting him for "his extravagant self-degradation, . . . the confusion he has introduced into American geology." Eaton deserved a lot better than that. His first major work was his *Index to the Geology of the Northern United States* of 1816, for which he claimed to have walked one thousand miles. The *Index* was in every way the successor to Maclure's *Geology* that Silliman had wished for. He was an early pioneer of using William Smith's methods for correlating strata. Eaton's geological map of the United States was directly based on Maclure's, but the level of detail was far greater. He followed up in 1821 with a survey of Rensselaer County, New York, and the following year he surveyed the geology of the Mohawk Valley and Erie Canal region.

Over his career Eaton published expansively. His books on American geology, combining direct observation with some often eccentric ideas about geological processes and the history of the earth, were enormously influential. One of Eaton's greatest achievements was his tutelage of two giants of American science, John Torrey and James Hall. The latter was his student at Rensselaer School, where Eaton spent the last eighteen years of his life. In turn one of Hall's finest pupils was John Strong Newberry, a professor at Columbia and director of the Ohio Survey. A line of scholarly genealogy links Mitchill, Eaton, Hall, and Newberry and the developing sense of an American geology in the first decades of the century to all the successes of the bone hunters of the second half.

Benjamin Smith Barton had been right in predicting that geology would be the first of the sciences to emerge in a unique American form. Through the work of Eaton, Mitchill, and especially James Hall, the stratigraphy of the Transition Series rocks (the part of the geological column that we now label Cambrian, Ordovician, Silurian, and Devonian) and fossils of New York State was not only deciphered but became a textbook example. And not just for a few fellow geologists. Across the eastern states, in the 1830s and 1840s, popular interest in geology and nature grew as well, as the public became fascinated not only with the thought of the mineral riches under the ground in the form of coal and metals, but also with the history of the earth itself. Popular lecturers titillated their audiences with news of an ancient and changed (perhaps even still changing) world of mountains being raised up and then ground down again, of ancient seas where now there is dry land, and of extinct creatures living in the Paleozoic seas—all of it seductively contrary to what the Bible said about the origins of the earth.

The successes of American geologists in deciphering the Transition Series, and the tantalizing problem of the vast horizontal beds beyond the mountains, helped further the growing realization at home and abroad that North America was potentially a huge open textbook for the discovery of the geological structure of the earth and its ancient inhabitants. The result was that in 1841 the greatest European geologist of his age (and, really, of all time), Charles Lyell, came to America to see for himself. Although the ideas of James Hutton had finally become the central theme of geology, it was only after they had been greatly extended by Lyell (another Scot, who lived and worked in London) that geology was placed on a thoroughly firm empirical and theoretical footing. Lyell's three-volume *Principles of Geology*, published between 1831 and 1833, written from the experience of his extensive travels around the world, made geology and paleontology a truly international science. *Principles* finally displaced Cuvier's *Essay* as the seminal text in geological science. Reading Lyell while on the voyage of H.M.S. *Beagle*, for example, convinced Charles Darwin that he should become a scientist, and it was as a geologist that Darwin had his first professional successes.

Lyell proposed to make an eight-to-ten-week trip to America, exploring the geology of New York State and "the country about the Falls of Niagara and Lake Ontario."[9] Lyell was in fact a most welcome guest, but there is always some danger in showing another expert the results of your research if you haven't fully analyzed and published it yourself. Some consternation was caused when James Hall later reported to Silliman: "Mr Lyell had made arrangements with Wiley and Putnam of N.Y. to publish an edition of his Elements [*Elements of Geology*, 1838] with notes and additions to American geology. You may well suppose that I was amazed, and can it be possible that Mr Lyell will take this course after all his repeated declarations that he should publish nothing till after the appearance of our Reports here? . . . piracy in its worst form . . . after having spent my time and money to explain to him the structure of the rocks of NY., in all of which I kept back nothing. . . . By a few weeks in this way he has learned what has cost us years of labor and which he is now to palm upon the Gullible American public as his own. Already the newspapers are lauding him in advance."[10]

It turned out that this alarm was based on a false rumor, but the incident demonstrates a particular difficulty in scholarship and foreshadows one of the greatest tragedies in American science. The entire world of ideas has a problem with priority and "ownership." If scholars and artists could not talk together and share ideas and experiences, progress would be very slow. But with communication comes the danger of what Hall angrily called "piracy." In geology the problem is compounded by the fact that the rocks cannot be hidden away in a studio or a desk drawer but are there for all to see (of course, the advice of a local expert is extremely useful in interpreting them). The episode may also show more than a little of the long-standing American dilemma with respect to Europe: Hall wanted to show Lyell the work he had done but still resented the fact that uninformed people might think they needed a European to explain the subject to them.

In the end, Charles Lyell came to America and was convinced: "We must turn to the New World if we wish to see in perfection the oldest monuments of earth's history, so far at least as it is related to its earliest inhabitants. Certainly in no other country are these ancient strata developed on a grander scale, or more plentifully charged with fossils. . . . [T]he order of their relative position is always clear and unequivocal." And Lyell was writing only about what he knew of New York State and New England. He had as yet no idea of the vast natural laboratory of geology represented by the American West, still only just being opened up to geological exploration and survey.[11]

Bad Lands
No Time for Ideas

We have recently received information from Mr. H. A. Prout, of his discovery of
the remains of a Palaeotherium in the tertiary near St. Louis.

AMERICAN JOURNAL OF SCIENCE AND THE ARTS, 1846

In 1841, toward the end of his first visit to America, Charles Lyell traveled to Philadelphia, where he met a young physician named Joseph Leidy who had already made a reputation both for his skill as an anatomist and microscopist and for his elegant and meticulous scientific drawings. Lyell pointed out to Leidy that although people like Eaton, Hall, Newberry, Say, Morton, and Vanuxem had used their knowledge of fossil *invertebrates* to make major discoveries in the stratigraphy of the New York Transition Series and the New Jersey Cretaceous, and despite the earlier discoveries of the mastodon and great-claw, Americans were not seriously studying fossils. There was much work to be done on collections steadily being assembled from the mid-Atlantic states; the great American West was still completely *terra incognita*. Lyell encouraged Leidy to take up the study of fossil vertebrates to fill those gaps. "Stick to paleontology. Don't bother with medicine. Stick to paleontology. That is your future."[1]

Joseph Leidy is one of the more intriguing characters in the story of American science. If he had never looked at a fossil, he would still be (or ought to be) famous as the discoverer of, among other things, the parasitic nematode worm that causes trichinosis. As a paleontologist Leidy was almost single-handedly responsible for describing the first fossil vertebrates to emerge from the West. He made possible the discoveries of dinosaurs and other scientific wonders that Thomas Jefferson guessed must be lurking beyond the Mississippi. But today he receives little credit for his paleontological contributions, his reputation having been eclipsed by others with a keener feel for self-promotion and, it has to be admitted, a greater sense of adventure.[2]

Although he trained as a physician, Leidy, a member of a fairly wealthy Lutheran family, never really practiced. Instead he found his vocation in teaching, in research at the microscope looking into the material causes and manifestations of diseases, and in documenting the finest aspects of the anatomy of the lower animals. It was his skill as an artist depicting his dissections that first brought him fame, but his genius was purely intellectual. The Academy of Natural Sciences became his spiritual home, and he was its president for many years. In 1845, at the age of twenty-two, he was appointed curator of the Anatomical Museum of the University of Pennsylvania. In 1852 he was appointed professor of anatomy at the university.

Leidy was a private man in an increasingly brash world, but he cultivated a broad range of scientific colleagues and correspondents. One of the most important of these was Spencer Fullerton Baird, formerly professor at Dickinson College in Carlisle, Pennsylvania, who in 1846 was appointed assistant secretary of the brand-new Smithsonian Institution. Under Baird and secretary Joseph Henry the Smithsonian eventually grew to rival and then eclipse the Philadelphia academy as the leading museum and research institution for the natural sciences in America. But throughout Leidy's career, relations between the academy and the Smithsonian were cordial and cooperative.

At the age of forty Leidy married, altogether to the surprise of his friends and relations. His wife, Anna Harden, was from Louisville, Kentucky, and little is known about her except that theirs was a loving relationship. Although childless themselves, they adopted a daughter, Allwina. Leidy was not particularly religious in later years but, striding the streets of Philadelphia with his gentle features and ample beard, he apparently took on a strong resemblance to romantic mid-Victorian representations of Jesus.

As his late marriage suggests, Leidy certainly avoided doing things hurriedly. After his meeting with Lyell, five years passed before he published his first work on fossil vertebrates, but the result was—typically— important. It concerned horses. Everyone knew that when the Spanish first came to the Americas, North or South, they found no horses living there, although the horses they brought with them flourished (another counterexample to Buffon's theory of degeneracy). It seemed self-evident that horses, asses, zebras, and their relatives were originally an Old World phenomenon; they had arisen and diversified in the Old World and never made their way to the New. Then Leidy studied a collection of fossils from Pleistocene riverbank sediments at Natchez, Mississippi, that Professor M. W. Dickeson (a scholar of Indian archaeology) had presented to the Academy of Natural Sciences. Dickeson thought that the prize of the collection was "the entire head and half of the lower jaw" of *Megalonyx*—Jefferson's great-claw—which was certainly a major discovery. But also present in the collection were a number of fossil horse teeth.[3]

In fact, the academy already had in its collections a number of other fossil teeth that were indubitably from one or more ancient species of American horses. And Charles Darwin had collected fossil horse teeth in South America.[4] There were even horse teeth in the Big Bone Lick deposits. Leidy drew the evidence together and established, once and for all, that the horse had originally lived in America but had become extinct in relatively recent times across the whole hemisphere, just as had the mastodon, saber-tooth, mammoth, and great-claw. The short paper describing his conclusions was typically terse and undramatic; he simply allowed himself to admit that the existence of Pleistocene fossil horses was "probably as much a wonder to naturalists as was the first sight of the horses of the Spaniards to the aboriginal inhabitants of the country, for it is very remarkable that the genus Equus should have so entirely passed away from the vast pastures of the western world, in after ages to be replaced by a foreign species to which the country has proved so well adapted."[5]

In the same year that Leidy wrote his first paper on fossil horses, a physician from St. Louis with the quintessentially American name of Hiram A. Prout published a short note about a fossil that had been found in a region of what was then called Nebraska Territory. In Silliman's *American Journal of Science and the Arts* for 1846, he described a fossil jaw that "a friend, residing at one of the trading posts of the Saint Louis Fur Company," had given him.[6] The specimen was a piece of a very large jaw with some teeth, the whole thing, although incomplete, being some fourteen inches long. Prout identified it as a species of *Palaeotherium,* the tapir-like mammal that had originally been discovered by Cuvier in the Paris Basin. It had come from an area known as the White River Bad Lands in what is now southwestern South Dakota and northeastern Nebraska. The term "badlands" comes from the description early French trappers had given of this region—*mauvaises terres a travailler.* It is an extraordinary landscape of deeply dissected

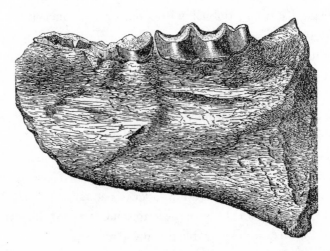

Prout's *Palaeotherium* jaw from the Bad Lands (from the *American Journal of Science,* 1847)

canyons and standing columns of rock. And the Mauvaises Terres turned out to be full of fossils of Early Tertiary mammals and turtles.

These Oligocene badlands encompass an area of some quarter million acres between the Black Hills and the White River. To reach them in the 1840s it was necessary to travel by boat up the Missouri to Fort Pierre and then take horse or mule overland westward, along the fur trappers' routes to Fort Laramie that followed the White or Cheyenne Rivers. Few people visited the Bad Lands, which had little forage for horses and even less water. But Indians (early maps show this as the country of the "Ohenonpa, Minikanye and Sichanga or Brule Dakota Sioux") traveled through and picked up things that they thought the white man might want to trade for. Fur trappers explored everywhere through the region, though in declining numbers because by the mid-1840s the fur trade was in severe decline, and mountain men like the famous Jim Bridger were turning their skills to guiding the ever-increasing flow of migrants westward through the Rockies to California.

True to form, this new American fossil was a giant. Prout at first stated that his new species (which Leidy later formally named *Palaeotherium prouti*) was "one half larger than the *P. magnum* [of Europe]."[7] In the second version of his paper, he went further: "In the largest Palaeotherium, hitherto described, the P. magnum, the [molar] teeth occupy a space scarcely one-third that of the Missouri animal." Even Leidy, in a burst of chauvinism, noted that Prout's *Palaeotherium* "must have attained a much larger size than any which the Paris Basin affords." When Leidy later determined that this animal was not a *Palaeotherium* but the first American representative of a distinct family of very large, quite weird, mammals, some sprouting horns, with small brains and all related not to the tapirs but to the horse and rhinoceros group, he gave it the modest new name of *Titanotherium!*[8]

The following year, a second strange fossil appeared from out of the Bad Lands. Joseph Leidy acquired this specimen through an extensive chain of connections. It found its way to Philadelphia from Leidy's friend Dr. Samuel Culbertson of Chambersburg, Pennsylvania, who had received it from his brother Joseph Culbertson. The Culbertson family included prominent missionaries and soldiers. Joseph Culbertson's son Alexander had joined the American Fur Company in St. Louis in 1829 and became its chief fur trader. His common-law wife Natawista was a Blackfoot Indian, and his easy relations with her people allowed him to play a prominent role as negotiator with most of the natives in the Upper Missouri country (including the resolution of the Blackfeet Treaty in 1855).[9]

Alexander Culbertson had collected the fossil during a trip he made through the Mauvaises Terres in 1843 with Captain Stewart Van Vliet (U.S. Army).[10] In the process of his travels he was also responsible for bringing many other Bad Lands fossils to Leidy. It is even possible also that Prout's original specimen had been collected during the same 1843 trip.

The new fossil from the Bad Lands turned out to be the skull of an ancient relative of the camels.[11] This second specimen was especially interesting because camels were yet another group, like horses, that had not been thought to have been an original member of the American fauna. Leidy named it *Poebrotherium*. Then, the following year, he received from Alexander Culbertson a specimen of yet a third new mammal, which he named *Merycoidodon* (and later renamed *Oreodon*). Over the following years, many hundreds of specimens of *Oreodon* and its relatives were collected in the White River Bad Lands, and it soon became clear that the camels and their relatives must have originated in North America and later become extinct there, like the horses, while flourishing elsewhere.

Leidy's descriptive accounts of these new forms, like his earlier works on American horse teeth, were brief and avoided flowery hypothetical elaborations. Having not seen the original field sites, and given scant information by the amateur collectors, he could say nothing about the geological context. He simply came to the point and then moved on, leaving the reader to draw the broader conclusions about how the original animals might have lived or why they became extinct. This aversion to theory is puzzling in so brilliant a man. Although this was very much the academic style of the day, Leidy took it to extremes. He is reputed to have said once, "I have not time for ideas or making money." It might be that he adopted his (literally and figuratively) bare-bones approach solely through the pressure of time, or because he was naturally averse to speculation (scholarly or financial). It is likely that he saw himself as very much part of an empirical tradition in American science that generally kept the accumulation of facts theory-free. "The most practical of geologists . . . have devoted themselves exclusively to the observation of facts, exhibiting even a fastidious avoidance of hypothesis."[12]

On the other hand, Leidy was certainly very interested in other people's theories. As to whether he had time, it is noteworthy that in the year 1847, in addition to his *Poebrotherium* and two more fossil horse papers, he published fourteen other studies in the *Proceedings of the Academy of Natural Sciences* alone. The subjects ranged from descriptions of new Protozoa and planarian worms, the mechanism by which the locust closes its wings, and the sense of smell in snails to a human cranium from New Holland (Australia). Each of them is marked by the same precision and economy, and all are descriptive rather than analytical studies. In his lifetime he published more than two hundred papers on fossils alone.

One reason for Leidy's reluctance to engage in speculative thinking about his fossils may stem from a bad experience he had as a young scholar. In 1853, he wrote a landmark paper on the "flora and fauna within living animals," based on his researches with the microscope.[13] In his introduction to this monograph on parasitology we see a very different Leidy from the reticent man of the next three decades. He tackled, directly and bluntly, three major issues: spontaneous generation, evolution, and the germ theory of disease.

This was before Louis Pasteur and many others had finally disproved the notion that living organisms spontaneously arose from water, although the idea was already largely discredited. In refuting spontaneous generation, Leidy established quite clearly that microscopic organisms had their own complex life cycles. To make his argument, that all life proceeded from preexisting life, he began by examining the very origins of the earth itself. Life did not exist at first on the early earth, which was far too hot and did not provide the "essential conditions." What, then, was the immediate cause? Remarkably, Leidy stated outright: "There appear to be but trifling steps from the oscillating particle of inorganic matter, to a *Bacterium;* from this to a *Vibrio* thence to a *Monas* (both are now known as bacteria), and so gradually up to the highest orders of life." This was evolution of the kind that had been espoused by Erasmus Darwin at the turn of the century when he projected that life had arisen from chemicals in the sea through an ancestral "filament." The conventional view was, of course, that God had created life in his own kind of spontaneous generation. Leidy was uncompromising, however. In the scientific view, "[special creation] can only be an inference, in the absence of all other facts; and if living beings did not originate in this way, it follows they are the result of natural conditions" (see Appendix B).

A further conjecture in Leidy's monograph concerns the causes of disease. This was a time when ideas about a "germ theory" were being debated as a rival to the idea that diseases were caused by noxious miasmas in the air. Interestingly, Leidy, while at the forefront of describing microscopic parasitic organisms, including those that cause disease, rejected the theory. Trusting contemporary technology, he said the idea that there could exist a class of "animalculae so small that they cannot be discovered even with the highest power of the microscope, . . . capable of giving rise to epidemics, but not discoverable by any means at our command, is absurd." Here he turned out to be wrong, of course, as the work of Koch, Pasteur, and many others later showed. And one of the classic discoveries in public health was that a species of *Vibrio* bacteria was the cause of cholera.

Dennis Murphy at the Academy of Natural Sciences has suggested that Leidy's statements on evolution got him in trouble with the Philadelphia establishment, and very nearly cost him the professorship at the University of Pennsylvania for which he was then competing.[14] If so, Leidy's subsequent reluctance to engage in speculative matters would readily be understood, especially when it was reinforced by his chagrin at having taken the wrong side on the germ theory debate.

By now Leidy had given up his medical practice, for which he had

never seemed very well suited. He took up teaching medical anatomy
and began to spend a major part of his time on fossil vertebrates. In
1848 and 1850 he made trips to England and Europe, where the great
men of the day—men like Owen, Darwin, Milne Edwards, Huxley, and
of course Lyell—encouraged him to continue his paleontological researches. But he did not give up his microscopical work and the study
of parasites. He remained, to his death, a man of remarkable breadth
and seriousness.

A steady flow of specimens came to Leidy from "Nebraska Territory," largely through the efforts of members of the Culbertson family
and other collectors associated with the American Fur Company, and
also from officers of the U.S. Army. For a variety of reasons, however,
Leidy never ventured out West himself until 1872, more than twenty-
five years after Prout's first specimen came to Philadelphia. In part this
surprising omission must reflect the simple problems of logistics. He
had many personal and academic commitments in Philadelphia. To get
to the Mauvaises Terres would have taken him at least four to six
weeks, traveling to St. Louis by the newly completed Baltimore and
Ohio Railroad, and then by river or horseback, or both. He would
have had to be away from Philadelphia for up to three months in order
to get three weeks of collecting in the field. But Leidy doesn't seem to
have had a very adventurous spirit anyway. Roughing it was not one of
his pleasures. He probably also felt that he already had a reliable
source of material. He could not possibly have known what enormous
numbers of specimens would be discovered when systematic pro-
grams of expert excavation in the Dakota Bad Lands replaced simple
surface collecting.

Whatever his reasons for not making his own research expeditions
to the West, and in leaving to others the task (and pleasure) of drawing
geological and evolutionary conclusions from his work, Leidy continued the pattern of making purely descriptive studies of fossil vertebrates in the style that had been set in place by other physicians in the
preceding decades. A different breed of scientist would be needed to
bring the study of fossil vertebrates into what we today see as the intellectual mainstream. And that would depend, first, on there being a more

ample supply of specimens, both in numbers of fossils and diversity of kinds.

David Dale Owen, as we have seen, was a member of Maclure's community at New Harmony to which so many Philadelphia scientists had repaired in the late 1820s. In 1847 the federal government commissioned him to make a second geological and mining survey of Wisconsin, northern Illinois, Iowa, and northern Missouri; his mandate was restricted to a study of what we would now call the Upper Midwest. The prime target was the discovery of coal and iron and the ores of other metals. Owen decided, however, to extend the work westward to the region of the Dakota Bad Lands. He explained this diversion of resources as necessary in order to "connect the geology of the Mississippi Valley, through Iowa, with the cretaceous and tertiary features of the Upper Missouri; a matter very important to the proper understanding of the features of the intervening country which it had been made my particular duty to explore."[15] Owen may also simply have been itching to see the place where the new fossil mammals were coming from.

Owen's plan at first was to reach the Bad Lands by traveling cross-country westward from the main expedition, but he was forced to abandon this idea because of supply problems. So in 1849 he sent his assistant Dr. John Evans by the more reliable route, up the Missouri River. Evans got a boost from the ubiquitous and ever-helpful American Fur Company, who took him and his horses by river as far as Fort Pierre. From there he followed the Cheyenne River to Sage Creek, where he collected Cretaceous fossil invertebrates before pressing farther west. The report of the journey later published under Owen's name colorfully described the arrival at the Bad Lands:

> After leaving the locality of Sage Creek . . . and proceeding in the direction of White River, about twelve or fifteen miles, the formation of the Mauvaises Terres proper bursts into view, disclosing . . . one of the most extraordinary and picturesque sights that can be found in the whole Missouri valley. . . . [T]o

the surrounding country . . . the Mauvaises Terres present the most striking contrast. From the uniform, monotonous, open prairie, the traveller suddenly descends, one or two hundred feet, into a valley that looks as if it had sunk away from the surrounding world; leaving, standing all over it, thousands of abrupt, irregular, prismatic, and columnar masses. . . .

From the high prairies, that rise in the background, by a series of terraces or benches, towards the spurs of the Rocky Mountains, the traveller looks down into an extensive valley, that may be said to constitute a world of its own, and which appears to have been formed partly by an extensive vertical fault, partly by the long-continued influence of the scooping action of denudation. The width of this valley may be about thirty miles, and its whole length about ninety, as it stretches way westward, towards the base of the gloomy and dark range of mountains known as the Black Hills.

Whoever wrote this section of Owen's final report made magnificent use of one of the most familiar analogies (close to being a cliché) of topographic description: the evocation of cliffs and ravines as the architecture of cities and castles. "The traveller threads his way through deep, confined, labyrinthine passages, not unlike the narrow, irregular streets and lanes of some quaint old town on the European continent. . . . [O]ne might almost imagine oneself approaching some magnificent city of the dead. . . . [T]he realities of the scene soon dissipate the visions of the distance. . . . [A]round one, on every side, is blank and barren desolation, . . . the scorching rays of the sun, . . . unmitigated by a breath of air, or the shelter of a solitary shrub."[16]

Now preserved as the Bad Lands National Park, the landscape has this effect on all visitors. The architect Frank Lloyd Wright in 1935 had an equally personal response: "I was totally unprepared for that revelation called the Dakota Bad Lands. . . . [W]hat I saw gave me an indescribable sense of mysterious elsewhere—a distant architecture, . . . an endless supernatural world more spiritual than earth but created out of it."[17]

If the landscape seemed barren and desolate to Dr. John Evans, however, his paleontological discoveries soon offset the oppressive, surreal
surroundings: "The drooping spirits of the scorched geologist are not
permitted to flag. . . . [F]inal treasures . . . embedded in the debris . . .
in the greatest profusion, organic relics of extinct animals. All speak of a
vast freshwater deposit of the early Tertiary period." Evans had discovered the mother lode.

In the formal report of his survey, Owen not only continued the development of powerful images of the great ancient lakes and seas of western America but also began the practice of describing the geology and
landscapes in dramatic terms that would appeal to the lay reader. And
his colorful evocations of the geological history of the region were in-
fused with patriotic (Jeffersonian) comparisons with Europe.

All the strata composing this formation have been a succession
of sediments or precipitates consolidated at the bottom of the
ocean. Alternating with these beds there are also other inter-
stratified, filled with the bones of quadrupeds which have perished on the banks and near the mouths of rivers, whence they
have been swept into estuaries and bays, and entombed. . . . At
the time these fossil Mammalia of Nebraska lived, the oceans
ebbed and flowed over Switzerland, including the present site
of the Alps, whose highest summits then only reached above its
surface, constituting a small archipelago of a few distant islands
in the great expanse of the tertiary sea.

[When] these singular animals roamed over the Mauvaises
Terres of the Upper Missouri, the configuration of our present
continents was very different from what it now is. Europe and
Asia were then, in fact, no continents at all, and up the valley of
the Mississippi as high as Vicksburg, was yet under water.
Mount Aetna . . . was yet unformed, and the fertile plateau [of
Sicily] was still deep under the tertiary Mediterranean Sea.

There was only one man in America to whom the fossils that Evans
collected could be sent for description: Joseph Leidy. The official

publication of Owen's survey included a long monograph by Leidy titled "Description of the Remains of Extinct Mammalia and Chelonia from Nebraska."[18] In this first of many such reports Leidy described eight species of mammals, all new, redescribed Prout's *Palaeotherium*, and added four new species of fossil turtles. His dry language—"the region of Mauvaises Terres of the United States appears to be as rich in the remains of Mammalia and Chelonia of the Eocene period as the deposits of the same age of the Paris basin"—contrasts strongly with the lively prose of Owen and Evans.

Even while the Wisconsin monograph was in press in 1850, Thaddeus Culbertson (younger brother of Alexander) was making his own long journey through the Upper Missouri under the auspices of the Smithsonian Institution. Culbertson, who had just graduated from Princeton Theological Seminary, was sent west in hopes that it would cure his tuberculosis. Spencer Baird, assistant secretary of the Smithsonian, arranged for the trip, advising him to "go to the White River country and collect fossils and send them to Leidy, because Leidy is the only person in the country capable of dealing with them."[19]

After traveling with fur company men, collecting many plants and fossil shells across the Missouri country, battling illness, and occasionally feeling energized and rejuvenated by the hot, dry western air, Culbertson made a three-week trip from Fort Pierre to collect in the Bad Lands. His guide from Fort Pierre was Owen McKenzie (son of the famous guide Kenneth McKenzie, "King of the Upper Missouri"), who had also accompanied John James Audubon on his expedition up the Missouri in 1843.[20]

As McKenzie and Culbertson rode into the White River Bad Lands, the topography conjured for them exactly the same imagery as it had for Evans. Culbertson wrote in his journal: "Never before did I see anything that so resembled a large city; so complete was this deception that I could point out the public buildings; one that appeared to have a large dome, which might have been the town hall; another, with a large angular top, suggested the idea of a court-house, or some other magnificent edifice for public purposes; and then appeared a row of palaces, great in number and superb in their arrangements. Indeed, the thought frequently

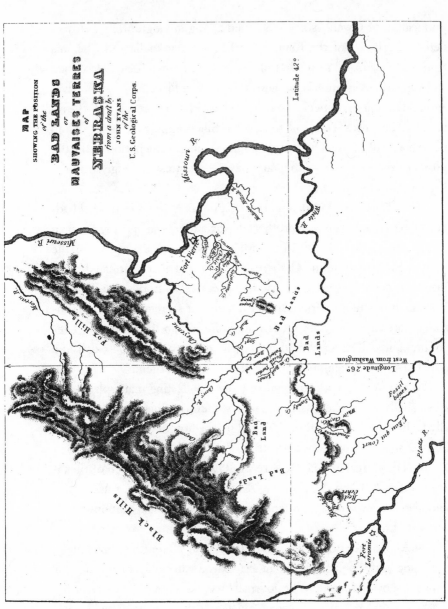

John Evans's map of the White River Bad Lands in the Dakotas, which at the time was known as Nebraska Territory (from David Dale Owen, *Report of a Geological Survey of Wisconsin, Iowa, and Minnesota; and Incidentally of*

occurred as we rode along, that we were approaching a city of palaces; with everything upon the grandest scale, and adapted for giants."

Like Evans, Culbertson found the landscape oppressive: "Fancy yourself on the hottest day of the summer in the hottest spot of such a place without water—without an animal and scarce an insect astir— without a single flower to speak pleasant things to you, and you will have some idea of the utter loneliness of the Bad Lands." But like Evans he also immediately saw the scientific potential: "We reached the place where the petrifactions most abound. . . . I was shown a number of ugly dark red misshaped masses, these my guide told me are petrified turtles. . . . It appears to me quite certain that slight excavations in some of these hills would develop very many perfect specimens."[21]

After collecting more fossils Culbertson boarded the steamboat *El Paso*, which had been chartered for the annual resupply trip to the trading posts. It had on board some hundred trappers and hunters and two hundred tons of ammunition and stores. The *El Paso* managed to reach a point up the Missouri "some hundreds of miles beyond Fort Union, and higher than any steamboat had ever gone previously," and then turned back. Culbertson returned on the steamer to St. Louis, writing in his diary: "I desired to feel grateful to Divine Providence for my safe return and restored health." But within weeks of his reaching his home at Chambersburg he was dead.

The new material allowed Leidy to revise and expand his Nebraska monograph in 1853 under the title *The Ancient Fauna of Nebraska*.[22] This work shows the extent of his network of informants and collectors. In addition to the specimens from the Owen-Evans expedition, Alexander Culbertson, and Thaddeus Culbertson's 1850 expedition, he also had access to the collection of "Professor O'Loghland of St Louis," and the "small but very excellent collection made by Captain Stewart Van Vliet, of the United States Army" (the fossils Van Vliet had gathered when he accompanied Alexander Culbertson through the Bad Lands in 1843). In the seven years since Prout published his first note on his specimen of *Palaeotherium*, the number of fossil mammals known from the Bad Lands had increased to fourteen species of three major types, plus the five kinds of turtle. "Every specimen as yet brought from the Bad Lands, proves to be of species that became exterminated before the mammoth and mastodon lived," Leidy wrote. Almost all the new forms were large (if only because a large specimen was more likely to be noticed and picked up by the traveler).

That same year, John Evans went back to make further collections in the Bad Lands. Other expeditions followed, and in 1869 Leidy expanded his monograph again, as *The Extinct Mammalian Fauna of Dakota and Nebraska . . . with a Synopsis of the Mammalian remains of North America.*[23] By now Leidy could tally 77 species of North American fossil mammals from all ages, including extinct beavers, camels, bears, and four different genera of horses, virtually all of which he had described and illustrated himself and many of which belonged to groups found only in North America. Among the new forms that Leidy described in this new work was also the first example of a saber-tooth "tiger" known from North America. Within two years the tally of mammals had reached 103 species.[24]

The diverse range of western material now being sent to Leidy for study was remarkable. As word of his work spread, more and more people began to send him specimens. In addition to his work on mammals, in 1856 he received from John Evans the first specimens of fossil fishes from what would turn out to be fabulously rich Eocene beds in the Green River region of Wyoming.[25] There were also fossil sharks from the Pennsylvanian period of Kansas and, although it did not seem an especially major landmark at the time, some new reptiles. In a brief note published in 1856, and later in an 1859 monograph on the Cretaceous-age fossils of the Judith River region of present-day Montana, Leidy described the first dinosaur remains from North America.[26]

Four different kinds of dinosaur were represented in the collection, all only in the form of isolated teeth. There was one obvious carnivore, which Leidy named *Deinodon,* and three apparent herbivores— *Troodon, Palaeoscincus,* and *Trachodon.* They had been collected by an energetic young geologist-explorer named Ferdinand Vandiveer Hayden, of whom we will hear much more in a later chapter. The association between Leidy and this dedicated professional explorer was to be the critical watershed in the discovery of fossil vertebrates in the West.

The second event of supreme importance was the formulation of a viable theory of evolution. Given his own early interest in evolution, it is not surprising that Leidy understood instantly the importance of Charles Darwin's theory of evolution by natural selection when it first appeared in 1859. Immediately on receiving a copy of *On the Origin of Species*, Leidy wrote to congratulate Darwin in unusually effusive terms, using the sort of language that was so notably lacking from his paleontological writings: "Night into day . . . I felt as though I had groped about in darkness, and that all of a sudden, a meteor flashed across the skies." In Europe, however, the opportunities that evolution presented for paleontology were not immediately appreciated. Darwin wrote back to Leidy, "Your note has pleased me more than you could readily believe: for I have during a long time, heard all good judges speak of your palaeontological labours in terms of the highest respect." Darwin continued: "Most Palaeontologists (with some few good exceptions) entirely despise my work. . . . All the older Geologists . . . are even more vehement against the modification of species than are even the Palaeontologists."[27] Louis Agassiz at Harvard remained opposed to Darwinian evolution all his life, insisting that the changing diversity of life on earth was the result of the working-out of a divine plan.

Evolution explained the changing patterns of origination of new forms and extinction of old ones that the fossil record so clearly documented. Leidy's careful descriptions of fossils from the West provided an increasingly important part of that documentation. Just before the publication of Darwin's book, however, perhaps Leidy's greatest personal triumph as a scientist came from a discovery closer to home.

TWELVE

Dr. Leidy's Dinosaur

Dr. Leidy states that the bones . . . were those of a huge herbivorous saurian. The animal was closely allied to the great extinct Iguanodon of the Wealden and lower Greensand of Europe; the genus is, however, different, and for it the name of Hadrosaurus is proposed.

WILLIAM PARKER FOULKE, 1858

From the late 1830s onward, a farmer named John Hopkins, while digging for marl (phosphate-rich rock) to spread on his land in Haddonfield, New Jersey (some eight miles from Philadelphia across the Delaware River), had occasionally turned up huge fossil bones. The bones seem all to have been vertebrae, plus possibly a shoulder bone, but visitors to the site had carried off what had been found. Then, in 1858, William Parker Foulke, a friend of Leidy who had a summer house near the Hopkins farm, heard about these fossils and tried to retrieve some of them. When that failed he got Hopkins to open diggings in the marl pit again. In Foulke's account: "It was no easy matter to find the pit itself; and after it had been found, many trials must be made to identify the exact place where bones had been discovered. At last success crowned the undertaking. In the west wall of the pit, under eight feet of surface rock, lay a thin stratum of decomposing shells, and two feet beneath this another, in and on which were found a pile of monstrous bones, enveloped in the rough, tenacious, bluish marl, from which they were carefully extricated with a knife and trowel."[1]

Leidy came out to the site, and it was soon clear to him that they were dealing with the skeleton of a very large reptile—a dinosaur, in fact. This was one of the first scientific exhumations of dinosaur bones, with "drawings and measurements being made of each bone where it lay, to prevent embarrassment in the story," as Foulke recounted. "Wrapped in coarse cloth and straw, they were dispatched first to Leidy's office at the University of Pennsylvania and then moved to the Museum in the Academy." The remains included bones of the pelvis, thigh, lower foreleg (ulna and radius), upper foreleg (humerus), fore and hind feet, plus vertebrae from the trunk and tail. There was no skull except for a jaw fragment, but (by a generous estimate) something like three-quarters of the skeletal structure was represented (if not from both sides, left and right).[2]

When assembled together, the remains of the skeleton turned out to belong to a completely new kind of reptile, some twenty feet long. Leidy gave it the name *Hadrosaurus foulkii*—not for Haddonfield but for *(H)adros,* meaning bulky, and for his friend Foulke—and announced that the animal was related to the herbivorous dinosaur *Iguanodon* that had been previously discovered in Europe. Leidy's *Hadrosaurus* was one of the most important finds of the century. It was the first more or less associated skeleton of a dinosaur, allowing the appearance of the whole animal to be reconstructed for the first time. From the arrangement of the pelvis Leidy decided that it was a bipedal animal that stood erect on long hind legs, with relatively short forelegs. This was a revolutionary conclusion, especially given the size of the beast. Later it would be shown that very many other dinosaurs, including Buckland's *Megalosaurus* and Mantell's *Iguanodon,* were also bipedal. *Hadrosaurus* was the first dinosaur to demonstrate what we now see as one of the two classic dinosaurian poses (the other being the elongated four-footed "brontosaur" type).

Leidy's descriptions of the pelvic structure also helped Thomas Henry Huxley demonstrate for the first time a close relationship between dinosaurs and birds, an idea that still reverberates in the paleontological community because of the implication that dinosaurs are not extinct after all, but live on in the form of modern birds. Typically, however, Leidy did not make this leap, leaving "ideas" to others. In this case, as it turned out, that would be at some great cost.

Leidy was completely aware of the sensational nature of his "discovery" of *Hadrosaurus,* as were the authorities at the Academy of

Natural Sciences, where the bones were put on display. They were not yet restored as a complete skeleton but were simply laid out, some in wooden boxes, for visitors to marvel over. A more dramatic disposition of the remains had to wait for ten years until Benjamin Waterhouse Hawkins arrived in America. Hawkins had sculpted dinosaur replicas for Sydenham in South London, when the Crystal Palace from Britain's Great Exposition of 1851 was moved there. Now, he had been commissioned to create a great Paleozoic Museum of fossil life that was to be installed in New York's Central Park.

Hawkins traveled about the country looking for suitable material for his artist's eye. When he came to Philadelphia he saw that *Hadrosaurus* represented a golden opportunity. He quickly obtained permission to make a restoration of the dinosaur as if it were the skeleton of some modern creature.

This would involve not only a great deal of work but much imagination and, not incidentally, a weight of iron to hold the whole thing up. Fossil bones, being made of stone, are much heavier than modern bones. The obvious tactic was to use the existing bones to model in plaster the missing elements from the opposite side, and then sculpt the rest according to the best zoological information available. This is what has been done with dinosaurs ever since. Hawkins's final mount of *Hadrosaurus* carefully preserved in a different color the distinction between the reconstituted parts versus the bones that had actually been found. The animal was posed in a lifelike manner, reaching for a rather unconvincing (iron) tree on which it was partly supported. The head was fabricated as a rough copy of an iguana head. This was the first time since the Peales that anyone had attempted such a reconstruction of a fossil skeleton on this scale. By modern standards it was clumsy, even crude. In contemporary terms it was splendid: huge and sensational. A grown person could walk under its rib cage. It completely captured the essence of the growing view of dinosaurs as lumbering brutes. When it went on view at the brand-new building that the Academy of Natural Sciences had just moved into, it caused such an uproar—it is said—that admission charges had to be instituted to keep the crowds manageable.

Part of Hawkins's deal with the academy was that he would keep his molds in order to make a replica for his Paleozoic Museum. He set to work in his studio on a number of projects, including a similar restoration of a smaller dinosaur named *Laelaps* that Leidy's junior colleague at the academy (Edward Drinker Cope, of whom much more later) had found more recently in New Jersey. Alas, Hawkins's great Paleozoic Museum project fell afoul of New York City politics and the notorious "Boss" Tweed. The studio was razed by vandals (vestiges of it may still exist buried under the grass of the park), but Hawkins managed to save the molds for *Hadrosaurus*. From these he made several plaster copies of his splendid mount. One was commissioned for the Smithsonian Institution, which exhibited it at the Centennial Exposition in Philadelphia in 1876. Later it was moved to Washington before being sold to the Field Museum in Chicago. It eventually fell apart around the turn of the century. Another copy went to Princeton University in 1874. In 1876 a third copy went to the Royal Scottish Museum in Edinburgh (built, incidentally, on the site of Charles Darwin's student digs), and that one lasted the longest, being still on display until World War I. It was the first mounted dinosaur skeleton on public exhibition in Europe (Buckland's *Megalosaurus* had been on view at Oxford's new University Museum since 1860, but only as separate pieces).[3]

In addition to its scientific importance, *Hadrosaurus* thus became the first dinosaur skeleton to be restored in an authentic life pose, the first to be copied and distributed to other museums, and the first to be displayed in Europe. In Philadelphia, the academy exhibited the original mount until sometime in the 1930s. Copies of the bones were remounted in a more modern pose in 1985, and the display was revised again in 1998.

THIRTEEN

Ferdinand Vandiveer Hayden

My explorations of the country west of the Mississippi began in the spring of 1853, prior to the organization of Kansas and Nebraska as Territories, and I have watched the growth of this portion of the West year by year, from the first rude cabin of the squatter to the beautiful villages and cities which we now see scattered so thickly over the country. We have beheld, within the past fifty years, a rapidity of growth and development in the Northwest which is without parallel in the history of the globe. Never has my faith in the grand future that awaits the entire West been so strong as it is at the present time, and it is my earnest desire to devote the remainder of the working days of my life to the development of its scientific and material interests, until I shall see every Territory, which is now organized, a State in the Union. Out of the portions of the continent which lie to the northward and southward of the great central mass, other Territories will, in the mean time, be carved, until we shall embrace within our limits the entire country from the Arctic Circle to the isthmus of Darien.

F. V. HAYDEN, 1871

The frontier had rapidly moved westward, and the discovery of gold in California at John Sutter's sawmill on the American River in 1848 produced a flood of westward migrants. The scientists followed, although not all of American science was conducted in the restrained and gentlemanly style practiced by the good Dr. Leidy. Two men in particular, James Hall and his teacher John Newberry, operated in a far more elbows-out manner. When Hall read Owen's Wisconsin report and saw Leidy's account of the vertebrate fossils of the Bad Lands, he determined that he would get a share of the action. Hall was then only forty-two but, like Leidy, had too many other responsibilities to be able to hazard an expedition to the Dakota Territory personally. Ferdinand Vandiveer Hayden, the man Hall first selected to collect for him in the

Bad Lands, really chose himself. It was a good decision. Hayden became one of the great scientific explorers of midcentury, both collecting fossils and using them to decipher the structure and geological history of the beds in which they were found.

After Hayden's initial work for Hall, most of the vertebrate fossils that he collected went to Leidy, with whom he contributed enormously to the early paleontology of the great deposits of fossil vertebrates in the American West. But he was in every way the opposite of the quiet Philadelphia professor of anatomy: a gifted hands-on collector and field surveyor, he became a skilled synthesizer and popularizer rather than a dry recorder of facts. Brilliant, calculating, thrusting, insecure (perhaps even paranoid), he managed harshly to polarize most of his wide range of acquaintances into friends and foes. And unlike Leidy he had unbounded energy for adventure. From his first tentative days in the Upper Missouri, his appetite for exploration and natural history never waned.

In 1853 Hayden was a young man, recently graduated from Oberlin College and penniless. He was born in 1829 to a rather shadowy family that lived first in Westfield, Massachusetts. When his parents divorced (or had they ever been married?) Hayden was sent off to live with an aunt in Ohio. Entering Oberlin College, Hayden at first seems to have read theology and then fell under the influence of the Natural Theology movement, which offered the chance to combine an interest in theology with a practical bent for natural history.[1]

After a spell as a schoolteacher, Hayden realized that he needed more training to advance in a profession, and the only real way to advance would be to obtain a medical degree. At that time a great deal of any medical curriculum was devoted to natural history, which suited him perfectly. He began studying in Cleveland, where he met two important scholars, Jared Kirtland and John Strong Newberry. He later enrolled at Albany (New York) Medical College and continued to build up a network of connections that might help him in his true avocation. In Albany he made sure that he met the eccentric but powerful James Hall, the New York State Geologist.

Hall has been described as "at times raving mad, vengeful, deceitful, and always suspicious, pugnacious," a man who "marched under the banner of self-righteousness."[2] For all that, he was a brilliant geologist, and Hayden, not exactly a shrinking violet himself, cultivated his friendship. Dissatisfied with medicine and looking for a way to develop his interests in natural history, Hayden had been pressing Hall and everyone else he could think of to help him find employment. As he wrote to Spencer Baird on February 16, 1853: "I could endure cheerfully any amount of toil, hardship, and self denial . . . to labour in the field as a naturalist. . . . I could live as the wild Indian lives . . . without a murmur." And on March 5, 1854: "My love for natural History is so great that I hardly feel any disposition for anything else."

Soon after asking Hayden if he would like to make the expedition west into the Bad Lands, Hall realized that he would need to send someone with more experience who was perhaps a little less wild-eyed. He chose his assistant Fielding Bradford Meek, a man with considerable experience in collecting fossils, plants, and invertebrates who had taken part in David Dale Owen's survey of Wisconsin. Given Hall's record for abusing his assistants, Meek must have been delighted to be able to leave town. He agreed to lead the expedition with Hayden as his assistant; Hayden was evidently keen enough to go west that he accepted the demotion.

Perhaps as a harbinger of events to come, when different individuals would rival each other for access to the good fossil-collecting sites, Meek and Hayden arrived in St. Louis only to discover that Dr. John Evans was already there and outfitting an expedition. He planned to explore the Upper Missouri on his way, eventually, to Oregon. Evans had been dispatched because Baird at the Smithsonian had learned that a party of German explorers was heading for the Bad Lands. But Baird had forgotten to tell Hall of this plan. The Academy of Natural Sciences was funding one-third of Evans's trip in exchange for specimens.

Something of a row developed about who was going to collect where. The standoff was partially resolved by none other than Professor Louis Agassiz from Harvard University, who was lecturing in St. Louis at the time (and perhaps also wondering whether it would be worth entering the Bad Lands stakes himself). In the end, the Evans and Hayden parties traveled together and were never really in conflict.

The "Germans" turned out to be the young Prince of Nassau with a companion and two servants. They left the boat early in the trip to collect fishes in the lower Missouri regions for Agassiz, and never went to the Dakota fossil lands.

Exploring from Fort Pierre to Council Bluffs, Meek and Hayden collected a large number of fossils. On their return, the collections were divided up. The vertebrate fossils went to Leidy to describe. The invertebrates (mostly shells), being most useful for the purposes of unraveling the layered sequences of rocks, or stratigraphy, of the region, were retained by Meek and Hayden. This collegial and effective division of labor remained in place for the next seventeen years. Oddly, Hall never published directly on the results of this expedition, although he read two accounts of them to the 1854 meeting of the American Association for the Advancement of Science.[3]

Now Hayden was really hooked and keen to go back out west. But he had neither a job nor any personal money, so he immediately started to float schemes for sponsorship. His first idea was that the Philadelphia academy, or even Leidy and other individual collectors, would stake him by advancing funds in return for the collections of fossils, animals, and plants that he expected to bring back. "Professor Leidy," he wrote, "Dear Sir, Pardon me for the liberty I take in addressing this to you, but I am anxious to obtain your opinion and counsel. . . . I think I could contribute much to Natural History by collections. . . . I would like to ask . . . if I could make some explorations for the Academy or do anything to defray a part or all of my expenses."[4]

But the Philadelphians were far too circumspect. So Hayden fell back to his second plan. The Indian agent at St. Louis, Colonel Alfred Vaughn, had offered to fund Hayden for two seasons of collecting, in return for which Hayden would catalogue all the collections and keep half. It was not a bad deal, and when it became clear there were no other options, Hayden accepted.

In spite of "the want of proper facilities for exploration, the wild and desolate character of the country, [and] the numerous bands of roving Indians," Hayden explored along the Missouri as far as Fort Benton and up the Yellowstone to the Bighorn River. He amassed a large collection of fossils, including many new vertebrates. It was during work at the confluence of the Judith River and the Missouri in modern Montana that Hayden found several localities with fossils that turned out—under Leidy's eagle eye—to be the teeth of dinosaurs. "I have also some vertebrate remains from the Upper Mo which I collected this summer which I think will interest you much," Hayden wrote to Leidy. "I find that the Bad Lands of the Judith are scarcely less interesting than those of the White river and I found some things which I hope will please you and contribute to science. They are mostly single teeth and vertebrae but will reveal a new feature in the geology of the Upper Mo."[5]

As already noted, Leidy described them and named them, but they remained an interesting sideline for more than twenty years while more exciting western discoveries were made of other reptiles (like mosasaurs) and many different kinds of mammals. Hayden did not find any more dinosaur remains on the Judith, and neither he nor Leidy could have had any idea of the wealth of dinosaur fossils waiting to be found in Wyoming and Colorado.

For Hayden, the work in the Judith River region was important in marking his real coming-out as an interpreter of field geology. It launched Hayden into the project to which he would become such a major contributor over the next twenty years: unraveling the structure and the history of the Cretaceous and Tertiary—those famous "horizontal" beds—of North America, and relating them to those of Europe. And although he was much the younger man, Hayden helped shape Leidy's career in one definitive way (apart from supplying him with fossils). Hayden's stratigraphy gave Leidy's paleontology its first scientific foundation and framework. At last, American students of fossil bones could catch up with their colleagues who studied fossil invertebrates and transcend old-fashioned natural history for natural science.

Hayden and Meek together helped give authority to a new and powerful image of the land between the Alleghenies and the Rocky Mountains. They documented that in the West those "horizontal" beds must have been laid down in a vast set of seas and inland lakes, waxing and waning over long periods of time and undisturbed by the kinds of convulsions in the earth that later produced the Rocky Mountains to the west. These ancient waterways were therefore symbolically the equivalent, and literally the origin, of the modern plains that appeared to so many travelers, from the early Spanish explorers onward, as a veritable ocean of grass, where the constant wind created waves tossing and rolling over a thousand miles of level terrain. There was nothing like this enormous level expanse, below the ground or at its surface, in Europe, or indeed on the East Coast of North America.

To be sure, they got a lot of things wrong, but Hayden showed a marvelous aptitude for pulling together a history of geological structure and change out of the multiple sections and exposures of the rocks. And he had an artist's eye for fleshing out his syntheses in dramatic terms. Hayden not only helped understand the geological history of all those level beds beneath the plains, whose vast scope and improbable consistency had so intrigued Barton, Maclure, and others; he also found the language to express the wonder of that history.

Hayden was also savvy enough to know that, while the geological and geographical surveying work was crucial to his job prospects and could attract a broad audience, the most glamorous side of it involved the fossils. The significance of western vertebrate fossils had been apparent from the moment David Dale Owen published his Wisconsin report, with its profusely illustrated section by Leidy documenting wondrous new kinds of ancient mammals from the American hinterland. By constantly amassing important collections and having Leidy describe them in special sections of his own survey reports, over the years Hayden was able to generate and sustain a level of interest in his work that would not have been possible with the rocks alone. Leidy therefore became extremely important to the continuance of Hayden's work and career, and Hayden became equally important for Leidy and for Leidy's successors. Almost all of Hayden's survey reports were accompanied by monographs from Leidy on the fossil vertebrates. These were of relatively lesser value in making detailed stratigraphic correlations of the American Cretaceous or Tertiary deposits with those of Europe, but they had the dramatic effect of revealing a diversity of ancient life in America unlike that seen (as yet) anywhere else in the world.

Later, Hayden would extend his popular work even further with ventures into "sun portraits"—photography. The distinguished photographer

William Henry Jackson was officially attached to Hayden's United States Geological Survey from 1870 to 1878 and made some two thousand images of western geology, including classic pictures of the Yellowstone region. Hayden considered himself the godfather of the movement to create national parks, such as Yellowstone, and Jackson's photographs were crucial in showing the glories of the area to a wide American public, most of whom had no chance of visiting in person. Hayden also published a wonderful pioneering book of scenes from the Rocky Mountains taken for the Union Pacific Railroad by another photographer: Andrew Joseph Russell.[6] Another lifelong interest was the languages and customs of the American Indians. In all of this, Hayden's acumen for self-promotion helped set a style that has not fully been abandoned by his modern successors, either the scientists themselves or the media writers who report on their work.

When he returned to the East in early 1856 with his half of the Vaughn collections, however, Hayden once again had no job and therefore little influence, although the Philadelphia academy did purchase many of his specimens. He began to write up scientific accounts of his discoveries with Meek, where his major asset was not just his evident ability as a field geologist but also his growing facility to describe his work in glowing prose for his audience. And this brought him to the attention of Lieutenant G. K. Warren of the Army Corps of Topographical Engineers.

By 1853, the eastern railroads had reached St. Louis, and that same year the Federal government authorized $150,000 for surveying possible routes through the contorted mountains and inland basins and deserts of the West. This appropriation was only the beginning of decades of work in which government surveys explored, mapped, and interpreted thousands of square miles of largely unknown territory (and untouched scientifically). The result was an understanding of geology that remains a landmark of field study, which depended in no small part on a hugely successful and influential set of partnerships between Hayden and the other surveyors, on one hand, and paleontologists like Leidy, Meek, and the paleobotanist Leo Lesquereux on the other. (A Swiss immigrant,

Lesquereux was America's first real expert on fossil plants. Failing to find employment in the New World as a scientist, he supported himself partially as a watchmaker while working occasionally for the New Harmony survey and Hayden and collecting for Louis Agassiz at Harvard.) The older geologists of the eastern states, like Hall and Newberry, began to find themselves outflanked by these newcomers and confined their work to the East and Midwest, where, to be sure, there was plenty to do.

With the exploding effort of western surveying being conducted in connection with routes for transcontinental railroads, Hayden had found the perfect berth for someone of his interests. Soon after returning to St. Louis in 1856, he was sent by Warren to assist in exploration of the Missouri from Fort Pierre to well north of the Yellowstone River, and along the Yellowstone itself. Hayden was ideal for such work as he could function both as expedition physician and naturalist. The next year he worked for Warren on an expedition to the Black Hills, traveling from Sioux City up the Loup Fork River and back to Fort Leavenworth, Kansas. Among other fossil discoveries were extensive Pliocene beds in the Loup Fork region and the Miocene beds of the Niobrara River region.

In 1858 Hayden and Meek were in Kansas together, exploring westward from Fort Leavenworth along the Solomon and Smoky Hill Rivers and the Santa Fe Trail. In the coal fields of the eastern part of Kansas he collected Pennsylvanian-age fishes that Leidy described. In Kansas Hayden and Meek identified outcrops with typical Permian age fossils and promptly got into a dispute with the State Geologist, George Clinton Swallow, over who said (and wrote) it first.[7] The following year Hayden was attached to Captain W. F. Raynold's expedition to the upper parts of the Yellowstone and the Bighorn Mountains, overwintering on the North Platte River. In 1860 they worked as far as Fort Benton and then back down the Missouri to Fort Union and then to Omaha.

The legend has grown up that the Indians would leave Hayden alone because they thought he was mad. The Sioux were supposed to have given him the name "he who picks up stones running." But while various Indian tribes might well have thought Hayden's behavior was crazy, the idea that he somehow thus acquired immunity from attack has no basis (it is, indeed, a story that is often told about fossil collectors in wild places).[8] Whatever the truth of the myth, the image of a man who, out in the field, picked up stones "running" conjures up someone of great energy (and perhaps also a healthy appreciation of the dangers of hanging about when Indians were in sight).

The partnership between Leidy and Hayden continued, to great mutual profit, for twenty years, interrupted only by the American Civil War, when most formal paleontology stopped and they both became military doctors. During the war years, however, Leidy finished another great monograph, *The Cretaceous Reptiles of the United States*.[9] This work was well received in the United States, although Leidy received a bitter blow from Europe in the form of a cruel review by the anonymous author "H," which read in part: "Altogether we must, while expressing our thankfulness for the memoir, such as it is, say that it is the least able contribution to palaeontology that we remember. Its best praise is that it contains no quackery; its worst condemnation is that it contains no science. It will always be valuable for its plates." If this were not condescension enough, the author reserved his most wounding remarks to the very end: "We look forward with hope, that remains so precious will some day be elucidated, and doubt not that the accomplished author of the Arctifera and discovery of Laelaps, will make available to scientific students the descriptions of his Philadelphian brother Professor."[10]

There could be no doubt to whom "H" was referring. The discoverer of *Laelaps* was a quarrelsome young associate of Leidy's at the Academy of Natural Sciences named Edward Drinker Cope, who was beginning to make a name for himself for his expertise with living fishes, amphibians, and reptiles, and some discoveries of fossil reptiles in New Jersey. At first Leidy, and many others, were sure that "H" must have been Thomas Henry Huxley, although it was not clear why Huxley would have wanted to act so uncharitably toward Leidy or so flatteringly toward Cope. In 1869 J. S. Newberry asked Huxley directly and was able to write to Leidy that although the review had produced "great surprise and regret among us when it was published, I can have the pleasure of assuring you that Professor Huxley did not write the article and knew nothing of it until it appeared and farther that he has no sympathy with the views or spirit of the article and condemns it as earnestly as we do. I have his authority for saying this to you. I suspect the avenues and the facts or assertions of the obnoxious article emanated from a source much nearer home."[11]

If "H" was in fact the initial letter of the reviewer's true name, then the possibilities are few. It seems inconceivable that "H" could have been Hayden, Leidy's closest geological colleague. Newberry's phrase "closer to home" would therefore appear to point to James Hall, but the hostility between Newberry and Hall requires us to be cautious about that conclusion. What hand young Cope had in the matter is unknown, but that he had no *direct* part is shown by the fact that he also thought Huxley was the author. Cope certainly liked the result. He wrote to his father: "If thee takes the Geological Magazine, London, thee will find a review by Huxley of Leidy's work on fossil reptiles, which is of the severest kind. Not handsome or Christian, yet I cannot help feeling some gratification, as it does not equal in unhandsomeness the manner in which both Leidy and Hawkins have treated me, and whom this article has dumbfounded. He takes occasion to excite Leidy's jealousy by complimenting me. However his strictures on L. are mostly deserved, and I am glad to see things estimated at their true value. This will become more apparent when my paper is printed."[12]

The complaint that Leidy's work was simply a compendium of facts with no analysis and no theory in part reflects the very different maturation of European and American science. Leidy had always seen his role with respect to the spectacular discoveries of the West as being to inform, not to interpret. In reviewing his *Hadrosaurus* for the *Cretaceous Reptiles* monograph, for example, Leidy did not develop his views of the similarity of the dinosaur and bird pelvis. In yet another summary of the fossil mammals of Dakota and Nebraska in 1869, no doubt in self-justification after the "H" review, he bitterly summed up his approach: "The present work is intended as a record of facts, in palaeontology, as the authors have been able to view them—a contribution to the great inventory of nature. No attempt has been made at generalizations or theories which might attract the momentary attention and admiration of the scientific community."[13]

If the state of knowledge of American fossils and their underlying geology had been as extensive as that for the Paris basin or English Wealden, the comments by "H" about the lack of analysis would have been appropriate. They certainly would not be unexpected if Leidy were writing today. But, for that time, the review by "H" was unfair. It was also personally devastating because, as previously noted, Leidy simply did not have that sort of mind. His view of his scientific role was to lay out the facts as accurately and simply as possible, undecorated by conjecture, and in this he may also have been reacting against the excesses of Harlan, Godman, Hays, and others, to say nothing of Koch, whose extravagant claims had brought paleontology into disrepute.

As a result, his reports rarely became out-of-date or controversial. He was a man, as he had said, who had "no time for ideas or for making money," and in this case was cruelly attacked for it.

After the Civil War, Hayden returned to western exploration, first under the auspices of the Academy of Natural Sciences, traveling through the Dakota and Nebraska country collecting from Tertiary formations. The geological history of the West (from the Missouri River to the Rocky Mountains) was quite different in Tertiary times (starting roughly 65 million years ago) compared with the Cretaceous that he had mostly been involved with so far. Toward the end of the Cretaceous, the Rocky Mountain uplift began, the seas regressed, and their marine sediments became overlain by newer deposits laid down principally in freshwater lakes and riverine environments. There was no single event in which salt seas were replaced with fresh; the geology of the Late Cretaceous shows a back-and-forth process, marine beds (and fossils) alternating with those of freshwater. This made the stratigraphy of the western Cretaceous difficult to correlate with beds of similar age in Europe and the boundary between Cretaceous and Tertiary difficult to judge (as previously noted, Leidy held out the possibility that the freshwater Judith River beds were Tertiary in age).[14]

A number of ancient (and modern) landlocked basins were formed among the mountain ranges of the newly arising Central Rockies. These basins filled with fine sediments from erosion of the nearby mountains and ash from volcanic eruptions. Because fossils were deposited there in quiet water rather than being transported by large rivers flowing rapidly to the sea, the state of preservation of fossil remains in these basins is outstanding. Immensely important fossil collecting areas came to be discovered as the ancient life of, for example, the Bighorn River, Wind River, Powder River, Green River, Bridger, Washakie, Uinta, North Fork, and Wasatch Basins was revealed in a whole series of waterworn badland topographies of buttes and canyons similar to those of the Dakota and Judith River areas. These gave access to formations from Paleocene to Pliocene ages that yielded an enormous number of new fossil vertebrates. Of these, collections from the Eocene-age basins were the most spectacular and scientifically important. Perhaps most familiar to the general public is the Eocene Green River Basin of southwest Wyoming, which has by now yielded literally millions of superbly preserved specimens of fishes alone. Hayden saw "hundred of thousands of perfect impressions of fishes . . . sometimes a dozen or two on an area of a square foot."[15]

Across the whole suite of basins, one finds recorded the history of the origin and early evolution of most of the modern groups of vertebrates, particularly the mammals, which radiated once the Age of Reptiles was over. The quality, quantity, and diversity of the preserved fossils—plants, fishes, even tiny insects—is unsurpassed anywhere in the world.

Hayden's assiduous cultivation of the great and the good—and particularly Baird and Henry at the Smithsonian Institution—finally paid off in 1867, when he was appointed (at a salary of two thousand dollars) to make a survey of "the territory of Nebraska" for the General Land Office. This was not a railroad survey but rather was concerned with the discovery of coal and other useful minerals. Hayden threw himself into this work and produced his usual detailed reports, in which he now modestly titled his operations the United States Geological Survey.[16]

The first Nebraska expedition was such a success that the following year it was extended to Wyoming, Colorado, and New Mexico. While these surveys were also principally concerned with economic geology, Hayden made a special point of looking out for fossils—invertebrates for Meek and plants for Lesquereux—that would continue to help sort

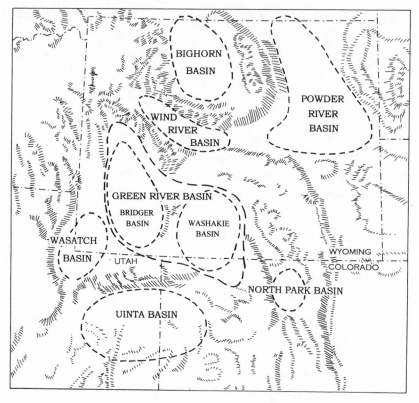

The principal Early Tertiary depositional basins in the Central Rocky Mountain region

out the stratigraphy. And, of course, he collected vertebrates for Leidy's description.

In three seasons, from 1869 to 1871, one of the regions where Hayden concentrated his efforts was the southwestern corner of Wyoming. With its arc of mountains (Gros Ventre, Wind River, Laramie, Medicine Bow, Uinta, and Wasatch Ranges) drained by streams creating the Green River as it flows south into Utah to join the Colorado, this had been one of the richest areas of the fur-trapping era. The tributaries of the Green River cut through the Eocene lake deposits of the central Rocky Mountain depositional basins, revealing—once they were found—one of the world's greatest treasure houses of fossils. Here

Hayden explored the Eocene Bridger Formation (which he named), the overlying Green River and Uintah Formations, and the Wasatch Formation beneath, all of which showed promise for producing magnificent fossil vertebrates.

In a monograph from 1873, Joseph Leidy would write of this region: "Fort Bridger occupies a situation in the midst of a wide plain at the base of the Uintah Mountains. . . . [T]he neighbouring country, extending from the Uintah and Wahsatch Mountains on the south and west to the Wind River Range on the northeast, at the close of the Cretaceous epoch, appears to have been occupied by a vast freshwater lake. Abundant evidence is found to prove that the region was then inhabited by animals as numerous and varied as those of any other fauna, recent or extinct, in other parts of the world. Then, too, rich tropical vegetation covered the country, in strange contrast to its present almost lifeless and desert condition. The country appears to have undergone slow and gradual evolution; and the great Uintah lake, as we may designate it, was emptied, apparently in successive portions and after long intervals. . . . [T]he ancient lake-deposits now form the basis of the country and appear as extensive plains, which have been subjected to a great amount of erosion, resulting in the production of deep valleys and wide basins, traversed by the Green River and its tributaries."[17]

The famous mountain man Jim Bridger had operated here. The main western overland trails, having crossed over the continental divide at South Pass, all led through the Green River lands before branching off to Oregon, California, or Salt Lake City. Black's Fork of the Green River actually passed through the Fort Bridger parade ground. Fort Bridger had been founded by Jim Bridger (after he gave up trapping in favor of guiding migrants) to supply the westward travelers. It was later taken over by the Mormons and then lost by them to the U.S. Army in the Mormon War. In 1858 it became an official army depot with an associated reservation of some five hundred square miles. After the Indian treaties of 1868 created the Wind River Reservation in southwestern Wyoming for the Shoshone and Bannock people, peace came to the region—another great boon for fossil collectors.

The English explorer Sir Richard Burton visited the fort in 1862, and he later wrote: "The fort was built by Colonel James Bridger, now the oldest trapper on the Rocky Mountains, of whom Mssrs. Fremont and Stansbury have both spoken in highest terms. He divides with Christopher Carson, the Kit Carson of the Wind River and the Sierra Nevada explorations, the honor of being the best guide and interpreter in the Indian country. . . . [W]hen an Indian trader [he] placed this post upon a kind of neutral ground between the Snakes and the Crows (Hasaroke) on the north, the Oglalas and other Sioux to the east, the Rapahoes and Cheyennes on the south, and the various tribes of Yutas [Utahs] on the southwest."[18]

As was usually the case in those early days, the fossils from this region had first come to light when Leidy received specimens discovered by some amateur collectors. The new Eocene mammals sent to Leidy came from two doctors at Fort Bridger. Joseph K. Corson (another Philadelphian—of course) was the army physician at Fort Bridger. James Van Allen Carter was a doctor in the town; experienced in western affairs, he acted as interpreter at the negotiations for the Treaty of Fort Bridger, giving the eastern Shoshone Indians the Wind River Reservation. In a coincidence of names he was also the son-in-law of Judge William A. Carter. (Joseph Corson later married another Carter daughter.)

Judge Carter, sometimes referred to as "Mr. Fort Bridger," was one of Wyoming's most prominent early businessmen. He had traveled to the region by wagon train in 1857 and never left.[19] Starting out as sutler (storekeeper) at the army fort, he branched out into cattle ranching, logging, and mining and acted as probate judge for more than thirty years. He essentially took over Fort Bridger in the 1870s, growing oats, barley, potatoes, and corn, although the climate (at seven thousand feet) was not suitable for the last. One source reported that Carter not only overcharged the government for supplies "but was selling much stuff from the reservation to other interests. . . . [A]n immense organisation was maintained by the sutler, who employed a hundred men."[20]

Carter was reputed to be worth more than $200,000. Whatever the truth about his honesty, he was a civilized man with a great library, fine wines, and a grand piano. Little happened around Fort Bridger without his knowledge and, preferably, his approval. Judge Carter, Dr. Carter, and Dr. Corson were pivotal in opening up the fossil beds of the region, and they started sending material to Leidy in 1868. For several years James Van Allen Carter collected for Leidy at Grizzly Buttes every Sunday; other favorite places were Bridger Butte, just west of the fort, and the badlands along Black's Fork. The first mammal materials that Leidy received from the Fort Bridger region turned out to include remains of a large unknown form that he named *Palaeosyops*.[21] Later he and others discovered that *Palaeosyops* was a member of a group that included a number of huge archaic tapir-like mammals related to his genus *Titanotherium* (from the Dakota Bad Lands). In Eocene times they had ranged in large numbers across the West.

Once Hayden had seen the sites for himself, he understood their immense potential: "There are indications that when this group is thoroughly explored it will prove second only to the 'Bad Lands' of Dakota in the richness and extent of the vertebrate remains." And he was right; in his 1873 report (which would turn out to be his last major paleontological monograph), Leidy could already identify and describe thirty-one species of mammals, eleven of turtles, four of lizards, and nine of fishes from the Bridger Formation alone, using collections made by his diligent friends from Fort Bridger and Hayden. A year later, other workers raised that number fourfold.

Hayden explored the Wind River Mountains, Green River, Bridger Pass, the Medicine Bow Mountains, and the Laramie Range. The haul in terms of fossils was enormous, and Hayden's reports once again contained major contributions by Meek, Lesquereux, and Leidy. He made a large collection of Green River fossils, including not only fossil fishes from the soon-to-be-famous "Petrified Fish Cut" (which went to Edward Drinker Cope for description) but also mammals and a single magnificent feather. This last he sent to New Haven for identification. It was "a true feather, as determined by Mr Marsh of New Haven; probably not a bird's feather, but belonging to some form of Archaeopteryx."[22]

Hayden's exploring and surveying (progressively more of the latter than the former) continued through 1878. He worked in Montana and Yellowstone Park in 1871, Montana, Wyoming, and Idaho in 1872, Colorado in 1873–76, and Colorado and Wyoming again in 1877 and 1878. Each

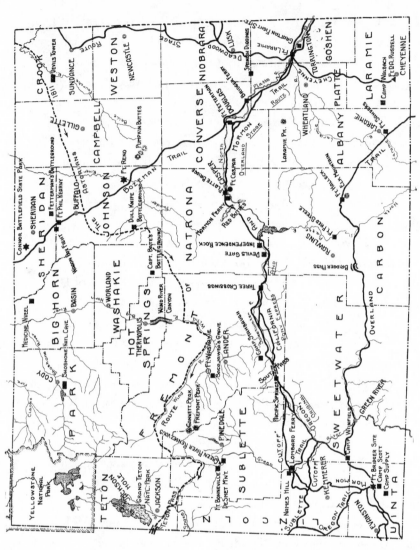

Map of Wyoming, including the major overland trails converging on Fort Bridger near the southwest corner (from Robert Ellison, *Fort Bridger, Wyoming: A Brief History*, 1931)

successive volume of his official reports was better illustrated than its predecessor and written in a more "accessible" language. His last report was twelve hundred pages long. Hayden had early on hit on the device of describing the geology as he observed it along his route—those routes being the major rivers into the Upper Missouri or the growing number of railroad corridors. The reports were thereby at once by readable by and useful to a wider audience, although the practice soon came under criticism.

By this time, Charles Lyell's prediction that the future of paleontology would be found in the American West had long since come true, especially in the long-standing themes in western geology involving the Cretaceous and its relationship to comparable periods in the eastern states and in Europe. In less than two decades, Hayden and Meek had established a basic structure of the Cretaceous and Tertiary periods in the West. It was a phenomenal achievement and their results (imperfect though many of their conclusions were) were being read with great interest in Europe.

One of Hayden's more controversial opinions was his assertion that, instead of the American West being geologically younger and less mature than Europe, it was older. "One instructive lesson," he wrote, "is derived from the mistakes of those eminent men that, in the progress of geological development, America was almost or quite one epoch ahead of Europe—that the fauna and flora of the Cretaceous period in this country was really more allied to those of the Tertiary Period in Europe, and that, geologically speaking, America should be called the Old World, and Europe the New."[23]

This was not an original thought. It seems to have arisen with the estimable G. W. Featherstonhaugh, who had floated the idea on the basis of a mistaken assumption that there were no rocks younger than the Coal Measures in the American interior; he probably meant the term "Old America" as a derogatory conclusion. The same theme had appeared in an essay by Louis Agassiz titled "America: The Old World," but Agassiz, like Hayden, doubtless meant it as a boast of American superiority.[24]

Hayden's letters to Leidy over these years are almost comical in their transparency. Every one started with the announcement that a new set of boxes had been shipped east and then quickly devolved into a request for Leidy to hurry up with his next manuscript, or make it longer, or better illustrated. For example: "Dear Dr., I have placed in the express this day a collection of bones . . . most of them are saurians bones I think, but there is one curious fossil, that looks like a part of a skull of a horned animal. I have sent all to you at once so that you might study, describe and figure for my forthcoming report at once. . . . Do not fail to write at once. As fast as I find anything I will forward it to you and you must keep all I send safely. . . . You will be much befitted I think. . . . Yours, F. V. Hayden."[25]

However, a significant change in the disposition of western fossils collected by the postwar surveys (both Hayden's and those of Clarence King along the fortieth parallel) occurred in the late 1860s. While, as before, all Hayden's materials were shipped back to Washington for the Smithsonian with the understanding that Leidy would continue to describe the vertebrates, Hayden was allowed to dispose of the specimens of which he had multiples. Hayden sold specimens to, among others, the academy in Philadelphia and to Professor O. C. Marsh at Yale University. In 1867, for example, Hayden sent Marsh four boxes of fossils "by cheap express," promising to deliver the rest in person (he also sent a photograph of himself). The Marsh correspondence shows that in May Hayden sent Marsh a receipt for "the sum of four hundred dollars, the balance of the account ($600) due for the collection of fossil vertebrates he purchased in 1866." Marsh also got collections from Major John Wesley Powell's surveys in the Rocky Mountain region, but these, exploring different territory and without the energy and interests of a Hayden, were far less productive of fossil vertebrates.

The second significant change was an addition to the roster of paleontologists contributing to Hayden's reports. Perhaps because of his impatience with Leidy's slow, deliberate habits and probably also because of the sheer volume of material and range of geological contexts, in 1870 Hayden started to send some of his vertebrates to be described by the young Philadelphia zoologist Edward Drinker Cope.

PART THREE

Giant Saurians and Horned Mammals

FOURTEEN

Kansas and a New Regime

Within the past two years settlers, in families and colonies, have spread westward, along the line of the Kansas Pacific Railway, and also on streams north and
south of the road, nearly to the one hundredth meridian. . . . It is safe to say that
the forces operating to throw population westward, taking into consideration
facilities of transportation, are three times as powerful as they were twenty-five
years ago. The result will be a gradual spread of people over the great plains, ar-
ranging their pursuits and modifying their habits to suit the capabilities of the
country and the necessities of their respective localities.

R. S. ELLIOTT, 1872

In the first five years after the American Civil War the rate of discovery
of fossil vertebrates in the West increased dramatically. In addition to
the efforts of Hayden and others, pushing their surveys deep into the
Upper Missouri regions of modern Colorado, Montana, and the Dakotas, many finds of fossil bones were now coming from people on the
ground. The population of the entire West was increasing rapidly both
from itinerants in the form of traders and army personnel and from the
permanent settlers in the new towns (especially those springing up
around the forts and along the continental railroad routes). In Utah, for
example, the population in 1850 was around 11,000; in 1880 it was
146,000. The burgeoning state of Kansas led the way in growth. In
1860 its population was about 200,000; in 1880 it was approximately
1 million. Eastern Kansas, with its superb farming land, had begun to be
well settled before the Civil War, while the western part of the state was
still essentially unpopulated for a long time and considered unfit for
agriculture, although it soon came to prominence for raising cattle.

In Kansas, not only was the land west of Fort Riley inhospitable to settlement, but so were the people, particularly the Cheyenne and Arapaho. The problem extended right across the southern plains. Rejecting the Medicine Lodge Treaties of 1867, under which the Arapaho, Cheyenne, Kiowa, Apache, and Comanche were given (but restricted to) their own reservation lands, war parties continued to harass settlers in western Kansas and eastern Colorado. The army in turn took its retributions. In a notorious set of incidents at Fort Wallace, George Armstrong Custer of the Seventh Cavalry was court-martialed for reckless behavior that had led to the loss of many men. He was reinstated a few months later and, a year after the first Medicine Lodge Treaty was signed, was responsible for the massacre of a Cheyenne village at the Battle of the Washita, in Oklahoma, inflaming an already tense situation.

Ironically, in Kansas in the 1840s and 1850s, one of the tribes making exploration and settlement of the western half of the state difficult was none other than the Shawnee. These were descendents of the people who, a hundred years before, had kidnapped Mary Ingles from her northern Virginia homestead and carried her to the banks of the Ohio. Since then, the Shawnee people had been steadily pushed westward, although their path was not a simple one. In the 1600s they had lost their homelands to other tribes and moved south into the Cumberland Valley of Tennessee, returning to eastern Pennsylvania, New York, and Ohio between about 1680 and 1730. By 1795 they had been pushed out again, this time by white settlers, into northern Ohio and Missouri. John James Audubon encountered groups of Shawnee at the confluence of the Ohio and Missouri Rivers in 1810. In 1811, under Chiefs Tecumseh and Tenskwatawa, they were defeated at the Battle of Tippecanoe. Between 1825 and 1845 they were pushed as far as Kansas (Topeka is the seat of Shawnee County). But the state was already the home of the Cheyenne and Pawnee, and most of the Shawnee eventually ended up in Oklahoma reservations.

Displacing the Indians was inevitable if Kansas was to prosper as part of the Union. The state was not only important in its own right for agriculture and mining, it was vital to whites who wanted merely to pass through. Good overland routes westward to Denver and southwestward to Texas and to New Mexico, along the three main tributaries of the Kansas River (the Solomon, Republican, and Smoky Hill Rivers), afforded a number of connections from the Missouri ultimately all the way to California. In 1861 a new daily stage route was established north and west from Atchison and thence on to Denver, Salt Lake City, and beyond, following the River Platte. A three-times-weekly mail stage later took a more southern route to Denver, passing almost due west across the state along the Smoky Hill River. The Santa Fe Trail led off into the southwest, with the Cherokee Trail splitting off northward at Pueblo to meet the Oregon Trail at Fort Bridger as the latter followed the North Platte River westward.[1]

Troops, traders, explorers, land agents, Indian scouts, buffalo hunters, pioneers, and farmers—all the varied cast of characters of the frontier—steadily pushed westward along these routes. Then the railroads came. A whole new territorial imperative was created when the Kansas Pacific Railroad (originally the Union Pacific Eastern Division) crossed the Missouri River, reaching Topeka in 1865, Salina in 1867, and Sheridan in 1868. The Central Branch of the Union Pacific took a somewhat more northerly route across the state and was only about halfway across by 1868. One of the first results of this development of the "transportation frontier" was the deployment of the army to a series of new forts along the railroad and stage routes.[2]

The army had a significant presence in the more remote parts of Kansas, and of all the officers serving in the West it was most often army surgeons who had the greatest interest in, and keenest eye for, fossils. One of the more prolific fossil collectors to be smitten with the beauty and strangeness of western Kansas was Dr. George M. Sternberg, an army surgeon posted to various sites in western Kansas between 1866 and 1870. In forays out from the forts he collected every kind of mineral, fossil, and Indian artifact and quickly acquired a reputation as something of an expert. He communicated with Baird and Henry at the Smithsonian, with Leidy (who described most of his vertebrate finds), and with Agassiz, Newberry, Hayden, and Lesquereux.

The discovery that truly established western Kansas as a mecca for fossil collectors was made by Captain Theophilus Turner, a graduate of Jefferson Medical College in Philadelphia. In 1866 he became the assistant surgeon at Fort Wallace, and his letters home give us a nice view of the land and its people. He liked neither. As he wrote to his brother, Fort Wallace "was within two miles of the Colorado line. It is placed on the south fork of the Smokey Hill River, a stream that is by no means a river." It was a hard posting with no fresh supplies ("I nearly went wild over an apple today. The first I have eaten in nearly two years"). There was little decent water and, in addition to the Indians, the constant threat of death from cholera.[3]

From its forts the army tried to protect travelers, railroad construction workers, and the assortment of men operating the staging posts. "You have no idea what a peculiar class of people is to [be] met with in this country," Turner wrote. "Nor can you realize till you understand how to conceive of an uninhabited desert, over which thieving murdering bands of Indians only roam except on the outland routes. It seems that there are men made for everything and it seems that a certain class were made to keep stage stations and be satisfied with sixty-dollars."

If Turner had a poor view of stage employees and the Indians, he was equally critical of the government. He wrote home, "You have doubtless seen by the papers that it is proposed to prevent Indians hunting or travelling between the Arkansas and Platte rivers, and that . . . includes nearly all of the best hunting grounds on the plains. The Indian thinks that they are going to starve to death and that they might as well die fighting. They have made a good commencement toward using up this stage road, having 'run off' all the horses and mules for over one hundred miles. . . . I think the army would rather whip the Indians into subjection than to secure peace by forming a treaty with them as they care nothing for treaties but continue depradations [sic] afterward without fear of punishment."

Although the Indians (mostly the Cheyenne) were a constant threat, Turner still managed to ride out into the country and collect mineral specimens. Late in 1867 he became a fossil collector when he discovered parts of a skeleton weathering out of the bank of a ravine some fourteen or fifteen miles north of the fort. He picked out two large vertebrae and gave them to John Lawrence LeConte, a young naturalist who at the time was making a long survey along a possible rail route from Sheridan, Kansas, to Albuquerque.[4]

LeConte took them back to Philadelphia, where he knew they would be identified and appreciated. But instead of passing them on to Joseph Leidy as might have been expected, LeConte gave them to Edward Drinker Cope. Cope was then twenty-eight years old and had already made a considerable name for himself as the volunteer curator of herpetology—the study of living amphibians and reptiles—as well as a paleontologist, at the Academy of Natural Sciences. His own (very successful) fossil collecting had thus far mostly been restricted to New Jersey. Cope quickly wrote to Turner asking him to collect the rest of the skeleton, at the academy's expense.

This was all a pleasant diversion for Turner. On December 20, 1867, he wrote: "On Friday last a party of us started on the prarie [sic] ostensibly for the purpose of hunting but in reality for the purpose of procuring the skeleton of an extinct monster which is embedded some fourteen miles north of here. . . . It was found to rest in a slate hill similar in appearance to those which are found on the road between home and Newton [New Jersey]." Eventually, the skeleton—"something over thirty-five feet of its vertebral column with about four inches of the anterior portion of its head with imperfect teeth, . . . a portion of a limb, a perfect bone eight or ten inches in length, . . . the whole . . . weighs about eight or nine hundred pounds"—arrived at the academy in Philadelphia. Cope ascertained that it was a huge plesiosaur, far bigger than the ones that Mary Anning had found at Lyme Regis. He named it *Elasmosaurus*. This Kansan monster was a great coup for Cope. It was just as dramatic a specimen as Leidy's dinosaur *Hadrosaurus* and even more complete.

Unfortunately, Turner was not destined to find more fossils. New problems with the Indians kept all the officers at Fort Wallace occupied, and then, in July 1869, he died suddenly of acute gastritis. Meanwhile, some of the first truly local (embedded, in the current term) scholars had entered the scene. Across the West, local governments and the burgeoning number of colleges serving the new towns that were springing up had begun to employ their own geologists and natural scientists. Again Kansas led the way, with one of the principal goals being to survey the state's soils and its mining potential. Among the first population of western scientists were several who were interested in fossils, among them Professor John D. Parker of Lincoln College (now Washburn University) in Topeka and his friend Benjamin Franklin Mudge, professor of mathematics and natural science at the State Agricultural College (now Kansas State University) in Manhattan, Kansas.

Benjamin Mudge at the time was forty-eight years old. Born in Maine and educated at Wesleyan University in Connecticut, he was a keen mineralogist but first practiced law before moving to Kentucky as a chemist for a coal company. An avowed abolitionist, he then moved on to Kansas, first getting himself appointed state geologist. The year he became a professor at the State Agricultural College, Mudge found fossil footprints north of Junction City. During summer vacations in 1866 and 1867 he collected plant fossils and invertebrates from the Smoky Hills that he sent to Meek. They turned out to be particularly useful in establishing the precise stratigraphic correlation of the Kansas Cretaceous with that of New Jersey and Europe.

As circumstances allowed, Sternberg, Parker, Mudge, and other part-time explorers began to range out along the stage routes that wound across western Kansas. Like Turner, they also found remains of "giant saurians" in the Cretaceous-age chalk hills and ravines of far western Kansas. But what they dug out were not plesiosaurs, like *Elasmosaurus*, but huge mosasaurs very similar to the great mosasaur from Maastricht that Cuvier had described more than fifty years earlier.

The "ichthyosaur" that Harlan had described in 1834 was also really a mosasaur, and the German naturalist Albert Goldfuss in 1843 had described a mosasaur from Lewis and Clark country that had been collected by "Major O'Fallon, Indian agent" at the Great Bend of the Missouri. Goldfuss named his animal *Mosasaurus maximiliani,* honoring his patron, the explorer and naturalist Prince Maximilian zu Wied-

Neuwied, who had made a scientific trip through the American West in 1832–34 and evidently obtained the specimen in St. Louis from O'Fallon. It seems very likely that Harlan's specimen and Goldfuss's were part of the same individual.[5] The prince was accompanied by, among others, the artist Karl Bodmer, whose paintings are an important early record of the western landscape.

The Lewis and Clark expedition may also have found mosasaur vertebrae along the Missouri, and mosasaur remains (mostly teeth) had long been known from the Cretaceous greensands of New Jersey. But these Kansas fossils were remarkable not only for their size, preservation, and abundance; soon it appeared that they were not at all difficult to collect. Entire seventy-foot skeletons were found right on the surface in the Cretaceous deposits of the Smoky Hill River region and could be extracted from the soft chalk with a knife. In harder strata, fossils had to be taken up as they were, still enclosed in large, unwieldy chunks of rock, and brought back to the laboratory, where the bones would be laboriously revealed with fine chisels and (nowadays) even acid. Needless to say, the Kansas mosasaurs were not only easier to collect, they were bigger and better than anything found in Europe.

Possibly the first of the Kansas mosasaurs was collected in 1867 or 1868 by "Colonel Cunningham and Mr Minor . . . in the valley of the Smoky Hill River," and was bought by Professor Agassiz for his Museum of Comparative Zoology at Harvard.[6] Around the same time, the Kansas Pacific Railroad's land agent in Topeka, William E. Webb, sent Leidy and Cope photographs (with measurements) of a specimen of *Mosasaurus missouriensis* that he had found "near the town of Topeka." Professor Parker said it was "seventy-five feet in length"; Webb claimed eighty. Webb also sent Cope a partial skeleton of a plesiosaur that Cope made the type specimen of a new form, *Polycotylus latipinnus*. By 1875 the remains of some five different genera and twenty species of mosasaurs had been described in Kansas alone.[7]

Benjamin Mudge also collected in company with Professor Merrill of Washburn College, Professor Felker of Michigan Agricultural College, Professor Warder of the Indiana Geological Survey, and students from his own college. During an expedition to the Republican and

Solomon Rivers in the summers of 1870 and 1871 he traveled as far west as Sheridan and Fort Wallace, "into the wholly uninhabited regions, the home of the bison and roving bands of Indians." There he found a rich trove of vertebrate fossils in the chalk. These fossils also did not go to Leidy. Instead, Mudge sent them to "a young and promising naturalist" in Philadelphia.[8]

That young naturalist was, once again, Edward Drinker Cope. A particular reason for choosing him may have been that Mudge had found remains of huge Cretaceous fishes, and Cope was far more of an expert on fishes than was Leidy. The new fish remains included the jaws, teeth, and vertebrae of *Saurocephalus*. This was the animal of which Lewis and Clark had collected a jawbone. Harlan had named it an ichthyosaur but now it could be seen to be a giant fish.[9] Over the following years Cope would show that there had existed a range of these and other monsters in the Cretaceous seas of Kansas.

The wider result of all this was that Cope, by promptly describing Mudge's material, established himself as a player in the western fossil stakes and, having already snagged the *Elasmosaurus,* became the leading authority on all fossil bones from Kansas. The well-worked story of the Cretaceous seas that had covered so much of the West now had a new cast of fossil characters—monstrous fishes and marine reptiles— and a new human interlocutor. Leidy could never have captured in words the drama of this sort of paleontological discovery. But eventually Cope did: "If the explorer searches the bottoms of the rain-washes and ravines, he will doubtless come upon the fragment of a tooth or jaw, and will generally find a line of such pieces leading to an elevated position on the bank or bluff, where lies the skeleton of some monster of the ancient sea. He may find the vertebral column running far into the limestone that locks him in his final prison; or a paddle extended on the slope, as though entreating aid; or a pair of jaws lined with horrid teeth, which grin despair on enemies they are helpless to resist."[10]

FIFTEEN

Entry of the Gladiators

Born in 1840, Edward Drinker Cope attended a prestigious and rigorous school—Westtown Friends School—but he did not attend university. The Copes, a Quaker family, had long been prominent in Philadelphia business circles; his grandfather Thomas Pim Cope owned a successful packet ship line. The young Cope should have been destined to continue in the family businesses, although he showed no interest in the world of commerce. He grew up on his father's rural estate just outside Philadelphia, and when he was sixteen his father, Alfred Cope, worried about his son's health, sent him to work during the summers on a series of family farms in the region. In 1860, his father bought the twenty-year-old a farm, hoping to establish him financially and give him more of an anchor in life than natural history.[1]

From a very early age, however, Cope had been passionate only about natural history. His interest was developed by living close to nature and reinforced by many trips to the Academy of Natural Sciences in Philadelphia. At school he was seen as having "an incessant activity of mind and body, usually highly amusing to an observer. People's attention was instantly caught by his quick and ingenious thought, expressed in a bright and merry way. His mind reached in every direction for knowledge, seizing upon everybody who came in his way, as a source or else a receptacle of information, which he was as ready to impart as receive."[2] In addition he soon turned out to be quite gifted as an artist.

For such a brilliant youngster, managing a farm was no more attractive a proposition than managing a shipping line. He eventually found it simpler to rent the farm and live off the income. After two years of fruitlessly pushing his son into directions he simply could not go, Alfred Cope relented and let him attend the University of Pennsylvania as a student of Joseph Leidy in anatomy and zoology. By this time Cope was already sufficiently skilled as a zoologist that he ensconced himself at the Academy of Natural Sciences and reorganized its collections of amphibians and reptiles on modern scientific lines. He was quickly recognized as having accomplished more as a relatively unschooled amateur than had any of his predecessors working in that august but conservative institution.

Like other midcentury naturalists on both sides of the Atlantic, the young Cope was caught in a trap: the natural sciences, no matter how intellectually attractive, did not easily produce career opportunities. Even at a great institution like the academy the curators were all volunteers. The best of a number of poor choices would have been to go into medicine, as Leidy had. Indeed, from Caspar Wistar onward there were many excellent examples to have followed. Similarly, in England, Charles Darwin had had the same problem, ultimately solved in the same way—with family financial support. Cope's father, like Darwin's, had to bow to the inevitable and acknowledge that his son was never going to be a productive part of the business community.

A more immediate difficulty was that Cope was a young man in a country engulfed in a civil war. The Quakers were, of course, pacifists, and there was no question of Cope actually being caught up in the fighting. Nonetheless, he was packed off to Europe on a sort of scientific Grand Tour. In London and Cambridge he met some, but by no means all, of the contemporary greats; he did not formally study there, although he spent a great deal of time in the British Museum studying the fish, amphibian, and reptile collections—just as he had at the academy in Philadelphia. He admired the museum's superb specimen of *Archaeopteryx* and also fell in love at least once.

After returning to Philadelphia in 1864 with his appetite for science whetted and his (always pronounced) energy, self-confidence, and ambition bolstered, Cope briefly held a courtesy appointment as professor of natural history and chemistry at Haverford College. Clearly this position was "negotiated" for him by his family. But his heart was not in teaching undergraduates; it was in research. And so he embarked on a life as an independent zoologist. In August 1866 he married a distant cousin, Annie Pim; the following year their only daughter, Julia, was born. Life settled down to a comfortable routine focused entirely on science. He was by no means poor; there was the rent from the farm that his father had given him, and his father's many generous "loans" usually turned into gifts. But until his father died in 1875, Cope never had the sort of ample funds that would allow him to move freely around the country or to hire as many field-workers to collect for him as he would have liked.

While a professor at Haverford, Cope began to travel modestly, as far as Ohio and Virginia. One of the areas of study in which he became expert was the strange blind fishes and crustaceans that develop in caves. He also began to extend in a different direction—time. He started to collect and study fossils, both in the Coal Measures of Ohio and the Cretaceous beds of neighboring New Jersey. One of Cope's early successes in the realm of fossils, and his first published paper in paleontology, was the description of a new Coal Measures (Pennsylvanian) fossil amphibian. The specimen had been sent from the Illinois state survey to Leidy, who passed it on to his former student. Cope named the new form *Amphibamus* (dual mover), to emphasize its froglike position between aquatic fishes and land tetrapods.[3]

As Vanuxem and Morton had shown, the Cretaceous beds of New Jersey were equivalent in age to strata in Europe that had already yielded a wealth of fossil material. Maclure and Thomas Say had long since described how, instead of massive layers of chalk, the New Jersey Cretaceous consisted largely of greensands and marl (marl was rich in phosphate, and farmers spread it on their fields to improve the soil). Exploring across southern New Jersey, Cope started to find mosasaur teeth, similar to those from the mosasaur found at Maastricht in Belgium.

In 1868, the same year that he described the great *Elasmosaurus* sent by Turner from Kansas, Cope had a great stroke of fortune. The superintendent of the West Jersey Marl Company wrote to tell him about some bones that his workmen had found in one of the pits near Barnesboro, New Jersey. These turned out to be the remains of a "gigantic extinct dinosaur," but unlike Leidy's *Hadrosaurus* of 1858, this was a carnivorous form and so was inevitably compared to *Megalosaurus* from England.[4] It was of a similar size—some forty feet. Cope named it *Laelaps aquilinguis,* and he remained fascinated by this creature for the rest of his life. Even six years later, for example, sitting in his tent in New Mexico, Cope sketched in his field diary images of what *Laelaps* might have looked like in life.

As far as fossils from the American West were concerned, at first, Cope followed Leidy in being content to receive specimens for study from contacts out in the field (like Mudge), rather than travel there himself. Between the end of the Civil War and the completion of the first transcontinental railroad link (1869), travel to the West still appeared too hazardous and time-consuming except for intrepid explorers like Hayden and the members of the numerous government surveys. Travel out west was also very expensive. There was much work to be done at localities that were closer to home and more readily accessible.

Although born to a Quaker family and retaining the old form of address "thee" until late in life, Cope was in many ways the least Quaker-like of Philadelphians. Where Quakers avoided personal aggrandizement and made their decisions slowly and carefully by consensus, Cope was impulsive, egotistical, and impatient of authority. He was careless with money instead of frugal. Tall and handsome, he distinctly had an eye for the ladies. Inevitably, he fitted in badly with Philadelphia styles and institutions. It was his misfortune that the Academy of Natural Sciences, where his intellectual heart was, and on whose collections of amphibians and reptiles he built his reputation as a scientist, was one of the slower institutions in the United States to accommodate to men like him. The academy was unwilling to take the steps that would be needed to make American science more professional and (intellectually as well as institutionally) more aggressive. The contrast between the calm, avuncular, steady Joseph Leidy and the young upstart Cope could not have been greater, and Cope did little to conceal his impatience with the ways of the "old guard" (ways that, of course, kept him from paid employment at the academy). Interestingly enough, this drive to professionalize science was one of the few things that Cope had in common with the man who became his lifelong rival, Othniel Charles Marsh at Yale.

Marsh had had a very different upbringing and education from Cope, but in many respects the men were cut from the same cloth. Marsh was born in 1831 on a small farm in South Danvers, Massachusetts. His mother died when he was young, and after that his father had an unsuccessful career variously in farming or keeping a variety of stores and was always poor. But Marsh had the advantage of an extremely rich and influential uncle, the financier George Peabody, who supported his late sister's son generously.[5] Peabody paid for Marsh to attend Philips Andover School, where he proved to be a brilliant student. A classmate remembered that "he stood first in every class every term without exception. He studied intensely, but tried to make the impression that he achieved his success without any work at all. In the debating club, he also took hold strongly, although he was at this time a slow and halting speaker, and never in his life was anything of a rhetorician. His superiority in managing practical affairs soon impressed all, and he became manager of the society and held the whole thing in his hands."[6] He also showed himself to be an adroit politician in student affairs.

Going on to Yale (Class of 1860), Marsh studied classics and developed his interests in mineral collecting and geology at large. He grew up to be an energetic outdoorsman, an excellent shot and keen fisherman, but shy. After graduating he became a member of the first class of the new Sheffield Scientific School at Yale, where he continued to study geology and mineralogy with Professor James Dwight Dana, Benjamin Silliman Jr., and George Jarvis Brush. While still a student Marsh published papers on the mineralogy of Nova Scotia and also on some intriguing fossil remains that he had collected on summer field trips to the Coal Measures there (probably in the summer of 1855).

When Marsh showed a Nova Scotia specimen to Louis Agassiz at Harvard, Agassiz promptly fired off an excited letter about it to the *American Journal of Science,* stating that "we have here undoubtedly a nearer approximation to a synthesis between Fish and Reptile than has yet been seen."[7] Nettled by this "poaching" (really an abuse of Agassiz's position), Marsh quickly published his own lengthy description of it, giving it the name *Eosaurus acadianus* and, in rather haughty language, concluding that Agassiz was wrong. "The highest authority" he said, would "establish a connection between these remains and the vertebrae of the Ichthyosaurus."[8] In fact, however, Agassiz had been more or less right: *Eosaurus* was a lower tetrapod, not an ichthyosaur. Perhaps the experience of being scooped by Agassiz fixed in Marsh the importance of controlling his own material; certainly in his later career Marsh fretted endlessly about establishing priority for the description and naming of new fossil species. The pressure always to be first became almost overwhelming in the face of the huge volume of new material to be described.

With the Civil War raging, Marsh, like Cope, headed off to Europe in 1862, first to London and then to study in Heidelberg and Berlin. In Berlin he briefly met Cope, who was on his own study tour. In 1864 Marsh moved on to Breslau, where he studied under Carl Ferdinand von Roemer, a geologist and paleontologist who had collected invertebrates in Texas as early as the 1840s, when a large German colony was established around Dallas. Among the lessons that Marsh learned from Roemer was that "the most inviting field for palaeontology in North America is in the unsettled regions of the West. It is not worth while to spend time on the thickly inhabited region."[9] He met most of the important paleontologists of the day (but apparently not Huxley). Like Cope he studied the collections of the British Museum, including *Archaeopteryx,* before heading home.

All this foreign travel and study also had one very particular focus. The entrepreneur and philanthropist George Peabody had proposed donating money to Yale for a new scientific museum (the one that today bears his name). With the endorsement of Professors Dana and Silliman, Marsh was being cultivated for a professorship at Yale to teach

geology and paleontology and to head the museum. Marsh duly returned to New Haven in July 1865 to take up his new career, although the
position was actually an honorary one; Peabody's personal allowance
substituted for a salary. Marsh began his teaching and wrote papers on a
variety of minor subjects (mastodon remains, fossil sponges, and the
Ohio Indian burial mounds) while also planning the new museum.
Among his recent paleontological interests were the Triassic footprints
from the Connecticut River valley that Edward Hitchcock had first described in 1836, and Marsh collected some of these for the museum.

Always a shy and private man, Marsh never married, and although
he was adept at cultivating the rich and famous, he seems to have had
few close personal friends.

Sometime in the late 1860s both Cope and Marsh decided to devote
their careers to the study of fossil vertebrates. Putting it in quasi-cynical
terms, both saw that fossils would provide a platform from which to
achieve great things. More generously, in the new scientific order created by the publication of Charles Darwin's *On the Origin of Species,*
and the new respectability of theories of evolution, fossils from the past
represented the science of the future for anyone interested in natural
history. Fact finding could be convincingly linked to theory making. As
Cuvier had predicted, the most dramatic and useful discoveries were
being made in the area of fossil vertebrates. Furthermore, in America,
the only serious worker in fossil vertebrates was Leidy—brilliant and accomplished but not interested in using fossils to create a politically powerful position in science (or, at least, only on his own rather reticent and
gentlemanly terms).

Apart from his early descriptions of *Eosaurus,* Marsh had little prior
experience working with vertebrates, but those studies showed that he
had an excellent mastery of anatomy and of detail. Study of mineralogy
had given him a keen appreciation for order and classification. Cope had
always concentrated on vertebrates, but his earliest interests had been in
the living forms, especially fishes, amphibians, and reptiles (herpetology).
His interests were first also in the area of taxonomy and classification.

Given their later head-on competition, it is highly ironic that both men's earliest ventures into the realm of fossils were descriptions of new Coal Measures fossil amphibians (although Marsh had insisted his was an ichthyosaur).

At first Cope and Marsh were reasonably good friends, or at least cordial colleagues, sharing in the pleasures of working with fossils and enjoying the fact of having common interests. Cope invited Marsh to join him for a visit to his sites in New Jersey, and they corresponded extensively.[10] Cope's second paper describing his dinosaur *Laelaps* from New Jersey also contains a short description of a new form of mosasaur, which he named *Clidastes,* and that the genus was "established on a species represented by a single dorsal vertebrae, which was found by my friend Prof. O. C. Marsh, of Yale College, in a marl pit near Swedesboro, Gloucester Co., N.J."[11]

It is not possible to pinpoint exactly when Cope and Marsh fell out, or why. But when they did, it was no small thing. They cultivated a mutual animosity with as much diligence (and sometimes more) as they devoted to the collection and description of their fossils. What started as a friendly professional rivalry eventually became a public scandal. Most likely the situation grew upon them gradually. But one deciding factor for Marsh may well have been that he had seen a point of vulnerability that encouraged him to be more aggressive. That weakness concerned Cope's *Elasmosaurus,* the plesiosaur fossil that Captain Turner had sent him from Kansas.

Unfortunately, instead of being one of his great triumphs—a magnificent monster assembled for display in the Philadelphia academy and even rivaling Leidy's *Hadrosaurus*—Cope's *Elasmosaurus platyurus* was to prove a great embarrassment. When he reconstructed the fossil he literally got it backward, mistaking the neck for the tail. He put the head on the wrong end. It was an understandable error, but plesiosaurs, as a group, were already quite well known in 1868; all of them had very long necks. Cope's reconstruction would have meant that *Elasmosaurus* was something never seen before: "different from Plesiosaurus in the enormous length of the tail, and the relatively shorter cervical (neck) region."[12] If he had been right, it would have been a most dramatic discovery.

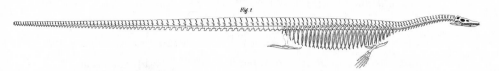

Cope's "wrong-headed" reconstruction of *Elasmosaurus*
(Ewell Stewart Library, Academy of Natural Sciences)

On a visit to Philadelphia in spring 1869, Marsh pointed out Cope's mistake, and a nasty argument developed. Leidy was brought in to adjudicate, but unfortunately for Cope, Leidy agreed with Marsh. One version of the story has the three men poring over the bones, laid out on a table, with Leidy silently picking up the head fragment and putting it down at the tip of what Cope had thought was the tail. Even worse for Cope was the fact that he had been so keen to get the word out about his brilliant new work that he had written of it in a paper due to appear in the *Transactions of the American Philosophical Society,* and had rashly had advance copies distributed. These had been sent to all his friends and colleagues, including Marsh. Hastily he had to try to get them all back before changing the main publication.

An uncharacteristic feature of this incident is that, rather than simply leaving Cope to correct his own mistake, Leidy published a short note of his own on the subject, thus rubbing salt into the wound. First Leidy read what could have been taken for some informal remarks on *Elasmosaurus* at the March meeting of the academy (Marsh was in the audience!). That seems harmless enough in principle, but the blunt remarks—"Prof. Cope has fallen into the error of describing the skeleton in a reversed position to the true one"—were reprinted in the *American Journal of Science*. And Leidy also claimed that *Elasmosaurus* was in any case not new; it was really the same as an existing plesiosaur named—by none other than Leidy himself—*Discosaurus*.[13]

It seems unlikely that Leidy would have done this unless there had already been a serious breach between him and Cope. That breach may

have occurred in connection with Benjamin Hawkins's plans to reproduce *Hadrosaurus* for the New York Paleozoic Museum—possibly
Cope's *Laelaps* had at first been excluded from the displays.

In a reply to Leidy's paper, and in his hurriedly revised description
of *Elasmosaurus,* Cope responded icily that he had been led astray by
none other than Leidy himself, whose earlier reconstruction of a different plesiosaur turned out to have erred in precisely the same way, something that Leidy had failed to acknowledge: "Prof. Leidy does not,
however, allude to the principal cause of this error, which was the similar reversal of the vertebral column in his descriptions of his genus
Cimoliasaurus," he wrote.[14] Both men had been deceived by the nature
of the articulations of the vertebrae, and Cope had also been assured by
Turner that in the field the head of *Elasmosaurus* had been found at the
tip of what turned out to be the tail.

No explanation would smooth over Cope's appalling solecism, and
the net result of the incident was both that he became even more es-
tranged from his major Philadelphia colleague—Leidy—and he had
made himself vulnerable to criticism by Marsh. Some new sides to the
personalities of Leidy, Cope, and Marsh were exposed in this episode.
Leidy turned out not to have been so saintly after all, revealing too
quickly and bluntly the extent of an evident distaste for Cope and his
methods; Cope was shown as working too quickly and too carelessly,
and taking criticism badly; and Marsh was revealed as competitive and
ruthlessly unforgiving.

It has been said that this event abruptly changed the personal equation
between Cope and Marsh; in later years Marsh claimed as much. But it is
unlikely to have been the single deciding factor. In fact, probably no single
factor precipitated their mutual animosity, and the descent of the Cope-
Marsh relationship into outright hostility does not seem to have been
abrupt. In retrospect, Cope and Marsh seem to have been destined to compete with each other rather than to cooperate. Given their very different
personalities and colliding ambitions, their estrangement grew inexorably
out of a hundred different small incidents and impressions. Even before
the *Elasmosaurus* affair, for example, Cope had had suspicions that "his"
collectors in New Jersey were being suborned to sell their finds to Marsh.

That these two giants of American science would come to excoriate each other as frauds and know-nothings might have been inevitable. But in a sense it was all madness. It destroyed them both and dragged down their friends. To put it dramatically, as the two former colleagues became ineluctably drawn into conflict their subsequent lives had every characteristic of a classic Shakespearean tragedy. It certainly is one of the defining narratives of American paleontology.[15]

One of the aims of this book is to avoid presenting the Cope-Marsh feud as the whole story of midcentury American paleontology. Before we turn to the legendary exploits of Marsh and Cope in the great collecting localities of the American West, therefore, there is another question to be asked. What was the greatest naturalist in the United States—Louis Agassiz, based at his great museum at Harvard—doing when news of all those fossil treasures started to come in from the West? Agassiz was never reticent in his pursuit of science. Since coming to America from Switzerland in 1846, Agassiz had made Harvard the center of American zoology. Why did he not compete directly with the youngsters Cope and Marsh? Why, indeed, had he not preempted them both and already used his undoubted influence to gain the most favorable position in the West?

Agassiz was fully familiar with all the fossil treasures that had come out of the West since Prout's and Leidy's first studies of the White River Bad Lands fossils. Captain George Sternberg had written to him from Fort Wallace in Kansas and sent him specimens. Agassiz had even been in St. Louis at the right moment to settle the differences between Evans and Hayden in 1853 as they set off up the Missouri. And he had made a western tour in 1868, with a party organized by Senator Roscoe Conkling of New York.[16] In the stifling heat of August, his group visited western Kansas, where Agassiz met Sternberg at Fort Hays and obtained one of the very first of the newly discovered mosasaur skeletons for Harvard, the one collected by Cunningham and Minor. At Fort Wallace he met Turner, who had only recently sent the *Elasmosaurus* skeleton to Cope.

Turner described him as "searching after bugs, fish and fossils. He is certainly a very funny old fellow and afforded us much amusement by his jokes and quaint manners." But Agassiz had not enjoyed that trip—made in the heat of August and September—and there had been genuine danger from the Indians because Kansas was "very unexpectedly in an horrible Indian war."[17] Typically enough, Turner was dismissive of the visiting eastern big-shots: "The present Indian troubles . . . are well calculated to infuse some little life into a man—or at least one would so infer from the fear expressed by all eastern people we have an opportunity to meet, . . . [who] . . . certainly showed more consternation than you would expect from the brave manner in which they discuss Indian affairs in Congress."[18]

Agassiz cast his professional eye on the scenery for evidence of glacial action (the subject that had made him famous as young man) and erroneously (and, to some contemporaries, comically) ascribed the entire topography to the effects of Pleistocene glaciation. Next the party traveled by rail tantalizingly close to where, within a couple of years, the great Eocene Bridger Formation collections would be made. There Agassiz met Hayden and collected, or purchased, some Green River Formation fishes. Hayden wrote to Leidy from "near Green River . . . I am far west at the end of the UPRR. Prof Agassiz has just left in high glee, he has found much of great interest. He has obtained a fine jaw and teeth of a species of Mylodon and several species of fishes with fossils. He says he now understands my work as he never understood it before."[19]

Agassiz did not return to the West. He may simply have felt that at more than sixty years of age he was too old for camp life and the privations of an extended collecting trip to Kansas or Wyoming. Whatever his reasons for not seizing the initiative by collecting in a region that obviously had a huge potential, Agassiz held back. He did not even send out an assistant. And then it was too late. An exchange of letters between Agassiz and Marsh gives a partial, after-the-fact rationalization of the issue. Agassiz claimed to have left the field to Marsh because Marsh had got there first—and was far better funded.

Early in 1873, after news of Marsh's first triumphs in the field ("your astounding announcement concerning Odontornithes") had become widespread, Agassiz wrote to him offering to have his expert staff in Cambridge make duplicate casts of some of Marsh's treasures, for study and display. Agassiz was a firm believer in the exchange of casts of important material between researchers as a supplement to published reports, far preferable to mere drawings. "If good casts of your specimens go forth, they will for ever be referred to in preference to any other publications."[20] Evidently this suggestion was not well received, partly because Marsh guarded his research material jealously, and partly because he had no interest in seeing scientific material displayed for the public. He may also have felt insulted by the implication that the illustrations in his publications were in any way to be considered inadequate.

Being so firmly rebuffed by the younger man, Agassiz wrote again, covering his tracks a little. "The matter is simple. From the beginning of the organisation of the Museum I have refrained doing in Cambridge what is done as well and may be better elsewhere, our means for scientific purposes not being, in any of our institutions, sufficient to do everything equally well. So when I saw how energetically you were pushing the investigation of our fossil vertebrates I desisted altogether in my efforts in that direction. . . . [T]here is no probability of these wonderful forms being found in sufficient number to supply several Museums."[21]

In fact, Agassiz eventually did employ private collectors out west for his own museum, too. And the supply of "wonderful forms" turned out to be almost inexhaustible. But even without Agassiz directly competing with them, the whole American West was, it seemed, not big enough to contain both Cope and Marsh without the two of them quarreling. Their mutual hostility fueled a frantic race to scour the West for fossils and, in the process, became one of the great legends of all science. Whether science as a whole gained or lost in the process remains debatable.

SIXTEEN

Riding the Rails

The decade of the 1870s was pivotal for the discovery of fossil vertebrates in America. From the many geographic and geological surveys, information about everything to do with the West—from soil conditions and timber stands to mining prospects (precious metals and, of course, coal)—was changing with ever-increasing speed. This fed dramatic changes in the economy. Between 1860 and 1870, wheat production in the north-central states more than doubled, and corn production increased threefold. Even more significantly, the amount of silver produced increased from 156,000 tons in 1860 to 36 million tons in 1873. In 1850 the country was already mining some 7 million tons of coal annually. But although emigrants to the West might have found it limitless in its geographical extent and personal opportunities, for fossil collectors like Edward Drinker Cope and Othniel Charles Marsh, when they began their own explorations westward, the land was already almost crowded and access to it immediately became contentious.

In 1870, apart from a number of important amateurs and the part-time collectors like Mudge's colleagues in Kansas, five men stand out in retrospect as having made serious professional contributions to the field of bone hunting: Hayden, Leidy, Mudge, Cope, and Marsh. Leidy had never collected out in the field, and Hayden was principally involved with the government geological surveys. Mudge tried to accommodate everyone and was content to remain principally a collector. The feud that subsequently developed between Cope and Marsh has become legend.

In the bitter thirty-year battle between these two giants, the primary issue was competition—for fossils, for access to the sites from which the fossils came, and for the prestige, authority, and even political position that came with describing these finds for science. Such rivalry was not original with Cope and Marsh. Even from the time of Hayden's first expedition to the Dakota Bad Lands with Meek, in 1853, when they found themselves literally in the same boat and with the same destination as Dr. John Evans, different groups of collectors, working for different patrons, had often found themselves in the predicament of competing for the same fossils in the same place. Cope and Marsh, however, took things to new lows.

Given the vastness of the West, why was there not room for all? That immensity was, of course, part of the problem: in such an extent of wilderness, where would one start looking? Part of the explanation therefore is that, as with any gold rush, once spectacular finds were discovered at a particular site, everyone wanted to go and collect there, or to closely neighboring regions, if necessary elbowing each other out of the way. As Leidy sat in Philadelphia steadily describing the western fossil vertebrates sent to him in the immediately pre- and postwar years, he was signaling to the rest of the world where the best fossil localities would be. As he apparently had no interest in following up those discoveries for himself, it became open season for others.

A related cause of the competition was that scholars like Marsh and Cope (the former to a greater extent) relied on the experience of their informants (often paid collectors) out in the field. These men really knew the ground and would offer their services to collect in places where they had previously found fossils. But those collectors also worked for more than one patron. For example, Charles Sternberg (younger brother of George M. Sternberg) collected for Cope in the 1870s, for Agassiz in 1881 and 1882, for Cope in 1883, and from 1884 onward for Marsh. Rival camps could therefore easily come to argue over who had the rights to, and priority for, given sites.

A further factor in all this was even more basic. The total number of productive regions, until the 1890s, remained small. And that in turn was because the fossil sites over which people squabbled were historically and geographically contingent. They were not simply the best sites out of hundreds found through careful scrutiny of the millions of square miles of the West; they were the handful of sites that had been discovered and developed *simply because they were accessible.*

Geography and history controlled the development of paleontology in the West. Even a cursory glance at a map of the United Sates between the Missouri and the Rocky Mountains will show the basic situation. The western mountains extend in a line (several lines, actually) running more or less north–south, and the rivers draining from them run (also or more or less) west to east from the mountains toward the Missouri River. At the northern end of this great region, the Missouri itself turns to run west to east as it drains the Rocky Mountains. The rivers feeding the Missouri, from the Arkansas to the Judith, have carved deep into the rocks to reveal thousands of square miles of badlands, principally Cretaceous and Tertiary in age, full of fossils. To get to these sites, the early scientific traveler had either to go up the Missouri and then head west overland, or to go wholly overland from hubs like St. Louis. Whether by steamboat or by wagon, travelers could take only a small number of routes that were largely defined by the rivers. Those routes were pioneered by the early emigrants heading to the Far West and the trappers and traders working for the fur companies.

At first, western travel meant using the rivers to get into the hinterland of the Upper Missouri country. The river route up the Missouri became particularly important after 1831, when the American Fur Company's paddle steamer *Yellowstone* showed that it was possible to reach points far upriver like Fort Benton. The Missouri steamboats principally served the fur companies, which used them to take trade goods upriver and bring down furs. They offered an attractive alternative to the more arduous overland route. It is no surprise therefore that the first vertebrate fossils were found in the vicinity of the fur companies' forts and trading posts. And it is no coincidence that many of the important fossil-bearing geological formations of the West are named for these early trading forts: Fort Benton, Fort Pierre, Fort Laramie, Fort Union, Fort Bridger, and so on.

An ever-growing network of wagon and stage roads had reached across the West before the Civil War. Many of these were originally Indian or buffalo trails that became institutionalized by simple dint of repeated travel. Later, the government improved them. These roads were the major route of travel for the gold seekers and farmers who walked behind their wagons from the east. Soon they were essential for the carriage of the overland mails all across the West and to California and Oregon. The most famous of these roads was the Oregon Trail, which led from jumping-off points in St. Joseph and Independence, Missouri, through Fort Kearney, Nebraska, along the North Platte River, to South Pass in Wyoming, and then down to Fort Bridger. The Mormon Trail, the Bozeman Trail, and the California Trail all split off from the Oregon Trail in western Wyoming. Other important trails were the Cherokee, Smoky Hill, and Santa Fe Trails crossing Kansas. All along these routes, more forts and settlements were established. In addition to the old fur companies' forts, there were now the stage posts, which tended to be set between ten and fifteen miles apart along the roads. After the Civil War, when Indian attacks on travelers were at their height, army forts were set up along the same routes. All these settlements attracted and concentrated people and, in turn, tended to focus the patterns of local exploration and the discovery of fossils. Such was the case with the classic sites of the Kansas Cretaceous, discovered by Turner, Sternberg, Mudge, and many others, along the Smoky Hill Trail from Independence to Denver.

After about 1870, the new continental railroads fundamentally changed the situation again. The Baltimore and Ohio had reached St. Louis in 1857, making that city even more of a focus for westward travel. Another early hub was St. Joseph, Missouri. Expansion of the rail network beyond the Missouri River and across the prairies began only after the Civil War, and because they needed the same flat elevations, the newly surveyed continental routes often followed the wagon roads. The Union Pacific route, for example, followed the Oregon Trail, taking the north bank of the Platte River between Omaha and Fort Kearney, while the stage road took the south side. The Kansas Pacific took its way west alongside the Smoky Hill River stage route from Topeka, through Fort

Hays to Fort Wallace. Therefore we find that another set of important geological formations were named in accord with stops on the railroads, and this has led to some confusion in identifying some of the old fossil localities. Specimens collected at "Monument Station, Kansas," for example, were probably not found near the railroad tracks but adjacent to the stagecoach station several miles away.

Once the transcontinental railroads had been driven through (and after the famous moment of linking-up of the Central Pacific and the Union Pacific on May 10, 1869), the paleontologists quickly followed. The reason that both Cope and Marsh naturally would want to collect at, for example, Fort Bridger, Wyoming, or Fort Wallace in western Kansas, was not simply that they knew interesting fossils had been gathered there: they now knew that they or their collectors could easily get there. In some cases, the beds were right next to the tracks. The famous Green River Eocene fossil beds, for example, which yielded a superabundance of fishes, reptiles, and mammals, just happened to be accessible via the Union Pacific Railroad, at Green River Station and Rock Springs Station. Until the route was straightened, the tremendous dinosaur site at Como Bluffs, Wyoming, discovered in 1877, was within sight of the Union Pacific tracks, and passengers would wave to the diggers from their Pullman cars as they passed by.[1]

As towns, some temporary and some permanent, grew up along the railroad routes, they very quickly became places where collectors could equip their expeditions with horses and supplies; there they could hire extra men and strike off on their own. Eventually more and more fossil sites would be found, progressively farther from the principal rivers, wagon roads, and railroads, but for a long time the exigencies of transport and access virtually dictated that some of the "hottest" sites would be close to the well-traveled routes and consequently visited by multiple collectors. Given these contingencies of access, it is not surprising that the pattern of discovery of fossil vertebrates was essentially coincident with maps of the means by which the country was made accessible: by water, road, and rail. Tracing out the parallel development of the transportation system and explorations for fossils allows us more completely to set vertebrate paleontology in its wider cultural context.

Of all the routes of access to the West, the railroads were the most important for the new science of American paleontology. They gave quick transport to the fossil digs; often the railroad companies, particularly the Union Pacific, would give free passage and freight haulage to the explorers. Without such largesse, Cope would never have managed to travel the country as extensively as he did. And because the routes were sometimes blasted through hills and mountains (instead of winding over and around), they produced valuable new exposures of rocks, revealing geological structures and fossil beds that had hitherto been hidden. Hayden, for example, in a letter to Joseph Wilson, commissioner of the General Land Office, wrote: "[I will] push on to Fort Bridger by way of the overland stage-route and returning along the Union Pacific Railroad, so as to construct a geological section of the route, making use of cuts in the road to give me a clearer knowledge of the different beds. My party consists of nine persons. We have a two-horse ambulance and a four-mule covered wagon, three tents, and four riding animals."[2]

Perhaps most important, the work of constructing the railroad routes significantly increased the number of people *in situ*. This included a whole new category of part-time naturalists: railroad men with expertise in surveying and construction and an interest in the landscape became geologists and fossil collectors. Exploring in their spare time many miles on either side of the tracks, these men discovered a number of fossil localities and made important finds. In describing the Eocene fish beds of the Green River in Wyoming, for example, Hayden gave credit to Mr. A. W. Hilliard, who "superintended the excavation along the line of the railroad [and] preserved from time to time such specimens of value as came his way."[3] The Como Bluff dinosaur beds were discovered by a Union Pacific station agent and the section foreman.

A fascinating aspect of the haphazard and opportunistic way that fossil sites were discovered and exploited is the fact that many sites were missed at first. The Union Pacific tracks had been driven past Como Bluffs, Wyoming, for example, almost a decade before dinosaurs were discovered there. Collectors had traveled past the site dozens of times en route to visit the Eocene beds of the Fort Bridger region, farther west. The country was simply too huge to be explored completely and more systematically.

The other side of the coin is that although the railroads gave increased access to the West, in 1870 it was a more hostile land than it had been even twenty years before. In the journal of his trip to the Upper Missouri with fur company traders in 1850, for example, Thaddeus Culbertson recorded only amicable interactions with Indians, as he wrote on April 10: "We discovered at a distance a company of Indians. . . . [T]wo scouts reached us at full gallop, and we accompanied them to their encampment. . . . [M]en, women, children, dogs and horses, all came out to look at us." On May 1: "On the opposite side of the river were pitched about two hundred Indian lodges. . . . [A] number of the Indians swam across the river, cold as it was, and something had to be cooked for them. . . . [T]hey were all anxious for horses, and two of them had the traders' receipt for thirty robes. They, of course, must be supplied, and soon one of them was capering around us on a fine bay horse, which he had selected. . . . [T]here appeared to be much good humour on all sides." Everywhere he and his companions went they were treated courteously by the Indians and were never under physical threat.[4]

The earlier history of the eastern states came to be repeated in the West, however, as more and more white men arrived to become settlers and miners, to claim land, to graze cattle, and to farm, and as the white men killed huge numbers of buffalo to feed their railroad workers, and the traditional migratory routes of the buffalo were disrupted. In short, when the white men became far more of a threat than a resource, conflict increased. Complicating matters, in the face of the waves of western migration, the eastern and midwestern tribes had also been forced farther west. The result was that Indians were fighting Indians. And all Indians were resisting the new emigrants from the East.

Only five years after Culbertson's trip, Hayden wrote to Leidy from the same country: "We have just heard of the murder of Mr Malcolm Clark, one of the factors in the fur company, killed by a war party of Sioux in ascending the Yellowstone. Several murders have already occurred and some hands have threatened to kill any white [men] they meet. In two days from this time I start for the 'Bad Lands' with a cart 4 horses and 2 men—There is no snow on the ground now and may be none this month and no danger is apprehended from Indians, whereas in the months of May and June it would be almost certain death for a small party, and much trouble for a large one."[5]

In the 1760s Ben Franklin and many others had worked to establish a boundary between the settlers of western Pennsylvania and the Indians of the Six Nations. As early as 1744 there had been the Lancaster Indian Treaty. In his retirement years, one of George Washington's priorities had been to try to find a solution to the problem of coexistence with the Indian peoples through granting them sole occupancy of large tracts of what is now the Midwest. Under President Andrew Jackson's Indian Removal Act of 1830, all the eastern tribes were "encouraged" to migrate west of the Mississippi. But an unstoppable flow of whites followed them. In 1834, responding to the same problem, now translated westward, a United States government report nominally established a zone between the Platte and Red Rivers that marked the "Western Boundary of Habitable Land." East of that line settlers would be protected by the army. But the exercise was essentially nugatory. Once again, the flow of people and the draw of natural resources created too strong a force to be resisted. The only solution seemed to be to confine the tribes to reservations, which meant that, since they lost their lands, the government had to support them and reacted brutally if they broke out of their designated areas.

After the Civil War, the situation became so bad that all parties venturing out west for the purposes of surveying and fossil collecting needed an army escort. And to obtain an escort, one needed an official connection to the government, or some influence. For paleontologists this meant that travel was greatly restricted. The army, occupied with protecting the emigrant roads and settlements, lacked the resources to supervise every group of fossil collectors who came out eager to explore, collect, and generally put themselves in harm's way. So this became another factor constraining the range of sites that collectors could explore. They had to content themselves with peaceful times or places, or be satisfied with regions where the army already was in place or was prepared to go. Only official United States Government Survey groups had a broad capacity to command escorts and explore widely.

Of course, no account of travel in the West would be complete without mention also of its desperadoes. Just as the tribulations of Mary Draper Ingles a hundred years before were due in part to white renegades, so, as the frontier moved west, horse thieves, stage and train robbers, cattle rustlers, and all those thugs who thought (usually erroneously) that thieving was an easier way to make a living than working, and who preyed on the hard graft and misfortunes of others, went with it.

There is little evidence that fossil hunters were ever seriously affected by the criminal element. They tended to travel in the more remote regions and were usually accompanied by military escorts. They worked with groups of a dozen or more men, had nothing worth stealing except possibly their weapons—and they were obviously just a little crazy. Marsh had a mild encounter with horse thieves in 1870. In 1878, an incident of attempted train wrecking (possibly involving Frank James, Jesse's brother) happened within a mile of Como Station and the famous dinosaur quarries. The gang removed a rail, but the problem was spotted before any train came along. One of the ringleaders was later killed while holding up a stagecoach heading from the Black Hills to Cheyenne. One was lynched at Carbon City, and another at Miles City, Montana.[6]

It was into this rapidly changing western world that the fossil hunters ventured in the 1870s, relying heavily on the experience of those who had gone before (particularly Hayden), but prepared to ignore anyone (particularly one another) who stood in their way.

SEVENTEEN

The First Yale College Expedition

We started from Fort McPherson [Nebraska] accompanied by a company of cavalry; for this was the country of the Sioux. . . . Across an unexplored desert of sand hills between the river Platte and the Loup Fork . . . the celebrated Major North, with two Pawnee Indians, undertook to guide us. . . . [T]he Indians, with movements characteristic of their wary race, crept up each high bluff, and from behind a bunch of grass peered over the top for signs of hostile savages. . . . Next in the line of march came the company of cavalry . . . and with them rode the Yale party, mounted on Indian ponies, and armed with rifle, revolver, geological hammer, and bowie knife. Six army wagons, loaded with provisions, forage, tents, and ammunition, and accompanied by a small guard of soldiers, formed the rear.

CHARLES BETTS, 1871

Before 1868, neither Cope nor Marsh had actually visited any part of the West. Cope was the more experienced in matters to do with Kansas, if only for the infamous *Elasmosaurus*. In his 1871 summary of collections from the Cretaceous rocks of Kansas (written in 1870), Cope noted that he would "be glad if his friends in the West" would forward to him in Philadelphia, "at his expense, specimens of bones or teeth which they may find."[1] Mudge had written to Marsh about his discoveries in eastern Kansas, and Professor John Parker sent Marsh some "saurian remains" from Kansas. Dr. George Sternberg wrote to him about the Smoky Hills Chalk, having found bones where everyone else eventually would, along the stage road between Fort Wallace and Monument Station. He reported to Marsh that the valleys of the Republican, Solomon, and Saline Rivers, north of the Smoky Hill River, "are probably equally rich but have never been explored. Chalk Bluff Creek, opposite Monument Station is also a rich locality."

Mudge had actually met Marsh years before; when Marsh was a schoolboy and Mudge a young lawyer they had been members of the same mineralogical club in Massachusetts. Mudge's early letters to Marsh from Kansas were somewhat obsequious, perhaps because he wanted to exchange information and fossils for copies of hard-to-get scientific literature. "I was much pleased . . . to learn of your present honorable position. I had seen some of your communications from Germany. I had also read in the Journ. of Science your article in the July No. [this was Marsh's account of an Ohio Indian mound]; I was much pleased with the detail and accuracy of your description. . . . I send you my Report on the Geology of Kansas. You will find it small but the appropriation was small. . . . I can send you a collection of fossils made in this immediate vicinity. They are small but interesting as being on the borders of the Permian and Coal measures. I have also a few fossils from the Cretaceous obtained in July. . . . My next excursion will be further S. and West. Some large portions of our State has never been visited by a scientific man."[2]

In July 1868, Mudge wrote offering to send "a fragment of a bone of a Mastodon," and between 1868 and 1870 he sent both photographs and specimens, including a slab of Pennsylvanian-age bird tracks from near Junction City. But Marsh was interested in neither footprints nor mastodons, and it was not until twenty-five years had passed that he published on them.[3] Meanwhile Mudge continued to explore westward into the valleys of the Solomon and Republican Rivers and sent his "saurian" material to Cope, while encouraging both men to come out in person.

Marsh moved first. Having seen the harvest of Kansas fossils that was being shipped to Cope in Philadelphia, he became determined to go straight to the source himself. He almost missed the opportunity to get in at the beginning of this "gold rush" of vertebrate fossils, but Cope hesitated just too long. In fact, Marsh's first trip west was not to collect

fossils but to attend one of the annual meetings of the American Association for the Advancement of Science, which in 1868 was held in
Chicago. A side trip offered at the August meeting was an excursion
courtesy of the Union Pacific to show off its newly opened tracks,
which then reached all the way to a railhead at Benton, Wyoming (near
Rawlins). One can only assume that the association and the railroad did
not pick Benton deliberately as a destination—they merely promised a
visit to the railhead, which was a constantly moving target. And the trip
west would take them through some glorious mountain scenery. As it
happened, the scientists must have had something of a shock, as Benton
lived up to every caricature of a western frontier town, and then, as
quickly as it had sprung up, became a ghost town.

The place existed for only three months (July to September 1868)
until the railroad moved farther west. At its height it had a population of
some three thousand desperate souls, twenty-five saloons, and five
dance halls, mostly in tents. Apart from the railroad, its principal indus-
tries were alcohol, gambling, and prostitution. In the largest of the
tented saloons the proprietors installed a magnificent mahogany bar, replete with fancy mirrors and the traditional paintings of nudes. It had
come from St. Louis and moved with the railhead, steadily west. In his
novel *UPR Trail,* the writer Zane Grey (not given to understatement)
wrote of Benton that at night, "every saloon was packed, and every dive
and room filled with a hoarse, violent mob of furious men; furious with
mirth, furious with drink, furious with wildness—insane and lecherous,
spilling gold and blood." In three months more than a hundred men
were shot to death.

Whether Marsh was titillated by all this, appalled, or simply ignorant because the train arrived in the daytime when the denizens of Benton were either working or hungover, we do not know. He had eagerly
participated in this field trip because he knew that at one of the stops
along the way—Antelope Station, in western Nebraska—supposed fossil human remains had been found. When the excursion train halted
there briefly, Marsh got the conductor to delay long enough for him to
locate the site (an excavation for a well) and then paid the stationmaster
to pick through the spoil heap and assemble some of the fossils for him,

which he picked up on the return leg. As he had suspected, they were not human fossils, but just as interesting to him: Pliocene remains of a pig, a camel, and—most exciting of all—a tiny species of horse, a creature standing only three feet or so high (something Leidy would dearly have loved to have had).

In later reminiscences Marsh rehearsed a vision of the western plains already familiar from Maclure and Hayden: "It was my first visit to the far West, and all was new and strange. I had a general idea of the geological features of the country I was to pass through . . . but the actual reality was far beyond my anticipations when I found myself surrounded [by the Great Plains] reminding me of mid ocean with its long rolling waves brought sudden torrents. It was in fact the bottom of an ancient sea, and not the petrified waters, that I then saw, and I was not long in deciding that its past history and all connected with it would form a new study in geology, worthy of a student's best work, even if it required the labor of a lifetime."[4]

One of the new and strange features of the West was the strongly alkaline Lake Como in southeast Wyoming, which had a population of curious "fishes with legs." In a neat invasion of Cope's intellectual territory, Marsh brought some of these creatures home. As Marsh no doubt suspected, these amphibians were suffering a developmental anomaly due to the chemistry of the lake water. In the more neutral water of the laboratory they metamorphosed normally into salamanders.

Marsh almost certainly hoped to begin his western bone hunting with an expedition in 1869, but was prevented by news of Indian unrest. This delay happened to work to his advantage, as a number of new localities came to light in the interval. Principal among these were the hugely productive Eocene deposits of the Bridger Formation in southern Wyoming, which he learned of from Leidy's 1869 paper on the first fossil mammals from Fort Bridger. The following year Marsh put together a group of students to accompany him on a field trip. It might seem odd that the group consisted only of students; evidently Marsh was a lonely and isolated man with few close friends and colleagues. So

he took twelve students. George Bird Grinnell, one of these participants, later wrote:

> Soon after I heard the rumor of the proposed western trip I took
> my courage in both hands and called on Marsh. . . . Within two
> or three weeks I saw him once more, was accepted as one of his
> party and only then discovered that as yet there was no party—
> except myself. In fact, he at once began to discuss with me the
> possibility of securing other undergraduates for his trip. . . .
> None of us knew or cared anything about the objects for which
> it was being undertaken. Vertebrate fossils meant nothing to us,
> but we all longed to get out into the uninhabited and then un-
> known, West, to shoot buffalo and to fight Indians.
>
> Marsh was possessed of considerable means and had a
> wide acquaintance. He had interested General P. H. Sheridan
> in his project and from him obtained orders directed to military
> posts in the West to provide the party with transportation and
> escorts. . . . [S]ome well-to-do businessmen had contributed
> funds to defray the expenses. . . . Some of these, being rail-
> road men, had given Marsh either free transportation for his
> party or at least rates lower than those usually in force.[5]

Marsh being Marsh, this expedition was no brief excursion. It
lasted for six months, during which the party traveled from coast to
coast. It would be the first of four such trips, in consecutive years, in
preparation for which each student was advised to buy and study a copy
of *The Prairie Traveller,* a handbook prepared by the War Department
as a guide for anyone making the transcontinental overland migration.[6]
It contained detailed instructions on a variety of useful topics, ranging
from lists of necessary clothing and "camp equipage" to the correct
way to harness a mule team and to build cook fires, and hints for parlay-
ing with Indians (including a useful lexicon of terms in various Indian
languages). It advised every traveler to go armed with Bowie knife, a rifle (at first the Henry .44-caliber was favored; the Sharp's .50-caliber
carbine was the weapon of choice in Marsh's time), and a revolver

(Colt, and later Smith and Wesson .36-caliber). Several of the students who took part in Marsh's four Yale expeditions later wrote up their adventures for the press.

One of these, by Charles Betts, was an account of the 1870 trip published in *Harper's New Monthly Magazine,* and it begins with a very familiar set of images:

> The peaks of the Rocky Mountains once projected as islands from a vast inland sea whose waves swept from the Gulf of Mexico to the polar ocean. In this era of the world, a tropical climate extended far beyond the arctic circle, and the tepid waters swarmed with sea-serpents and other reptilian monsters. At the close of this period, known to geologists as the cretaceous, a slow upheaval drained this ocean from the continent, and left behind great lakes, whose shores and waters teemed again, in tertiary times, with new forms of tropical life. Rhinoceros, crocodiles, and huge tortoises backed upon the banks or lay beneath the shade of gigantic palms; and as the ages rolled away prolific nature brought upon the scene the mammoth, mastodon, and horse. During the tertiary period mud and sand accumulated in the lakes to a depth of many hundred feet and entombed the bones of all these animals. Then came a time when all was dry, and torrents from the mountains wore through the deep accumulations. . . . To the region of these eroded basins Professor O. C. Marsh, of Yale College, had long contemplated a geological expedition.[7]

The plan was to use the railroads to leapfrog from one major collecting area to another, picking up army escorts at each stop. As a result, during that one expedition, the party covered more ground and collected more fossil bones than all previous explorers put together had accomplished in the previous twenty-five years. The Union Pacific Railroad, as usual eager to promote itself, first gave the group free transport west to North Platte, Nebraska, from where they went overland to nearby Fort McPherson to collect an army escort, horses, mules, and wagons. Judging from accounts of the trip, the men of the fort were none too pleased to be baby-sitting a group of eastern kids, but the senior officers were soon impressed by Marsh himself, then a fit young man (if already a little portly) and a keen outdoorsman but not, it turned out, much of a horseman.

The initial outing in the field was a two-week circuit north from Fort McPherson to the Loup Fork River, and the first sign of the adventure that the students sought came with news that a party of antelope hunters had just been attacked by Indians. William F. Cody—later better known as Buffalo Bill—who was working as a guide at Fort McPherson, "brought in the moccasins and some trinkets from an Indian boy who had been killed in the skirmish, at which the newcomers from New Haven stared in wonder," in Grinnell's account.[8] Undaunted, the students, none of whom had yet fired a gun in anger or (some of them) even mounted a horse, set off north on captured Indian ponies into the Sand Hill country explored a decade earlier by Hayden and Meek.[9]

"Just before dark, water was found and we camped for our first night out of doors," Grinnell later wrote. "That night at the camp fire Professor Marsh talked to us and to an audience of soldiers about the geological changes that had taken place here in past ages and about the discoveries of unknown animals that we hoped to make. Buffalo Bill, who had ridden out with us for the first day's march, was an interested auditor."[10]

After grueling, often waterless marches, following creek beds and the occasional Indian path, the party collected along the Loup Fork River, finding Upper Pliocene fossil rhinoceros material and some six different species of fossil horses. They wanted to go farther north to the Niobrara River, but news of fresh Indian troubles stopped them. According to Grinnell, "The Sioux and Cheyenne occupied the country of western Nebraska and to the north and northwest, and they objected strongly to the passage of people through their territory, and when they could do so—that is, when they believed they had the advantage—attacked such parties."[11] So the Yale men circled around southwestward back to Fort McPherson, using an old portion of the Oregon Trail to travel east along the North Fork of the Platte River and marveling that the wagon tracks were still visible.

Both Marsh and his students had to learn not only how to get around and survive in the wilderness but also to find the fossils. Unlike the East, where fossils turned up in excavations like marl pits, ditches being dug, roads, and the banks of streams, in the West nature had done the initial work. Collecting meant that the group (watched over carefully by the soldiers) would divide up, each man prospecting over a segment of territory, scrutinizing the surface, and in washes, gullies, and canyons, for signs of bone. Often that meant spotting a mere trace of bone at the bottom of a bank and digging down to where the main body lay. Thousands of years worth of erosion had exposed fossils at the surface, and in those early stages of searching it was not necessary to make excavations and quarries to get a good haul of material. Often a knife was all that was neccesary to get the bone out of the ground. But everything depended on the prospector getting a practiced eye for distinguishing bone from the shades of brown in the sand and rock. It was hard, boring work and not at all what the students had expected. But there were enough fossils in the ground to encourage even the most un-enthusiastic spirits. In Charlie Betts's memorable account:

> The soldiers not only relieved us from all fear of surprise, but soon became interested and successful assistants; but the superstition of the Pawnee [guides] deterred them for a time from scientific pursuits; for the Indians believe that the petri-fied bones of their country are the remains of an extinct race of giants. They refused to collect until the professor, picking up the fossil jaw of a horse, showed how it corresponded with their own horses' mouths. From that time they rarely returned to camp without bringing fossils for the "Bone medicine Man." . . . Our researches resulted in the discovery of the remains of various species of the camel, horse, mastodon, and many other mammals, some of which were new to science.[12]

Moving on by rail to Fort D. A. Russell, near Cheyenne, Wyoming, the men discovered a new set of exposures of Oligocene badlands between the North and South Platte Rivers, collecting turtles, bird and rodent material, and many specimens of Leidy's *Oreodon* and *Titano-therium* (formerly Prout's *Palaeotherium*), together with the signature snail for the Bad Lands strata, *Helix leidyana*. Marsh now had his own "mauvaises terres" locality to match Leidy's, triumphantly sending a report back to the *American Journal of Science:* "This interesting series of fresh-water Tertiary strata lies almost horizontal, dipping apparently, but very slightly, toward the north-east. It probably forms the southwest border of the great Miocene lake-basin, east of the Rocky Mountains, which is so remarkable for its extinct animal remains."[13]

At Antelope Springs the group did some serious digging and collected more of the fossil horse material that Marsh had seen at the railroad station two years before. Then, in September, they were off again—west, overland and this time across very rough terrain to northern Utah and to Fort Bridger in southwest Wyoming. Here they found the remains of what appeared to be a huge extinct mammal that seemed to be an elephant relative with horns on its skull. "South of the Fort were great washed deposits of greenish sand and clay of Eocene age, and here we found great numbers of the extraordinary sixhorned beasts later described by Marsh as Dinocerata," Grinnell wrote. "It was from this locality too that came *Eohippus,* the earliest horselike animal."[14] Along the valley of the Green River they collected from "an Eocene deposit" where "petrified fishes abounded; and we found a small bed containing fossil insects."

Up to this point there was not much chance of Marsh trespassing on anyone else's territory. For as long as paleontologists and other naturalists have been exploring new lands and recording their finds, they have generally followed an informal but firmly observed protocol of honoring the territorial rights of colleagues who came first. When someone works in a particular area, that area is their turf until they relinquish it—usually to a student or other protégé, very rarely by simply walking away. Respecting the ownership of a site or an area is crucial to maintaining any kind of civility or order in the profession.

The situation was somewhat complicated by the fact that Hayden was a government employee and the information he produced was public property, but—at least until Marsh's party reached the Green River

area—a gentleman's agreement had generally held. Marsh had cooper-
ated with Mudge in Kansas; the Loup Fork beds had been discovered
by Hayden but were not being actively worked; Marsh had obtained
Antelope Station material two years before, so that was "his." The sticking point was the whole Fort Bridger region and its complex of sedimentary basins, for Hayden and Leidy had been there first and staked
their claim to digging in the area. Whatever Marsh's right to collect
there, however, there is no doubt that his work in southwestern
Wyoming, in this and subsequent years, significantly shifted the whole
focus of collecting fossil bones in North America.

In Middle Eocene times the environment in the Green River Basin
and nearby basins had been warm-temperate to subtropical. Huge shallow lakes that waxed and waned with fluctuations in the climate had
been surrounded with cypress swamps. At higher elevations, first oak,
maple, and hickory formed forests, and then pines, spruce, and fir grew
near the mountaintops. In this rich environment a huge variety of animals and plants had flourished, while in the lakes fishes abounded. The
environment was diverse enough to support vast herds of the curious
horned mammals that so intrigued all the early collectors there, and also
populations of the first primates (relatives of monkeys) to be collected
in America, and many birds including even the nests of shore birds.
The fossils were preserved in sediments eroded from the higher ground
and in huge deposits of ash from volcanoes, probably those in the modern Yellowstone region to the north. In 1870 the territory around Fort
Bridger was still green meadowland, excellent for grazing and watered
by branches of Black's Fork—a welcome oasis for travelers in a landscape that otherwise was a semi-desert of dry prairie grass and sagebrush.

The students' accounts of the trip conveniently omit any mention of
Hayden or Leidy's prior work on fossils from the Bridger area, but Hayden, who was still actively exploring and surveying there, openly resented Marsh's arrival. Marsh could have approached the situation
tactfully. Instead he offended Hayden with his arrogant behavior, his

insistence on going wherever he pleased, and the fact that his small army of collectors was picking the fossil fields clean. This was no small matter. In Marsh's expeditions not just the students but the Indian scouts, the local men he hired, and the soldiers all collected the fossils. When Marsh cavalierly ignored Hayden's priorities, Hayden wrote anxiously to Leidy: "I [will] start a box or two of vertebrate remains to you . . . in order that you may . . . see if there are any new species [and] have a chance to describe them before Prof Marsh returns. He has been ransacking the country with great success and claims important discoveries."[15]

Hayden also warned Leidy: "I advise you to publish your species as soon as described. Marsh is still down on Green River. I may see him. I will send everything to you as soon as I get it, so do not fail to attend promptly to those boxes when they arrive."[16] A month later Hayden complained again: "He is raging ambitious, Marsh . . . claims that I have interfered with him near Bridger."[17]

Behaving in this manner would have been self-defeating for Marsh if he had not had the support of the local men, the same men who had earlier been sending specimens to Leidy. It is clear, for example, that Marsh had carefully cultivated the influential Judge Carter, who essentially sponsored Marsh's 1870 trip to Fort Bridger. "You will find no more interesting and satisfactory field for investigation than this portion of Wyoming Territory. One that is entirely free of the dread of hostile Indians," Carter wrote to him. "In coming here you will have to stop at Carter Station, and the Commanding Officer, Major Lamothe, will aid me in transporting yourself and party."[18] What Carter made of the apparent conflict of interest with respect to his supposed friend Joseph Leidy is unknown. Perhaps a professor from Yale was more impressive to Carter than someone like Leidy, who refused to go west himself (or, indeed, a government surveyor like Hayden).

Perhaps because of the conflict with Hayden, Marsh decided to trek from Fort Bridger through the Uinta Mountain region to where the Green River was joined by the White River, a distance as the crow flies of some ninety miles, but longer overland, as they first followed Henry's Fork to the Green River before heading south. "No exploration of this region had ever been made; but hunters and Indians had brought back fabulous stories of valleys strewn with gigantic petrified bones," Betts wrote. This was brutal country for mules and wagons, however. Soon the party had to abandon the wagons and cache their stores of grain in order to proceed. "To this geological paradise the shortest route lay across the Uintah Mountains, the altitude of whose lowest pass is eleven thousand feet; but we could find no guide through these rugged defiles, and were obliged to follow the circuitous course of the rivers."[19] The group rounded the eastern end of the Uintas and looked off to the southwest, and again Betts marveled:

> After crossing an extensive lake table a grand scene burst upon us. Fifteen hundred feet below us lay the bed of another great Tertiary lake. We stood at the brink of a vast basin [the Uinta Basin], so desolate, wild and broken, so lifeless and silent that it seemed like the ruins of a world. . . . [The land] was ragged with ridges and bluffs of every conceivable form; and rivulets that flowed from yawning cañons . . . threads of green across the waste, between their falling battlements. Yet, through the confusion could be seen an order that was eternal, for as age after age the ancient lake filled and choked with layers of mud and sand, so on each crumbling bluff recurred strata of chocolate and greenish clays in unvaried succession.

The prospectors were exhausted and the collecting was slow. Betts continued: "Though we found none of the gigantic bones of which we had heard so much from hunters and Indians, yet, as we ascended the [White] river, the fossils increased in number, until from one point of view we counted eleven shells of Pliocene tortoises."

A Shoshone showed them a route back over a beautiful forested country to Henry's Fork at the southern end of the Bridger Basin, where it turned out that a band of on-the-run horse thieves had appropriated their grain store. They caught up with the brigands, and Marsh rather bravely faced them down and took the wagons back north to Fort Bridger. After all these exertions Marsh gave everyone a rest, and they took the Union Pacific to Salt Lake City and then went on to California, before returning to Green River, Wyoming, to collect more fossil fishes.

It was November, but still they had not finished. On the advice of George Sternberg that Indians would be absent from western Kansas at the end of the year (he was wrong), Marsh now took his party by train to Fort Wallace, Kansas, and started to collect in the Smoky Hill Chalk. Here, in Betts's account, "remains of cretaceous reptiles and fishes were collected in great quantities. One trophy was the skeleton of a sea-serpent, nearly complete, and so large that we spent four days in digging it and carrying it to camp."[20] With this, Marsh was now treading on the toes both of Hayden and Cope; apparently he also did not include Mudge in this work. At least, when Marsh said that he was going to Kansas, Hayden deliberately stayed away: anything to keep the peace. Hayden reported to Leidy: "Marsh has gone to Kansas after reptiles. At his request I do not go, leaving the whole field open to him."[21]

As the various memoirs of Marsh's four western trips show, these expeditions were very hard work, especially for such novices. Given that the students were not prepared for the rigors of fieldwork and camp life, it is remarkable that the four ventures were carried off without loss of life or even serious injury or illness. A great deal of the credit must go to their local guides and army escorts. But life as a fossil collector was not nearly as exciting as the students had hoped. "Most of us had gone out mainly for the sport," one of them recounted. "We expected to see something of the Wild West, and I know that I also hoped to learn some geology. Most of our time was in point of fact spent in the bad lands."[22]

There were only two serious defections: during the first expedition, James Wadsworth, "who knew nothing about fossils[,] and another fellow finally broke away from the digging job and went off on their own, joined a troop of cavalry chasing Indians with Buffalo Bill as scout (they had a very rough time of it and came back thoroughly battered but happy)."[23] Marsh was not always a popular or communicative leader. The students found that "Marsh was rather a figure of fun to the youngsters altho' they liked him," as Wadsworth's son later recalled. They had little respect for his horsemanship, however. "My father always referred to the Professor as 'Whoa Marsh' and explained that the boys gave him that nickname because of his nervousness as a rider—constantly say 'Whoa!' to his horse."[24]

At least one student (from the fourth expedition, in 1873) later remembered: "We found it difficult to get any information from Professor Marsh on what we were doing. I cannot recall that he ever gave us even a cursory lecture on the geological formations on which we were working. Or the possible significance of what we were finding. If we asked him questions, he was very apt to give a few of his characteristic grunts and return a noncommittal answer. At that time his bete noire was Professor Cope of Philadelphia, and I always thought he was afraid that if he told us anything it might possibly leak back to his antagonist."[25]

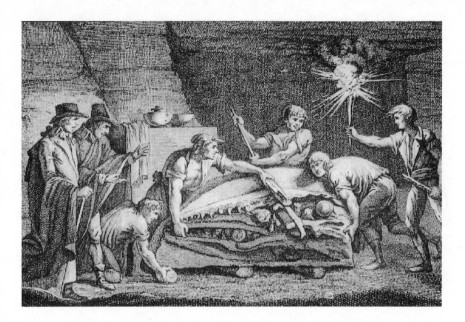

Excavation of the Maastricht mosasaur in 1780 (from Barthélemy Faujas-de-St.-Fond, *Histoire Naturelle de la Montagne de Saint-Pierre de Maestricht*, 1799)

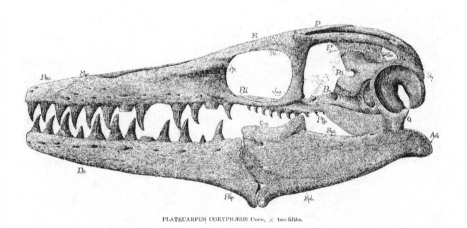

PLATECARPUS CORYPHÆUS Cope, × two fifths.

Skull of the mosasaur *Platecarpus coryphaeus* from the Cretaceous of Kansas (from Edward Drinker Cope, *Synopsis of the Extinct Batrachia and Reptilia of North America*, 1873)

View of the White River Bad Lands of South Dakota (photograph, Kevin Walsh)

Monument Rocks, Logan County, Kansas (photograph, John Charlton and Kansas Geological Survey)

Joseph Leidy with tibia of *Hadrosaurus*, 1859; half of a stereoscopic pair (Academy of Natural Sciences)

Ferdinand Vandiveer Hayden in camp, Yellowstone, Wyoming, 1871 (photograph, William Henry Jackson, National Parks Service)

Benjamin Waterhouse Hawkins and his mount of *Hadrosaurus*, half of a poorly retouched stereoscopic pair, circa 1868 (Ewell Stewart Library, Academy of Natural Sciences)

Edward Drinker Cope, 1876 (Ewell Stewart Library, Academy of Natural Sciences)

Othniel Charles Marsh, carte-de-visite by Moulthorp and Williams, New Haven, 1862
(Peabody Museum of Natural History, Yale University)

Street scene in Benton, Wyoming, 1868 (Wyoming State Archives, Department of State Parks and Cultural Resources)

Petrified Fish Cut, Green River, Wyoming, 1869 (photograph, William Henry Jackson, United States Geological Survey)

Captain Theophilus Turner, M.D., carte-de-visite by I.G. Owen, Newton, N.J., circa 1866 (Ewell Stewart Library, Academy of Natural Sciences)

Dr. James Van Allen Carter, 1864 (Wyoming State Archives, Department of State Parks and Cultural Resources)

Dr. Joseph Corson, 1869 (Wyoming State Archives, Department of State Parks and Cultural Resources)

Judge William Carter in an undated photograph (Wyoming State Archives, Department of State Parks and Cultural Resources)

Badlands of Black's Fork, Uinta County, Wyoming, 1870 (photograph, William Henry Jackson, United States Geological Survey)

The Uinta Mountains, near the head of the West Branch, Black's Fork, Summit County, Utah, 1870 (photograph, William Henry Jackson, United States Geological Survey)

Marsh's student party of 1870, near Fort Bridger; Marsh is standing sixth from the left (Peabody Museum of Natural History, Yale University)

Benjamin Mudge in an undated photograph (Kansas State Historical Society)

Arthur Lakes, circa 1880 (Peabody Museum of Natural History, Yale University)

William Harlow Reed, circa 1880 (Peabody Museum of Natural History, Yale University)

Sam Smith of the Rocky Mountains, undated photograph (Peabody Museum of Natural History, Yale University)

Morrison Formation "hogback" formed by a protective cap of sandstone (*right side*), near Cañon City, Colorado, 1898 (photograph, I. C. Russell, United States Geological Survey)

Fault in the Judith River beds, "just below the Benton–Cow Island trail," Blaine County, Montana, 1905 (photograph, T. W. Stanton, United States Geological Survey)

Watercolor painting by Arthur Lakes, showing William Harlow Reed (*left*) and Edward Kennedy at Como Bluff, Quarry 10, the site of the excavation of the Yale *Brontosaurus*. On the reverse of the image Marsh has written, "An uncovered stenosaur [probably *Morosaurus*] most of the bones too badly rotten to ship, Como Wyoming." (Peabody Museum of Natural History, Yale University)

Como Bluff seen from the south in a modern photograph, sites of Quarries 9 and 10 in foreground

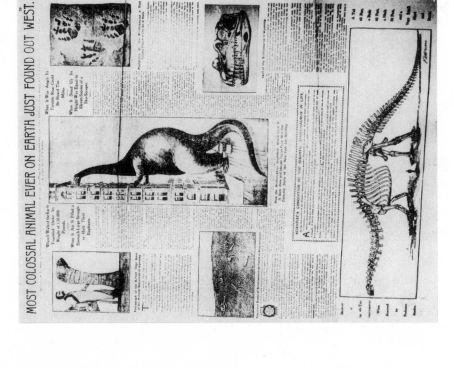

Flammarion's fanciful reconstruction of an *Iguanodon* five stories tall (from Camille Flammarion, *Le Monde Avant la Creation de l'Homme*, 1886)

(*right*) Discovery of a *Brontosaurus* announced in the *New York Journal and Advertiser*, December 11, 1898 (Peabody Museum of Natural History, Yale University)

EIGHTEEN

The Competition Begins

When everyone had returned to New Haven, the number of fossils that had been collected by the first Yale expedition was enormous, but Marsh was particularly intrigued by something that he had found on the very last, bitterly cold, day of collecting near Fort Wallace. With night falling and his soldier escort getting increasingly nervous, Marsh persisted in stopping to look at yet another bone. "After sunset . . . I saw on my right, about a dozen feet from the trail, a fossil bone. . . . [I]t was hollow, about six inches long and one inch in diameter, with one end perfect and containing a peculiar joint that I had never seen before. . . . I cut a deep cross in the gray chalk rock . . . so that I could be sure to find the spot again."[1]

Back in the laboratory in New Haven, comparisons with European material showed him that it had come from a flying reptile similar in anatomy to the birdlike creature from Solnhofen, first collected almost a century before. Except that the size was all wrong—it was from an animal with a wingspan of at least twenty feet, monstrous compared with the pigeon-sized pterodactyls of Europe. It was "a gigantic dragon even in this country of big things."

For his second western expedition, in 1871, Marsh took ten students (one of whom, Harry Zeigler, was a veteran of the first trip).[2] They headed immediately to Kansas and Fort Wallace, where one of the students, describing the landscape for the *Hartford Courant*, echoed Captain Turner in writing disparagingly of the "belt of sand, which was called the Smoky River, but which required a stretch of courtesy to be so regarded. . . . [T]he whole country was destitute of water."[3]

Marsh quickly found the spot he had marked the previous winter and collected more material of the creature he was to name *Pteranodon*. This huge flying reptile is notable not just for its size but for its huge beak and the curious keeled crest protruding from the back of the head. Marsh later remembered: "My journey from New Haven was amply repaid, but greater rewards were to come, for during the month that I spent at hard work in this region, other dragons came to light, even more gigantic and much more wonderful than I had before imagined. For, unlike all other dragons, living or extinct, known to science or mythology,

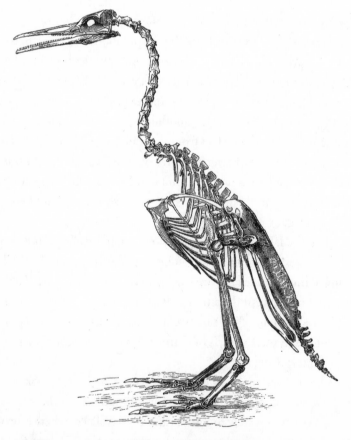

Marsh's reconstruction of the toothed bird *Hesperornis,* from the Cretaceous of Kansas (from O. C. Marsh, *Odontornithes,* 1881)

these reptiles, notwithstanding their gigantic size and vast spread of wings . . . were without teeth and, hence, comparatively harmless."[4]

A second discovery was possibly even more dramatic—the skeleton of "a large fossil bird, at least five feet in height." It was a giant, almost wingless and with strong legs—evidently a diving bird. Marsh later named it *Hesperornis regalis.* Unfortunately, however, this first specimen lacked a head. Once again, Marsh had not included Mudge in this collecting trip, possibly not wanting to share in the discovery of the *Pteranodon* material that he had been eagerly anticipating all winter long.

Instead of the biting cold that had affected the group the previous year, this time the group had to contend with the brutal summer heat, so they stayed in Kansas for only a month before heading off by train to Denver and then to Cheyenne, Wyoming. Once again the expedition was fitted out at Fort Bridger and spent six weeks in the Eocene Green River Basin. "The country consisted of desert plains, too sterile to produce grass, but thickly overgrown with sage brush," in the words of a biography of Marsh published some seventy years later. "The plains were occasionally varied by buttes or eminences which were entirely bare. The streams of water were more abundant than in Kansas, and their banks were fringed with cotton-wood trees which formed an agreeable relief from the gray expanses either side. The views presented of the Uintah Mountains were magnificent."[5]

The Yale party again found rich pickings in the Fort Bridger region, collecting a huge variety of material, including more kinds of fossil horses. Almost all of what they found was on the surface, and most of it was in the form of teeth and portions of jaws. Marsh also gathered more pieces of the strange giant mammal that had been found the previous year. He had decided it was a new species of the animal from the Dakota Bad Lands that Leidy had named *Titanotherium,* and therefore he named his species *Titanotherium anceps.* Of all the specimens collected in Wyoming, perhaps none would turn out to be so contentious and so important for Marsh's (dwindling) relationships with his putative colleagues than the large "sixhorned" mammals, apparently related to the elephants, found by the field parties of 1870 and 1871. Once again

Marsh was straying into Hayden's territory, but unlike Hayden, who stayed in any one place only long enough to collect samples, Marsh's group collected in earnest. Inevitably, that put him ahead of Leidy in the race to describe new species.

Almost certainly on the recommendation of Judge Carter, Marsh employed several local men as guides for his party. When Marsh left, he commissioned them to continue collecting for him. He also persuaded James Van Allen Carter to keep an eye on them for him, including running a tally of their days worked and disbursing their wages. Two who worked most consistently for Marsh between 1871 and 1873 were B. D. Smith and John Chew.[6] Chew, judging from his rather infrequent but neatly penned letters to New Haven, was more educated. Marsh was very erratic about paying them, although employing these two was a wise investment on his part and established a pattern that he followed for the rest of his career. Smith and Chew became skilled collectors, knew the country, and kept a steady flow of boxes coming to New Haven through the winter and following spring.

After leaving Wyoming, the Yale party traveled all the way to central Oregon, where a keen geologist, Professor (and Reverend) Thomas Condon of Oregon State University, had found a set of localities in John Day County with abundant mammal fossils (elephant, rhino, horses) in lake deposits of volcanic ash, clay, and sand exposed in the deep canyons cut by the river. Condon had already sent fossils to Leidy and wanted to keep the best specimens for his own collection, but he had told Marsh he was willing to send for study "without reserve everything that shall seem to me at all characteristic from this field and will designate those you may retain. . . . The basin of the Columbia River has never been submerged since its emergence from the Cretaceous Ocean. A vast lake system reaching geographically from the Rocky Mountains to the Cascades, and geological, from the Eocene to the Surface soil has here its record of vertebrate life."[7]

This could be considered yet another trespass into an area that Leidy might have thought was his. Interestingly, in later years Condon became very disillusioned with Marsh and eventually welcomed Cope and his collectors to "Fossil Lake." A major point of contention was the fact that Marsh refused to return the specimens he had been loaned. Condon wrote to Marsh asking for them in 1877 and again in 1880; even as late as 1890 he had not received them. "You can have no use now for those fossils of mine," he complained. "Every point they reached has been more than covered by your more recent find. . . . I need them too." The stern tone of this letter is neatly balanced by the sarcastic way Condon signed off—"With the keenest sense of the annoyance and hindrance such request must be to you, I am yours respectfully, Thomas Condon."[8]

To get to the John Day deposits the Yale party had to take a train to Kelton, Utah, travel by stage some six hundred miles to the John Day River, and then go downstream to Canyon City. They returned by steamer from The Dalles to Portland and San Francisco. Some then went cross-country by rail, arriving in New Haven in mid-January; Marsh and others took the longer route back via Panama. While in San Francisco, Marsh fired off to the *American Journal of Science* a very brief note announcing the discovery of his *Hesperornis* from Kansas.[9] As in the previous year, altogether the trip had taken six months. Again he and his men had collected dozens of crates of fossil material. But this time the expenses (some $15,000) had come directly from Marsh's own funds.

Edward Drinker Cope must have squirmed at the news of Marsh's first Yale trip. Both the students' accounts in the popular media and Marsh's broadside for the scientific community in the *American Journal of Science* rubbed it in. Although Cope had gained the initial advantage through his contacts in Kansas, he had let Marsh get there first. Worried that he might now have left it too late, Cope set out for the West himself in August 1871. Over all the years of his fieldwork he kept up the practice of writing home to report on his travels for his father, his wife, and his beloved daughter Julia. For Marsh we have a huge number of the letters that he received, but few that he sent. By contrast, many of Cope's own letters from the West survive, giving us a quite different perspective from the one we get from Marsh's trips.

Typically, Cope set off by himself, having taken the precaution of getting an order from General John Pope (commanding the Department of the Missouri) for a military escort when he got to western Kansas. He would pick up civilian assistants when he got there. Although Cope was by popular reckoning a wealthy man, like Marsh he depended on family funds. Possibly it was a careful eye to finances that had helped keep him from heading out west earlier; certainly this first western trip was a quite modest affair. In fact, newly discovered letters show that Cope was seriously short of money.

He began his trip west in August with a visit to Wyandotte Cave in Indiana. The adaptations of animals for living in caves, completely away from the light, had long fascinated him, and Wyandotte had a unique fauna. In addition to the humans who had sheltered there in prehistoric times, there were interesting animals including the Indiana bat, a dozen or so insect species, spiders, and a blind crawfish. In Indiana Cope also searched for local collectors of Indian artifacts who might sell items to his patron William Sansom Vaux, an avid collector of minerals and ethnological objects and a generous supporter of the Philadelphia academy. "I have been looking out for thee here and have seen some nice things," Cope wrote.[10]

Running out of cash, he suggested to Vaux that the academy should reimburse him for a collection of alcohol-preserved fishes, reptiles, and mammals from Costa Rica (the Van Patten Collection) that he had recently purchased for seventy-nine dollars. He wrote from Indiana, "Manhattan [Kansas] will be my headquarters for some time. I am quite anxious to recover part of the money paid Van Patten as it left me at a bad time."

Three weeks later Cope was writing to Vaux again, and sounding a lot like Hayden in his early letters to Leidy: "Did you get my letter from Indianapolis . . . ?" And two weeks after that he wrote, even more plaintively, from Fort Wallace: "I have written three times . . . [about] the very low state of my finances owing to the purchases I made of van Patten for the Academy. . . . I am now on the fossil beds and can obtain for the A.N.S. any amount of these things if I have means. I have already truly surprised myself at my success in getting saurians & fishes, but am restricted in my means. I am fitting out an expedition, which will be out some time & will cost me considerable for subsistence etc. . . . Philadelphia is behind in these matters & it seems to me that the interests of the Academy should be aided by a little of the necessary. Can't you pay your part of the subscription to Van Patten's colln. to me here?"[11]

On the way to Topeka Cope had his first look at the prairies, but for someone who often wrote so vividly, his reports home were as featureless as the landscape itself: "They [the prairies] are wonderful to me and look more like the ocean than anything I have ever seen. . . . [T]he whole country from Indiana to this place is as flat as New Jersey."[12] The next day he wrote slightly more enthusiastically to his sister (perhaps he didn't want Annie to think he was having too much of a good time). "Thee . . . would be refreshed by a little breeze from . . . the most interminable prairies. . . . The Flowers we have often heard of, but I did not suppose they were so tall; as high as a man's head. . . . Sunflowers, Cornflowers and various composites, flax, sage, euphorbia, endless verbenas, etc., cover the great expanse in every direction. The air is delightful, and it is impossible not to be taken with the spirit of the push for the West. The plains are said to be alive with buffalo, so much as to stop the trains."[13]

By the time Cope got to Kansas, Marsh and his party had long since departed for Wyoming. Arriving in Manhattan to meet at last with Benjamin Mudge, Cope found him unhappy, even angry, with the way Marsh had treated him. Cope described Mudge as having been "abandoned" by Marsh: "Prof. Mudge wanted to accompany Marsh and Marsh wouldn't let him go! I'll let him go!" This is an episode that has been much referenced by writers on Cope and Marsh. In fact, in a letter written from Topeka on September 6, 1871, Cope said that one of Marsh's "guides is at Ft. Wallace, left behind, and in want of a job." While many authors have taken this to mean Mudge, Mudge was presumably then not in search of paid work, and from the writing it seems clear that the guide and Mudge were two different people. Furthermore, by September 6, Cope had not yet got as far west as Fort Wallace.

Cope and Mudge got along well. Mudge showed him all his collections at the State Agricultural College, including his specimens from Fort Wallace. Typically, Cope immediately sent off to the American Philosophical Society a short paper describing them.[14] It would be nice to know what Mudge thought of that, as Cope had evidently not been interested in offering his host co-authorship of the work.

Cope, probably with Mudge, then spent a week or so exploring with a small party along the Saline and Republican Rivers. In Topeka he met the enterprising land agent William E. Webb, who had previously sent him the mosasaur and plesiosaur specimens. "My friend W. E. Webb the land agent here I find to be a delightful young fellow, used to the plains, as a surveyor of land all over the state, very familiar with localities of fossils, which turn out to be very numerous and gigantic. . . . I will . . . come out in the 11th month & spend a month on a special expedition with Webb. . . . Such an opportunity is very fine, with a man who knows the ground. Professor Marsh has been in that country for 3 weeks but has no such chance and it will cost nothing. . . . I am coming home soon and will return the last of the 11th month . . . to make an exploration with Webb and a small party."[15] Cope seems not to have made this second trip, however, and it seems reasonable to assume that he simply couldn't afford it.

Happily, Vaux came through with twenty dollars, and in September Cope was able to continue to Fort Wallace for seventeen days, courtesy of a pass for the Kansas Pacific Railroad. On arrival he presented General Pope's letter to Captain Butler, commanding the fort. As Marsh had already been there earlier, the officers based there were used to the needs of paleontologists. Butler assigned to Cope's party Lieutenant Whitten, five soldiers, and a six-mule wagon with tents and provisions. "The men took army rations and I laid in as follows: 5 cans of oysters at 21c, 1 box of sardines at 40c, 1 box condensed milk, 2 cans of peaches, 2 loaves of bread, 1 bottle of lime juice. These I shared with Lieut. W. . . . They tried to lend me a military hat and revolvers, but I left them behind as I hate the sight of them." Reflecting the sporadic state of Indian hostilities, Cope later said that his "exploration in Western Kansas was made during a state of hostility of the Cheyenne Indians, and in a region where they were constantly committing murders and depradations." In another letter he wrote: "The Indians are peaceable and nowhere in this region at present."

Although Marsh and his party had already collected from around Fort Wallace, Cope made an impressive haul of fossils from three localities along the Smoky Hill River, camping "five miles south of Butte Creek," on Fox Creek, and Russell Springs.[16]

The finds included lots of mosasaur material, huge turtles, and fragments of the monstrous flying reptile that Marsh had already described as *Pteranodon*. Cope claimed that he had found two species of these last creatures, one being "the largest Pterodactyle as yet known from our continent, the end of the wing metacarpal exceeding in diameter that of the species described by Professor Marsh from the same region." In the same publication he described a second specimen (based on a few bones from the wrist), under the new name *Ornithochirus harpyia*, adding the cutting note that it might really have been the same as Marsh's *Pteranodon ingens*, but "this cannot, however, be definitely ascertained, as his species is imperfectly described."[17]

During this trip, one of Cope's escorts (Martin Hartwell) "observed the almost entire skeleton of a large fish, furnished with an uncommonly powerful offensive dentition." In his letter to the American Philosophical Society, Cope named it *Portheus molossus*. It was yet another of the giant Cretaceous fishes related in some way to *Saurocephalus*, most of which had been described from very incomplete material. Eventually it turned out that Cope's *Portheus* was the same as a fish previously described by Leidy (on the basis of a tooth from New Jersey) as *Xiphactinius*. Setting aside the problem of names, however, Cope's party had found really good material of the kind of fish that was abundant in those Cretaceous seas. Jefferson would have loved this monstrous predator up to twenty feet in length. It became an essential part of any museum display of fossil vertebrates, especially when the specimens contain prey fish four or five feet long inside them, as at the Sternberg Museum of Natural History in Kansas.

NINETEEN

Buffalo Land
Who Was Professor Paleozoic?

The next twenty-four hours constituted a regular field-day for the Professor, being distinguished by an event which, from a scientific stand-point, was among the
most important of our entire expedition. This was the discovery of a large fossil
saurian, which we came upon while exploring quite in sight of Sheridan, and not
more than half a mile from its eastern outskirts. . . . Of the countless millions of
saurians then existing, capricious Nature had seized upon this one, to transmute
it into an imperishable monument of that extinct race. In those ages of roaring
waters and hissing fires, she had clothed the bones in stone, that they might with-
stand the gnawing tooth of time, and thus handed them down to the wondering
eyes of the Nineteenth Century.

WILLIAM E. WEBB, *BUFFALO LAND*, 1872

The new railroad era of fossil collecting in the West started in Kansas. As
the Kansas Pacific Railroad pushed west from Topeka, across the Permian formations of the Flint Hills and into the open country of Cretaceous hills and valleys, it gave access to dozens of fossil sites from which a
wonderful array of mosasaurs, the plesiosaur *Elasmosaurus,* and a later
host of other forms such as birds with teeth and huge flying reptiles were
discovered. It was in Kansas that both Cope and Marsh had their early
successes, aided by a small but active (and growing) number of Kansas
residents like Mudge, who were in a position to explore the local terrain
and who wanted to use their collections to advance their personal ambitions by becoming associated with the big guns back east. By the end of
1871, all the major players except Leidy had visited Kansas in person;
Hayden (with Meek), Agassiz, Marsh, and Cope had all collected in the
Cretaceous deposits of the western half of the state.

Although a great deal has been written by, and about, the early settlers in the West, there is not a lot of published work that can tell us
what people like Cope and Marsh experienced day-to-day—the landscapes, the people, the difficulties of traveling or of setting up camp and
staying in one place. The accounts by students in Marsh's Yale College
expeditions give a somewhat romanticized glimpse of life in the dissected landscapes of the Smoky Hills among the Indians and rattlesnakes. Cope's letters home tell us a little more. Captain Turner's
letters have not the slightest tinge of romance about them; he found
Kansas to be a hard and hostile place. For later years, Charles Sternberg's autobiography is a wonderful (heavily dramatized) source.[1]

And there is also a single novel.

Starting around the 1850s, a new literary genre, the western novel,
gripped audiences both in America and abroad. These novels offered
adventure and dramatic settings in the endless plains and soaring
mountains; the handsome young adventurer or brave pioneer family
would battle not only the elements and the land but also the Indians and
renegade whites. The distinction between hero and villain was always
crystal clear. From James Fenimore Cooper's tales of western New York
State to Mayne Reed's novels of the Mexican War (*The Rifle Rangers* of
1850 and *The Scalp Hunters* of 1851), and the dime novels of "Ned
Buntline" (Edward Judson) and Erastus Beadle, not much had changed
since the real-life saga of Mary Draper Ingles: the setting had simply
moved west. The hardy pioneer, the deceitful savage, the unforgiving
land, the eventual triumph of virtue—none of it was original, all of it
was new. Soon the fictional romance and the real adventures began to
coincide, overlap, and feed one on the other. Whether the stories were
true or not scarcely mattered; everything in them contributed to an image of the evolving spirit of America, the rugged individualism of its
people, and its destiny of westward expansion. Even the Indians eventually became "noble" savages of the kind beloved of European romantic writers, but only after their lands had mostly been taken and the
buffalo killed.

Very few of these romances feature any kind of scientist, and certainly no paleontologists decorate their pages; scholars and heroes of

derring-do are normally at polar opposites. But one story does feature fossil collecting, and the paleontologists it portrays are far from heroic. The book is William E. Webb's *Buffalo Land,* published in 1872.[2]

Modern readers will find its style overblown and the humor somewhat heavy-handed, but it is the only genuinely funny book written about fossil collectors anywhere. A contemporary reviewer also said it gave "the first really correct and satisfactory idea of the Plains country."[3]

The central character of the novel is Professor Paleozoic, characterized as a man who "ordinarily existed in a sort of transition state between the primary and tertiary formations. He could tell cheese from chalk under the microscope. . . . [A] worthy man, vastly more troubled with rocks on the brain than 'rocks' in the pocket. . . . [L]earning had

THE PROFESSOR—A REMARKABLE STONE.

"Professor Paleozoic," as depicted in Webb's *Buffalo Land*

once come near making him mad." Professor Paleozoic is on his first fossil-collecting trip to the West, and the description of him and the other characters in the book shows that its author both knew his Kansas landscape and was familiar with the scientists who regularly trekked west in search of fossils and other scientific treasures.

Evidently they often looked extremely foolish to the locals.

In the late 1860s and early 1870s, William Edward Webb was the agent for the National Land Company in Topeka. This was the time when the Kansas Pacific Railroad was pushing steadily westward, opening up and selling section lands that would eventually become beef farms and magnificent wheat fields. (At Topeka Station of the Kansas Pacific one could also book buffalo shooting trips along the railroad tracks.) Webb was a multifaceted character, as any entrepreneur in the West had to be: a speculator and businessman, a local politician, an amateur photographer, a hunting guide who took groups out to shoot buffalo, and also a part-time fossil collector. He had already collected fossils of mosasaurs in western Kansas in 1868 and had sent specimens and photographs to Leidy and Cope. It seems reasonable to assume that he finished writing the book in the winter of 1871.

Webb knew and traveled with William Cody. The man who became Buffalo Bill had worked briefly as a Pony Express rider and, when still only twenty-one, was employed to shoot animals to feed the workers building the Kansas Pacific line. He claimed to have killed more than four thousand buffalo. In 1868 he became an army scout for the Fifth Cavalry. How many of his exploits were real and how many were invented is still a matter of debate: the New York journalist Edward Judson met him in 1869, and in articles for the *New York Weekly* and some five hundred novels written under the pseudonym Ned Buntline he made sure of that. Webb also knew Wild Bill Hickock, sheriff at Fort Hays, Kansas, then a frontier town to which Webb's fictional party of hunters and collectors found its way. Both legendary characters appear in *Buffalo Land*. When Webb wrote the book, he did not have to invent a story—he merely described life as it was, albeit embroidering it somewhat when it came to his cast of visiting characters.

In large part, *Buffalo Land* was written as a frank advertisement of the delights of western life, aimed to lure eastern emigrants out to settle in Kansas. Although the main story is about an expedition to collect fossils and hunt buffalo, should the descriptions of the landscape and western life not sufficiently have enticed the emigrant (or the lurid tales of the Indians repelled them), Webb added a long salesman's appendix in which "Additional facts concerning the natural features, resources, etc., of the Great Plains and contiguous territory" were laid out, with sections on the geography, climate, stock-raising, "trees and future forests," and fuel resources ("coal in immense quantities").

Webb has an enormous amount of fun in his tall tale, with his cast of hapless easterners in all their pretensions and naïveté. In addition to Professor Paleozoic, the book features Dr. Pythagoras, whose favorite theory was "development" (a common name for the theory of evolution in those days), a politician in the form of the "New York Alderman," the obligatory phlegmatic, tweed-wearing Englishman, named Genuine Muggs, together with an entomologist and his son intriguingly named Colon and Semi-Colon. Webb, the narrator, takes all of them on a hunting and "fossicking" expedition, with Buffalo Bill himself as their initial guide.

The party set out from Topeka, following the Kansas River to its junction with the Republican River. "Above that point, under the name Smoky Hill, it stretches far out across the plains, and into the eastern portion of Colorado. Along its desolate banks we afterward saw the sun rise and set upon many a weary and many a gorgeous day. . . . Here Fremont marked out his path towards the Rocky Mountains and the Pacific." Along the way Professor Paleozoic explains the geology, but Webb has him state that the "vast plateau lying east of the Rocky Mountains . . . was once covered by a series of great fresh-water lakes" and expound a theory that "man existed on this earth when . . . the waves lapped against the Rocky Mountains." The evidence for this last observation was the existence of supposed fossil human footprints near Bavaria, Kansas.

At Fort Hays ("remarkably lively and not very moral"), the travelers met both Wild Bill Hickock and Buffalo Bill Cody, and also General Phil Sheridan. "Being then the depot for the great Santa Fe trade, the town was crowded with Mexicans and speculators. Large warehouses along the track were stored with wool awaiting shipment east, and with merchandise to be taken back with the returning wagons." From there they struck off overland toward the northwest and on to the Saline River valley.

In these adventures across the plains, Webb portrays Professor Paleozoic as something of a cross between an earnest young Louis Agassiz and Bob Hope in the old movie *Paleface*. Dr. Pythagoras is presented as the archetypal nerd. And in the mishaps of the group, Webb manages to defang the image of a dangerous West full of savage Indians and poisonous snakes. If these bumbling characters can survive, Webb is telling his readers, then you—who must be far more savvy—can too. This leads to some inconsistencies. It was necessary for dramatic purposes, for example, to depict the Indians as a constant threat and to tell some gruesome tales of butchery, but when the party met up with some Cheyenne the Indians were presented as something of comic book characters.

"In White Wolf we had found as fine-looking an Indian as ever murdered and stole upon his native continent." Professor Paleozoic was prevailed upon to smoke a peace pipe with them, and then "the council broke up and in an incredibly short time thereafter many of the Indians were reeling drunk." After that (in one of the oldest jokes known to man), having failed to steal anything, the Indians stole away. Later on in the narrative, the party observed (from a safe distance) an epic daylong fight between Cheyenne and Pawnee war parties. Taking up his shovel, Professor Paleozoic intoned, "Let our task be to bury the dead. It is extremely problematical whether any of these red men will go out of the valley alive." However, after charge and countercharge, daring horsemanship, and nonstop firing, the casualty list stood at three dead ponies and a couple of braves injured.

The book follows the wayfarers as they travel westward from Topeka, and most of the narrative is taken up with hunting buffalo and other game, and more tales of pioneers and old Indian fights. The fossil collecting began when the party reached the region of Sheridan, Kansas. "We all stood beside the huge fossil. It lay exposed, upon a bed of slate, looking very much like a seventy-foot serpent, carved in stone. . . . 'This fossil, gentlemen,' said the Professor, 'is that of a Mosasaurus, a huge reptile which existed in the cretaceous sea.'" From this point onward, the professor begins to sound much more like a real scientist, and toward the end of the book his homilies about the geological history of the land they are traveling are supplemented by a long extract of a scientific paper by "my friend Professor Cope of Philadelphia" on the Cretaceous of Kansas. In this work the basically marine nature of the Cretaceous seas of western Kansas is explained, and there is no further reference to freshwater lakes, or Cretaceous humans and their footprints.[4]

The more technical style of these sections is a striking contrast to the racy flippancy of the rest of the book, but it provides a valuable literary device for Webb by giving an authoritative backing to what he has so far been treating lightly. This in turn gives credibility to the blatant sales pitch of the Appendix. Altogether, not only is *Buffalo Land*—Webb's only book, apparently—well written and a downright good read, it is a surprisingly good reflection of the rustic charms of life in Kansas, literally and figuratively at the end of the line from the East.

Why Webb chose to pillory the geologists and paleontologists who had started coming west each season is not hard to guess: the easterners Hayden, Meek, Lesquereux, LeConte, Agassiz, Marsh, and Cope all passed through western Kansas (riding the rails), and all must have seemed faintly absurd, especially compared with local paleontologists like Mudge or Parker. Webb seems to have read Mark Twain and learned well.

But who were the real-life models Webb used as the basis for his particular characters? Webb would have known, and no doubt positively relished the fact, that readers would try to find clues to the identity of the New York Alderman, Colon and Semi-Colon, Genuine

Muggs the Englishman, and not least Professor Paleozoic. All of his characters have just a little too much believability about them to be pure invention; the reader feels at once that they are people he might know.[5]

Young Dr. Pythagoras's preoccupation with "development theory," or evolution, suggests Hayden, Parker, or LeConte. Louis Agassiz seems at first unlikely to have contributed much to the depiction of Professor Paleozoic, but he had visited the Kansas fossil fields in 1868 with a party of the great and good organized by Senator Roscoe Conkling of New York, who might easily have been the model for the alderman. With his devotion to glacial theory, Agassiz is more likely than anyone else to have held the view that the Kansas Cretaceous was of freshwater origin. The first great mosasaur found in Webb's narrative was discovered by "Frenchman Louis" (a possible reference to Agassiz), a man who collected fossils because his prime work as a watchmaker did not bring in enough money (this recalls Leo Lesquereux, who collected for Agassiz). That great mosasaur was said to have ended up in the collections of Harvard University, as was the case with the real first mosasaur from Kansas, which Agassiz purchased in 1868. Also the entomologist Colon is said to have been from Boston.

Professor Paleozoic serves Webb's narrative purposes ideally, and over the years conjectures have grown about the model for his central character. Several attempts have been made to identify Professor Paleozoic definitively with Cope, but the evidence is mixed.[6] Professor Paleozoic is depicted as young and thin (either Cope or Hayden would fit). He signed a letter to a class of his students "H" (Cope did not have students, but Hayden did, although this might also have been a reference to the "H" review of Leidy's 1865 paper). He was fond of the ladies (definitely Cope). He was indifferent to rocks in the pocket (which even suggests Leidy). But Professor Paleozoic has his geology wrong.

The other prime candidate is Marsh. Buffalo Bill Cody definitely accompanied Marsh's first Yale expedition, if very briefly. But there is no evidence that Webb went on any of Marsh's trips or that Buffalo Bill ever went with Cope. In the end, nothing much about Professor Paleozoic suggests Marsh, except that the description of the buffalo hunting and a buffalo stampede match those of the Yale expeditions, but by then the experience of the buffalo hunt was becoming commonplace, as was the condemnation or disdain for it shown by many eastern visitors.[7] Railroad tracks like those of the Kansas Pacific cut across the traditional migratory routes of the buffalo. As a result they tended to gather at the crossing places, where they were exceptionally easy prey for railroad passengers, who simply emptied their guns through the windows as the trains—which slowed down in case any animal should leap onto the tracks—passed through. Even at the time this seemed stupid, mindless slaughter and a long way from "sport." For people like Webb, it was the downside of the emigration that he was otherwise encouraging; hence the emotive descriptions in *Buffalo Land.*

In his book *The Bone Hunters' Revenge,* David Rains Wallace concluded that Professor Paleozoic was based at least in part on Marsh because of Webb's very first dig at the professor.[8] "While in Kansas some years since, he penetrated a remote portion of the wilderness, where, as he was happy in believing, none but the native savage, or, possibly, the primeval man, could ever have tarried long enough to leave any sign behind." There, he found "an upright stone, with lines chiselled on three sides and on the fourth a rude figure resembling more than any thing else one of those odd fictions which geologists call restored specimens. On a ledge near were huge depressions like footprints. They were the foot-prints of birds, no doubt. . . . Both specimens were forwarded to, and at the expense of, noted savants of the East." The denouement is that Professor Paleozoic had found and removed "a stone telling in surveyors' signs just what section and township it was on." As for the tracks, "Whether the bird-tracks had a common origin, or were hewn by the hatchets of the red man, is a point still under discussion."[9]

Wallace believed that this little gaffe was a reference (in reverse, as it were) to the Onondaga Giant affair—an early fossil hoax that Marsh had helped unmask. The Onondaga Giant was a ten-foot limestone sculpture of a human figure, supposedly excavated near Cardiff, New York, in 1869. P. T. Barnum tried to buy it, and none other than James Hall proclaimed it real. Scholars rushed to say that it was some kind of ancient classical relic, possibly Phoenician, indicating important connections between the New and Old Worlds. Marsh looked at it, saw the recent tool marks, and bluntly pronounced it a fake.

In truth, the model for the bumbling Professor Paleozoic was in part none other than Webb himself. Two years before the book was published, Webb had written to Cope from Topeka: "I have two large sandstone blocks, a fragment of one of which I send you. On one is a perfect impression an inch or so deep, resembling . . . a human foot. . . . On the other are two impressions, much deeper, & one very fair affinity of a foot. I will answer for the impressions being genuine, & not sculptured, & in the opinion of all who examine made at the time of formation of the stone. . . . The stones were found upon a sandstone ridge. . . . I will try to send you photographs tomorrow."[10]

At the bottom of Webb's letter, now in the archives of the academy in Philadelphia, is a penciled note in Cope's hand, undated and evidently directed to an academy staffer: "Please tell him the tracks are undoubtedly sculptured by the Indians & are common all over the United States. Done by medicine men & represented [to] be tracks of the deity. Thank him. Edw. D. Cope."

When, in the book, Webb had the party find the human footprints, he even said that "many scientific men, among whom is Professor Cope, affirm that they must be the work of Indians long ago." But Webb, the narrator of the tale, still preferred the opinion that he put into the mouth of Professor Paleozoic, that they were "imprints . . . of human feet" and that they had been made in Cretaceous times.[11]

One final note can be added to this incident. In answering Webb's letter, Cope comes across as surprisingly expert on Indian culture for a man who had then never ventured west. However, he had had some help. Charles Lyell had discussed such carvings as early as 1846.[12] The minutes of the March 18, 1870, meeting of the American Philosophical Society records that: "Prof. Cope exhibited three photographic pictures of figures of the human foot incised in upper cretaceous red sandstone, near Topeka, thought by western men to be fossil impressions. . . . A discussion of the use of the foot in aboriginal picture writing followed."[13]

So Professor Paleozoic, like everything and everyone else in the book, was evidently a composite. Webb, an experienced salesman, was far too canny to risk portraying any single person too realistically when his purposes could be served far better by parodying several of them. He must have known that his book would have been read avidly by fossil collectors, their friends, and especially their enemies. He would not have risked offending the first or encouraging the last. His whole career as a land agent and outfitter of expeditions depended on generating good will, not bad press.

Webb's book, however, may have stung Cope a little, because contact between the two men seems to have ended after its publication. And the copy of *Buffalo Land* belonging to the library of the Academy of Natural Sciences in Philadelphia has a notation on the flyleaf: "This book was taken away from the library by the late Professor Cope and returned by Mrs. Cope, June 5th 1901."

TWENTY

1872

The Year of Conflict

Joseph Leidy finally traveled west to collect for himself in 1872. It was the first time he had seen any of the prairie landscapes, the mountains, and the great fossil basins and badlands from which others had been sending him specimens for twenty-five years. He had never had a monopoly on the Dakota Bad Lands region, but it was imperative for him to get into the field in person if he was to have any chance of enforcing a claim to the Eocene vertebrates of the Bridger region. Hayden had long ago warned him that he was vulnerable to being preempted. Marsh had demonstrated the same by charging into Fort Bridger with his Yale expeditions and collecting huge amounts of material. Marsh would probably be going back to Fort Bridger yet again, and Leidy also knew that Cope was planning to visit the area. Leidy had no choice but to accept a long-standing invitation from the Carters and Dr. Corson and perhaps repair some of the damage—although it turned out that in many respects he was already too late.

Leidy went west on the new transcontinental train to Denver, Salt Lake City, and then on to Green River and Fort Bridger, Wyoming, where he was treated as a distinguished guest. His fieldwork and a second visit the following year allowed him to compose a large monograph on the Bridger Formation and its fossils. Now Leidy was finally able not only to collect fossils but to write a firsthand narrative description of the scenery and the geology in what was becoming a popular genre—a prose evocative of the "feel" of the western landscapes, so foreign, so hostile, apparently so romantic to easterners, and—to paleontologists—so promising. In his own restrained way, he indulged in evocative passages, such as one describing the Green River Bad Lands.

I was astonished at the appearance of the country extending from the horizon in the north to the snowy-peaked Uintas on the south. An utter desert, a vast succession of treeless plains and buttes, with scarcely any vegetation and no signs of animal life. Everything parched, abundance of river courses without water, the stones at my feet baked in the sun. On ascending the butte to the east of our camp, I found before me another valley, a treeless plain, probably ten miles in width. From the far side of this valley butte after butte arose and grouped themselves along the horizon, and looked together in the distance like the fortified city of a giant race. The utter desolation of the scene, the dried-up water-courses, the absence of any moving object, and the profound silence which prevailed, produced a feeling that was positively oppressive. When I then thought of the buttes beneath my feet, with their entombed remains of multitudes of animals forever extinct, and reflected upon the time when the country teemed with life, I truly felt that I was standing on the wreck of a former world.

From the lower plains the neighbouring terraces . . . appear like vast earth-work fortifications, . . . frequently the terraces are so extensively eroded and traversed by narrow ravines that they appear as great groups of naked buttes. . . . [N]othing can be more desolate in appearance than some of these vast assemblages of crumbling buttes, destitute of vegetation and traversed by ravines, in which the water-courses in midsummer are almost completely dried. . . . [Amid] these assemblages of naked buttes, often worn into castellated and fantastic forms, and extending through miles and miles of territory, . . . it requires but little stretch of imagination to think oneself in the streets of some vast ruined and deserted city.[1]

In this rehearsal of the by-now familiar analogy between the cliffs and ravines and a deserted city, he seems to have written with both Owen's 1850 report (with its identical descriptions of the White River Bad Lands) and the journal of Thaddeus Culbertson's trip directly in front of him. This is not unlikely, as he had had already quoted the relevant passage from Culbertson's journal in his 1853 report.

Edward Drinker Cope had already arrived at Fort Bridger before Leidy in 1872, but because of difficulties with men and equipment he did not get out into the field until July 14, when he headed out for Cottonwood Creek and eventually into the Washakie Basin—well away from where Leidy was collecting. Leidy arrived at Fort Bridger the following day and quickly set off on a short trip with the local doctors, James Van Allen Carter and Joseph Corson, to the valley of Dry Creek, forty miles east from Fort Bridger. They camped for three days and explored among the buttes for signs of fossils. With all the inevitability of this continuing classic tragedy, even though they prospected in a different region, they found the same sorts of materials that Cope would collect later that summer. Then Marsh, who had probably not meant to go into the field at all in 1872, turned up at the end of October, no doubt to help secure his own interests in the area. And he collected more.

It was during this first camping excursion with Carter and Corson that Leidy made one of his most important discoveries. Everywhere the three went they found turtle material—it was abundant enough to be a nuisance—and what Leidy thought were at least two species of the large tapir-like mammal *Palaeosyops* that he had already named. Soon the collecting improved. "We were fortunate in obtaining the remains of two of the largest and most extraordinary mammals yet discovered in the Bridger tertiary deposits," Leidy reported. "One of these was a tapiroid animal exceeding in bulk of body and limb the living Rhinoceros, though the head seems to have been proportionately small. Dr. Carter discovered many fragments of a skeleton, including a whole humerus, portions of jaws, and a much crushed and distorted cranium." Then the next day, "if not the most interesting, the most exciting incident of

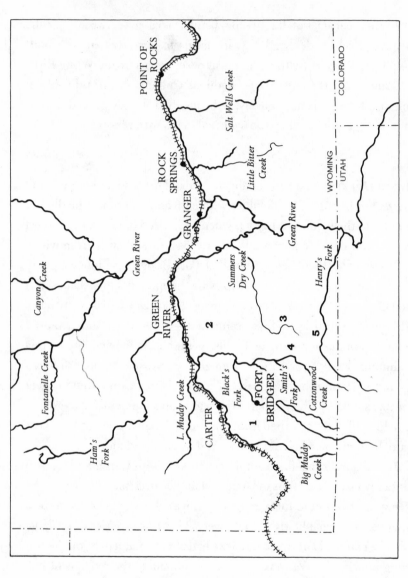

The Fort Bridger region of southwestern Wyoming, showing some of the historical fossil localities. These include: (1) Bridger Butte, (2) Church Buttes, (3) Big Bone Buttes, (4) Grizzly Buttes, and (5) Lone Tree.

our exploration of Dry Creek Buttes was Dr. Corson's discovery of the upper canine teeth, apparently of the most formidable to Carnivores, the enemy of the Uintatherium, Palaeosyops, and other peaceful pachyderms. The teeth resemble those of the Sabre-toothed Tiger. The more perfect specimen consists of nearly nine inches of the enamelled crown." (One gets the impression that Carter and Corson were rather better at spotting fossils in the field than Leidy, who had never tried it before.)

Leidy might simply have had these bones boxed up and sent home with his other fossil discoveries to be described later. Instead he showed a surprising turn of speed and perhaps even competitiveness. While still out at Fort Bridger and only days after first seeing the specimens, he wrote a quick description of them (dated July 24) and sent it to Philadelphia for publication by the academy. In this brief note, Leidy named the new genus and species *Uintatherium robustum* for the skull and limb material. He assigned the tooth to a second genus, *Uintamastrix,* with the species *Uintamastrix taro.* (Looking back it seems impossible to us that he would not have known that the remains all belonged to one animal, but it would be unfair to judge Leidy, who did not have a whole animal to compare with but merely saw two very different kinds of teeth. The advance "pamphlet" of his paper was issued on August 1, 1872. Even for the standards of the day, that was phenomenally quick. As with many other works (for example, his note on Cope's *Elasmosaurus*), Leidy also sent an almost identical version of his paper to the *American Journal of Science*.[2]

There was a reason for this haste: Leidy knew he was going to be in an all-out competition with Marsh and Cope to make the most exciting finds and describe them first. And the subject they would contest was bound to be mammals, such as these strange horned creatures. Marsh had signaled his anticipation of a general rivalry earlier that spring by suggesting that, to avoid arguments and recriminations about the critical issue of who published what, and first, each would send the others copies of their papers the moment they came off the press. Marsh "agreed with each of these authors in March, 1872, that we should send to each other, on the day of publication, any papers . . . we might issue, the date of publication to be either printed or written on each pamphlet."[3] Marsh's plan, however, had exactly the opposite effect of reinforcing the competition and creating a situation guaranteed to energize even the reticent Leidy.

Typically, on both this trip and one the following year, Leidy did not confine himself to studying the landscape, geology, and fossils but also made notes on the local plants and Indian artifacts. In his field excursions he also spent a good deal of time on freshwater biological studies, and perhaps significantly, when Leidy reported on the first trip, fossils were almost an afterthought: "I have returned from Fort Bridger, . . . [where] I spent five weeks delightfully. . . . [T]he streams about Bridger contain very little of interest, as water from the snows of the Uintas are too cold for most forms of animal life. I have specimens of a curious parasitic leech from Henry's Fork, which I have described and figured. . . . I visited the shores of Salt Lake. I had not the means of examining the water as it should be done. I collected some dipterous insects, and also three algous plants from near the shore. . . . I also examined the algous plants of the hot spring near Salt lake. These plants I shall describe. At Bridger, with the aid of Drs Carter and Corson, I collected many vertebrate fossils. Some of these further illustrating my former descriptions I will have figured, and may require perhaps some five more plates. I shall try to have my report ready for the printer as soon as my other duties will permit. I hope you may be able to bring me some interesting material to add to my report to you."[4]

In his previous solo trip to Kansas, Cope had discovered that it was going to be difficult and expensive to work in the West. Apart from the military escorts, he needed local help for guides and collectors, and there was the constant problem of competing for localities and the same fossils with Hayden and Leidy on one hand and Marsh on the other. Lacking the funds—though certainly not the nerve—to forge ahead on his own, the only obvious way around the problem was to join forces with Hayden. Therefore on his return to Philadelphia in 1871 Cope devoted part of the winter to badgering Hayden for an official appointment to the survey. Cope had influence in Washington (at the National Academy of Sciences and with Baird and Henry at the Smithsonian), and he used it. Hayden agreed to take him on formally (if without salary) as paleontologist to his United States Geological Survey.

For the next few years, this afforded Cope the way to collect in the West and to publish on his finds. In addition to producing huge numbers of short papers, it meant that he could periodically gather his data into long monographs, dealing with everything from fishes to mammals, that Hayden would publish for him. His style of dramatic writing ideally suited Hayden's need to make a popular impression with the official reports of his surveys. And, although Hayden might have been apprehensive about having Cope as part of his team, he had no doubt that these reports would be invaluable. Interestingly, he had previously asked Marsh to join him, but Marsh was too canny to give up his independence. Cope needed to give up his. Whoever joined the survey, however, it was bound to be at the expense of Joseph Leidy, who to this point had been the primary recipient of Hayden's collections.

Cope set off on his first foray on behalf of Hayden's survey in June 1872 full of optimism. With Hayden's reluctant agreement he had decided that he had to go see the superbly productive Eocene lake beds of Wyoming for himself. Later in the year Hayden wrote to Leidy, trying to rationalize this and to smooth over some of the resulting unhappiness. His letter reveals that a rift had developed between him and Cope: "I asked him not to go into that field that you were going there. He laughed at the idea of being restricted to any locality and said he intended to go whether I aided him or not. I was anxious to secure the cooperation of such a worker as an honor to my corps. I could not be responsible for the field he selected in as much as I pay him no salary and a portion of his expenses. You will see therefore that while it is not a pleasant thing to work in competition with others it seems almost a necessity. You can sympathize."[5]

Cope took his wife, Annie, her sister, and his daughter Julia with him as far as Denver, where they rented lodgings. Hayden promptly reported back to Leidy, who was about to set off for Fort Bridger himself, to try to keep the two men apart. "Cope is here with his family and several ladies, has taken a house for the summer. He will work around here and by and by go to Bridger. I hope you will not disagree. If you go to work around Bridger at once you will have completed the examination before Cope gets there. He is going to operate between Cheyenne & Colorado Springs. . . . It seems to me a good deal might be done in the vicinity of Church Buttes. There are some remarkable bones and teeth there. . . . Meek is sick at the Hatten House. When you come here, go and see him."[6]

Cope also went to see Meek. Apart from the Christian duty of visiting a sick man far from home, he also had a paleontological mission. During the previous field season, near Black Buttes Station on the Union Pacific line, "fifty-two miles east of Green River, and near the Hallville Coal mines," Meek had found some large bones and he had suggested that Cope should take a look. Cope found the site and "succeeded in recovering sixteen vertebrae, including a perfect sacrum, with dorsals and caudals; both iliac and other pelvic bones, those of one side nearly perfect; some bones of the limbs, ribs and other parts not determined." Cope saw that it was a dinosaur and subsequently named it *Agathaumas* (great wonder) *sylvestris*. If he had had any of the head he would have discovered that it was something like *Triceratops*, a member of the group with a big bony frill around the neck and rhinoceros-like horns on the nose. In the event, Cope thought it resembled *Hadrosaurus*. It was the certainly big: "if the reader would compare the measurements . . . he will observe that those of the present animal exceed those yet described from North America."[7] *Agathaumas* also conclusively proved, in Cope's opinion, that the coals of the Bitter Creek Basin of Wyoming were Cretaceous in age.

From Black Buttes it was only a short journey along the Union Pacific past Green River, Church Buttes to Carter Station and then south a few miles overland to Fort Bridger, where Cope arrived in late June. At this point he was not sure what Marsh's plans were, but he knew that Hayden was still busy in the area and that Leidy had also announced he would be heading for Fort Bridger soon. Sure at least of his support from the survey, Cope wrote home: "I will have every facility furnished by the Interior Dept.: expenses paid, orders for men, wagons, beasts, provisions, etc." He did not suspect that Marsh had enlisted the two Carters and Dr. Corson against his interests. He also could not have guessed that Hayden himself would sabotage him. When Cope got to Fort Bridger he found that nothing was available. He complained snippily to General E. C. Ord: "On reaching this post Capt. Clift in command informs me that Dr. Hayden's first party have deprived him of all animals, bridles, saddles, etc. essential, to the outfit of this expedition. I risk reporting what is already known to yourself, in adding that the remaining teams are all employed in furnishing wood to Camp Douglass. . . . [T]he men on duty tomorrow will number only fourteen."[8] Dr. Carter reported to Marsh: "Hayden came through sometime ago and crippled Prof Cope's prospects for transportation etc."[9]

By the beginning of July, Cope, who didn't need a military escort in (then) peaceable Wyoming, had cobbled together a small outfit at his own expense: "One wagon with four mules for $500; hired another team with driver at $180. per mo.—$1.00 per day each for men, 4 mules and wagon. Hired one teamster, a packer and guide, and a cook, so that the party consists of five men. I brought with me three young men from Chicago who wished to be benefited by the chance to study." Cope soon fired the oldest of these three, a man called Garman who "began to make ridiculous and unreasonable demands of pay for services, time, etc., and poisoned the two boys with mutinous ideas . . . his whole scheme was to get up an expedition of his own. I am glad to be rid of him at the outset."[10]

There was another serious problem concerning the men Cope employed: two of them—namely B. D. Smith and John Chew—were Marsh's collectors. Adding to the confusion, in his letters home Cope mentions his mule packer "Sam," who was another Smith—Sam Smith, who sometimes referred to himself as "Sam Smith of the Rocky Mountins [sic]." Some authors have assumed that B. D. and Sam Smith were the same person, but the handwriting of the two men in their letters to Marsh is quite different. B. D. Smith was possibly an older man; he addressed his letters to "Friend Marsh," whereas Sam Smith wrote more respectfully to "Professor Marsh Sir." James Van Allen Carter, in his letters to Marsh, always referred to B. D. Smith simply as Smith, while Sam was always "Sam Smith" or "Sam," presumably to distinguish between the two. And, in a letter to Leidy the following year, Dr. Carter referred to "Sam Smith, that's Cope's own Smith."[11] Possibly B. D. and Sam were related. It is evident that B. D. Smith (and probably Chew as well) stayed with the party from mid-July at least until the end of August. Cope may not even have known that they were in Marsh's employ, although it seems likely that either the Judge or Dr. Carter would have told him. In any case, it was B. D. Smith's and Chew's choice to take Cope's dollar.

When he found out, Marsh was furious. B. D. Smith wrote to him claiming that it was just a small matter and tried to demonstrate his fidelity to Marsh by pointing out that he had turned down an offer to take part in a military expedition (for "Captain Jones of the Engineers") just weeks before Cope arrived (a fact known to Dr. Carter). He offered the ingenious excuse that he wanted keep an eye on Cope and guide him away from the best places for fossils. That may even have been true, and it seems that Smith actually took two trips with Cope. First Smith wrote to Marsh, on July 5, 1872: "I am going out again the morrow and will stay out one month I don't know weather I will be able to go any more or not I am afraid it is injuring my eyes them other bone pickers is at the fort yet cant get anyone to go with them they have ben waiting a week for me to come in and wanted me to go they oferd me 50 a month and board and pay me for the use of my team."

Later that month, however, he wrote from Carter Station, admitting that he had gone out with Cope again after all: "Sir, as I came threw the post this morning I found I had to either go with them or have them follow me They want to go up green river if they can get a man to go with them if not they are going to Henry's Fork and I think that is the best bone place there is around and I don't want them to go in there so I thought I'd better go up green river with them for a few weeks They only intend to stay hear about 4 weeks more they have been hear 3 weeks already and done nothing and I doubt that they will do much at bones."[12]

On August 5, he followed up by writing: "Friend Marsh . . . I was with Prof, Cope a few days but have left him my eyes was sore and I thought I would rest them a few days. I got your letter and despatch on the 20 I am going to start for Pine Bluffs in the morning to collect for you I think Cope has heard of the place and will go there but I don't intend he shall get ahead of me and if he does get there I don't think he will be much in the way for he don't understand collecting very well and he has fell out with his party and is alone he has one team hired at 6 a day and one that he bought."[13]

Cope would have been overjoyed to find skilled collectors to help him. Perhaps Smith and Chew just wanted to make extra money, but another clue to why they went with Cope is given in a letter from Chew to Marsh. On July 16, 1872, Chew wrote saying that Smith "is with another party I [will?] continue the work until I hear from you hope you will let me know immediately whether you want me longer or not. I am alone now and cannot go as far away as I would like to but will do the best I can in the vicinity of Church Butte until I hear from you."[14] B. D. Smith followed up in his letter to Marsh: "The man Mr Chew that is with me rote to you but has no answer yet." If, as was often the case, Marsh had been careless about making firm commitments with them, that might explain why, with the chance of making some money, they went off with Cope.

With Leidy collecting with his friends in the Bridger Basin and with Marsh still absent, Cope ventured farther afield, setting off on a cripplingly rugged journey. For two months he took his party over some two hundred miles of brutal country and into areas that no one except Hayden's surveyors had yet explored. Cope first headed north and east of Fort Bridger, following Black's Fork all the way to the Green River. This was his first exposure to the badlands of Wyoming. It was some of the same country that Marsh had collected over. After just three days he wrote to his father from Cottonwood Creek, on July 17, 1872: "I have had great success and in two days have found 25 or 30 species of which 10 are new; one of the latter a kind of flesh eater with flat claws. I found three turtles, three tapirs, and one Palaeosyops in one place!"

After two weeks in the well-watered region of the heads of Ham's Fork of Green River, Cope worked down toward Church Buttes on the Union Pacific line and then east to the town of Green River. "I have 20 sp. mammals (8 new)," he reported.[15] He then headed south, exploring "Mammoth Buttes, which form the water-shed between South Bitter Creek and [the headwaters of] the Vermillion and examined the Bad Lands of the Washakie Basin carefully. . . . [I]n reaching this point we crossed a portion of the Cretaceous formation, and I took especial pains to determine the relations of the strata at these points."

It was in the Bitter Creek region, "a howling wilderness where water is scarce and bad, grizzly bears plenty," that Cope made his own finds (critically important, as it turned out) of the apparently ubiquitous giant mammals with horns on their heads. "I spent three weeks on the head waters of Bitter Creek and only left because the spring at which we camped gave out, after being for some days so impure as to make several of us unwell. . . . [A]t the same time my teamsters began to do wrong; one got to complaining and used the grain I had set aside for the team, to the great injury of the latter. The other (having chased after straying mules) spent three days in a state of intoxication. In the meantime the mules starved. When he returned he . . . stole $20 or more worth of provisions . . . ran the wagon into a ditch and started out for another spree. I discharged him at once and came on to Green River by R.R. while the team came after."[16]

Cope now headed all the way north again, crossing back over Black's Fork of the Green River and making his way across country with a much reduced party, traveling some seventy-five miles to the confluence of Fontanelle Creek and the Green River. Again they found the giant horned mammals: "I found two skulls one nearly perfect of the species I call E. cornutus. This was a monstrous animal, and Elephantine in size and proportions. . . . Since coming up here I have had good success. I found 30 species, several of them new. . . . Altogether I have 50 new species, vertebrates so far. . . . There are remains of three species of the last, and over 13 individuals. Six I found entombed near together."[17]

As the group headed up into the Salt River Range, well above seven thousand feet, the going became so rough that Cope left his outfit behind and "took my packer and one mule, packed with bedding and provisions." They made camp at the top of one of the tributaries of the Fontanelle called Willow Creek, with Cope acting as cook: "Here we camped, picketed our animals, and built a fire. The fire was needed for the night was frosty and the blankets none too many. Our supper of bacon was soon cooked, the beans warmed, and with bread and fried rice made a good meal. . . . I brought back from the bluffs, fossil insects of interest. That afternoon we made 10 miles towards the Fork [Ham's Fork] and reached it night before last after a rough and beautiful ride. . . . Coming down from the mountains I had a splendid view of the Great Eocene lake basin before me, with Uintah Mts. for its southern and Wind river Mts. for its N.E. shores. . . . I expect to be in Fort Bridger in 4–5 days to join my small but excellent family."[18]

Cope's father was not the only one he told about his discoveries of horned mammals from Bitter Creek. Cope may have known that Marsh had a strange new giant mammal in his collections from around Fort Bridger in 1871 (although Marsh's note on that species had yet to be published in the *American Journal of Science*). He may have known about Leidy's paper on *Uintatherium* and *Uintamastrix;* he certainly had heard from the Carters, B. D. Smith, and Chew about Leidy's discovery of new specimens. So, even before he returned to Fort Bridger in September, whenever Cope found promising new specimens, he wrote short papers about them—as fast as he could—and sent them off for publication in Philadelphia. Before the month was out Cope had sent the complete manuscripts of three short papers (respectively the "Description," "Second Account," and "Third Account of New Vertebrata from the Bridger Eocene of Wyoming") to the *Proceedings of the American Philosophical Society.* Before the field season was completed he had sent off additional works on new turtles of Wyoming, coal and fossils of Nevada, his new dinosaur from Wyoming, and two papers on fossils from Bitter Creek. And, as detailed in the next chapter, in one soon to be notorious case, he even used a telegram to try to establish priority for a name for a new genus of mammals.[19]

None of this was easy. To start with he had to solve the not insignificant problem of coining names for all the new species. Only a working taxonomist can truly appreciate how difficult it is to come up with descriptive combinations of Latin and Greek words and, moreover, ones that have not been used before. Cope's field notebooks show him experimenting with various roots and prefixes—all this while living in a tent somewhere in the back of beyond. Excited by his discoveries, he was all the while wondering whether Marsh and Leidy might be producing names and descriptions ahead of him. The stage was set for an almighty dispute over priority for naming new species. But the seeds of antagonism had been planted simply by the fact that Cope had gone to the Bridger region that year.

It has been said that Cope was reckless in the field. Certainly he worked himself and his crews very hard—to the point of exhaustion, in fact. In September he came down with a fever and for the first time realized the dangers of working, essentially alone, in the wilderness. On the 18th he limped into Fort Bridger, delirious and running a high temperature. Annie (with little Julia in tow) had come from Denver to be with him. Now she had to nurse him at their grim boarding house. She wrote back to her parents that Cope had "some inflammation of the brain for a day or two producing great restlessness, and no sleep, only dozing to frightful scenes, making it most roilsome and wearisome. . . . Dr Corson seems to understand his case, and is most attentive."[20] A month later Cope had recovered enough to write to his father: "Nothing is left of the fever. . . . [T]he most depressing and distressing is . . . carbuncles. I have two huge ones on the back of my neck with two medium and five little ones. . . . [M]y nights are positively happy under the influence of an opiate. . . . [D]uring my fever I had terrible visions and dreams, and saw multitudes of persons, all speaking ill of something. I had many other nervous states, all of which caused great suffering."[21] Probably he had developed a systemic staphylococcus infection from sores on his neck.

Despite all its triumphs, in addition to the physical hardships, Cope's 1872 trip was accomplished at considerable professional cost. It triggered intense rivalry with Marsh over the credit for being the first to describe the strange six-horned "elephants" from Wyoming. He had (perhaps unwittingly) crossed Marsh by employing Smith and Chew. Marsh also charged that Cope had been sneaking a look at the specimens Smith had collected for Marsh and that Cope actually had gained possession of some of his material from both Kansas and Wyoming. The worst of it was that this last charge was true, if only again inadvertently. When it came to eventually packing up specimens and shipping them back east, B. D. Smith sent some of Marsh's material to Cope in Philadelphia by mistake. As soon as he got back home and found them, Cope sent the fossils on to Marsh, but irreparable damage had been done to their relationship.

Marsh evidently wrote in a fury to Cope, who replied: "I wish you had mentioned to me about missing specimens from Kansas, Wyoming, etc. when the first suspicion crossed your mind that I knew anything about them. It is far more irritating to me to be charged with dishonourable acts than to lose materials, species, etc. I never knew of any losses sustained by you or specimens taken by any one till those were sent to me that you now have. Should any such come to my hands I will return them, as I did the last. . . . All the specimens that you obtained during August 1872 you are to use. Had I chosen they would all have been mine. I allowed your men Chew and Smith to accompany me & at last when they turned back discouraged I discovered a new basin of fossils, showed it to them & allowed them to camp and collect with me for a considerable time. By this I lost several fine things, although Smith owed me several days work." (Cope is referring to the Bitter Creek work.)[22]

Marsh replied in part: "In regard to Smith, let me remind you that I had spent no little time in teaching him to collect fossils; had entrusted to him valuable information about localities (including those east of Green River); had given him an outfit, and engaged him for the season at his own price. When in June he had a good offer to go on a military expedition, he declined, saying that he could not go without my consent. You, however, enticed him away, even before he had shipped his specimens as directed, and they were then delayed with great loss to me. I would not have done this to you for all the fossils in Wyoming. This act of yours created a strong prejudice against you at Fort Bridger among both officers and civilians. These separate parties promptly informed me of it, and denounced you."

Marsh spent a lot of the summer of 1872 in the laboratory and may not have intended to go into the field at all. Among his many publications that year was a series of four papers in July on his Wyoming collections of the year before. Here he described no fewer than fifty-four new species, many of them representing new genera. They included "large pachyderms," tapirs, hoofed animals, relatives of the dogs and foxes, two different kinds of bat, several kinds of small insectivores something like hedgehogs, and what turned out later to be several kinds of very early horse relatives. Most of these new taxa were based on isolated teeth rather than complete specimens, which is why so many of them were given names—*Centetodon, Centracodon, Harpalodon*, for example— that use the Greek word *odons*, meaning tooth.[23] In September of the same year Marsh brought out an even shorter of pair of notes. It was in the first of these that he created the genus *Tinoceras* by renaming the material that he had, the year before, called *Titanotherium anceps;* the second note created the species *Tinoceras grandis.*[24]

In addition to all this, he had another major discovery on his hands. Earlier in the year Benjamin Mudge had made a small collection of bones that he first started to send to Cope and then fatefully changed his mind, dispatching it to Marsh instead. Marsh opened the box and found himself looking at part of the skull of a bird. At first it seemed that Mudge had sent bits of two animals, a bird and a reptile. Then he put them together—it was a bird with teeth, something never seen before. All modern birds lack teeth (hence the phrase "as scarce as hen's teeth"), but here in the chalk of Kansas was something major, notable not only for its novelty but also representing something predicted by the growing field of evolutionary paleontology. Birds with teeth were obviously some kind of direct link to their reptilian ancestors. (Although the famous *Archaeopteryx* had been discovered in Germany in 1861, there was at this point no specimen with a skull, and therefore it was not yet known that *Archaeopteryx* also had had teeth.) From material collected in Kansas the previous year Marsh was pretty sure that his big, loonlike bird *Hesperornis* also had teeth. He now had two examples of these toothed birds: the large one that he called *Hesperornis* and Mudge's smaller one—a bird with well-developed wings, which he called *Ichthyornis.*[25]

Marsh was well aware that both Leidy and Cope had plans for summer fieldwork. Dr. Carter, concerned over the intense jockeying for position and the conflicting pulls on local loyalties, had written trying to broker a deal between him and Leidy that would, by implication if not explicitly, exclude Cope. "I am sincerely sorry to learn of another disappointment to you by the action of Prof. Cope. This is the first I knew of his having been over any of the fields explored by you. . . . I [am] fully prepared to estimate how provoking it is—not to put it stronger—to be thus anticipated [Carter seems to be referring to work in Kansas in 1871]. . . . [W]ould it not suit you to come out alone this summer and enjoy a while with us. . . . I would be delighted to have you two together here, and my faith is that it will work beneficially to all in this way. Dr Leidy is a genuine friend of yours and I believe a most generous man."[26]

Carter was also worried that Marsh's large parties of students were becoming a matter of contention. "Can't you come alone—that is not with an 'expedition.' I think a great deal of you Professor, but with due respect I don't 'go a cent' on the Yale boys as helpers to science." Marsh ignored this advice, but he did decide to go west in 1872 after all—in October. Perhaps because of the lateness of the season (and also the cost of the 1871 trip), possibly because of Carter's comments, he took only a small party of four students. Once more Marsh went first to the Cretaceous beds of Kansas, and again the results were spectacular, one of the many highlights being the discovery of more fossil toothed-bird specimens. He collected more material of *Hesperornis,* including another almost complete skeleton. Then it was on to Wyoming where Marsh, no doubt guided by Smith and Chew, who were able to show him the sites that Leidy and Cope had found productive, collected his own examples of the strange giant horned mammals that seemed to have the dominated Eocene landscape. It was during this late trip to Fort Bridger that two of Marsh's students found remains of yet another—and perhaps the largest of all—of the giant mammals of the Eocene. This creature, which Marsh gave the modest name *Brontotherium gigas,* had a curious dished shape to the skull with a pair of horns on either side of the nose. Eventually it would turn out to be a creature that had roamed the west in huge numbers, as had Leidy's *Palaeosyops.*

By this time Marsh's very short note naming *Tinoceras* had been published in New Haven and, like Cope, he had realized that these giant elephant-like mammals could represent yet another major coup. Some were related to the creature Leidy had already named *Uintatherium*, but *Tinoceras* seemed to be of a different kind and new; he later created a whole new order to contain such mammals—the Dinocerata. For now, however, it seemed that Leidy and Cope had gotten ahead of him, something guaranteed to bring out the worst in a man like Marsh.

TWENTY-ONE

The Case of the
Great Horned Mammals

Among the many extinct animals hitherto discovered in the Tertiary of the
Rocky Mountain region, none perhaps are more remarkable than the huge mammals that have recently been described from the Eocene beds of Wyoming. . . .
[T]hese animals nearly equalled the elephant in size, and had limb bones resembling those of Proboscideans. . . . [T]he skull, however, presents a most remarkable combination of characters, . . . long and narrow and supported three
separate pairs of horns.

O. C. MARSH, 1873

I found two skulls one nearly perfect of the species I call E. cornutus. This was a
monstrous animal, and Elephantine in size and proportions. Its skull is three ft.
long and the hips 5 ft. across. The head of the femur is as large as the top of my hat.

EDWARD DRINKER COPE, 1872

By the spring of 1873 Cope and Marsh, who had once been colleagues
and even friends, had become bitter enemies. While the divorce had
been a long time brewing and doubtless was inevitable, the immediately
precipitating events concerned their discoveries of the strange horned
mammals in Wyoming. The dispute tells us a great deal about the state
of paleontology at the time and about the personalities of the two men.
It also marked the beginning of the end for Joseph Leidy as a major
player in this great game.

Leidy, Cope, and Marsh had each discovered specimens of gigantic
fossil mammals, some with as many as three sets of horns on their skulls, in
the Eocene beds of the Bridger Basin, Green River Basin, Uinta Basin, and
Wasatch Basin. These creatures were early equivalents of the large hoofed

animals (deer, cattle, horses) we know today and had apparently roamed the West in large numbers like modern buffalo or the elephants of the African plains. From their size it seemed not impossible that some of them were in fact related to "pachyderms" (elephants)—although it was hard to see where a trunk might have been located. Cope insisted on sketching in a trunk when he attempted to reconstruct their appearance. And while the fossils were referred to as having six horns, they actually had none: the horns were bony protuberances from the skull, not true horns.

It took a while to work out that several different groups of these monsters were involved: using their later names, they were the Dinocerata and Amblypoda (very distant relatives of the elephants) and the Titanotheria (equally distant relatives of the tapirs, horses, and rhinoceros). The Titanotheres included the horse/tapir-like forms that Leidy named *Palaeosyops*, which would be joined by the huge, grotesquely horned animal Marsh would call *Brontotherium*. The Dinocerata (including animals named by Leidy as *Uintatherium*) were the "sixhorned" creatures that immediately became so contentious.

Whatever group or groups they belonged to, and whatever their purely scientific significance, it was obvious from the first that the giant mammals being discovered in Wyoming had major dramatic potential. They were the sort of animal that Jefferson would have loved—huge, grotesquely ornamented, and, even if they were (yet again) herbivores rather than ferocious carnivores, they were fantastical enough to fill the viewer (including any Frenchman) with a hearty respect. Their names were chosen deliberately to convey a sense of size and importance, being based on the classical word roots for giant (*mega-*), stretched (*tino-*), terrible (*deino-*), thunder (*bronto-*), and a kind of ancient god (*Titan*). Whoever could make these unique creatures his own would greatly enhance his scientific and popular reputation.

All scientists, working largely out of sight of the public, are driven to be recognized favorably by their peers; the only way to do that is to get important scientific results published. In the mid-nineteenth century just as today, scientists also craved attention in a more popular domain. Their results had not only to be scientifically important but impressive, even spectacular, to the general public. Paleontologists have a natural advantage in this respect, although the fact that they some-times work with spectacular animals popular with the public also works against them, causing them to overreach in their rhetoric and even to concentrate on trivial (even frivolous) subjects.

The professional reputation of scientists rests on a foundation of their works—their papers—published in reputable journals. This is a matter not just of simple communication but also of establishing "terri-tory." When it comes to discovering and describing new kinds of fos-sils, what matters most—to some workers it matters even more than getting the facts right—is being first. The first discoverer gets to give the animal its scientific name, and from that point the name of the discov-erer and the animal are indissolubly linked. Whenever Leidy, Cope, and Marsh worked on material from the same area (either collected by themselves or found for them by others) they were immediately and in-evitably caught up in a race for priority. When they found spectacular material like giant horned mammals of the Eocene or (later) the even larger and more dramatic dinosaurs of Wyoming and Colorado, the in-tensity of the competition became almost unbearable. The case of these giant mammals is an object lesson demonstrating the confusion and bad will that can be created by any people (and Cope and Marsh in particu-lar) who try to work, both literally and figuratively, in the same field. And it will also demonstrate how scientific names should be (or actually were) established.

As a reminder, the scientific names of animals are a combination of three parts. Humans, for example, are *Homo sapiens* Linnaeus—the name codifies the genus, always capitalized (*Homo*), the species, always lowercase (*sapiens*), and the person who first defined the species. The name of the genus must be unique but the name of the species can be used in different genera. *Passer domesticus* is the house sparrow; *Gryllus domesticus* is a cricket. If a species named in one genus is later found to belong in a different genus, its specific name is maintained, thus retain-ing a historical trace of the original naming. If either a genus or species have been named for material that clearly had been given a name al-ready, the earlier names prevail and the new ones are sunk. Such a fate befell the giant fish from Kansas that Cope named *Portheus molossus* in

1872; it turned out to be the same as the fish that Leidy had described as *Xiphactinius audax* two years earlier.

In a more complex case, among the new mammals collected from Wyoming by Marsh in 1871 was the creature he gave the name *Titanotherium anceps.* That is to say, he declared that this was a new species— *anceps*—in the existing genus *Titanotherium.* The genus *Titanotherium* had been created by Leidy to contain the very first of the fossil mammals from the Dakota Bad Lands: *Titanotherium prouti* and this animal had first been named by Hiram Prout as *Palaeotherium prouti.* (When Leidy realized that it was different from Cuvier's *Palaeotherium,* he had to create the new genus.) Later Marsh moved his species *anceps* out of *Titanotherium* and into a new genus, *Tinoceras.* Even Cope's beloved *Laelaps* had to be renamed *Dryptosaurus* because the former name had been used before (for a beetle!). This is the sort of thing that can give the observer a blinding headache and almost bring the participants to blows.

Although both Cope and Marsh collected extensively in the Cretaceous beds of Kansas, they managed at first to come to something of an understanding over "territory." Although they sparred uneasily over the material, it must have been reasonably clear that Cope had the prior claim on the fishes and reptiles, and that Marsh had the field of toothed birds and pterosaurs more or less to himself. Mudge, who was the man-on-the-spot and also man-in-the-middle, as far as Kansas was concerned, was able to play a role in keeping the two men from mortal combat by judiciously directing material to the "right" place, Philadelphia or New Haven. When it came to the fantastical Eocene-age horned mammals that were unearthed in the Bridger region of Wyoming between 1869 and 1873, however, all bets, it seemed, were off.

Everything in the ensuing dispute between Marsh and Cope depended on the details. Leidy published his descriptions of *Uintatherium* and *Uintamastrix* on August 1, 1872. Marsh, with the issue of priority obviously weighing heavily on his mind, in a postscript to his four *Preliminary Descriptions of New Tertiary Mammals* from his Wyoming collections of 1871, took the unusual step of listing the publication dates for the four parts. The addendum (dated August 19, 1872) established publication dates for July; it noted also that the four parts had been distributed "in pamphlet form" on June 21.[1]

Earlier in 1872 Cope published a description of a new mammal with "a remote affinity . . . to 'Titanotherium' " that Hayden had collected in the "Wasatch group, near Evanston, Utah," from "beds . . . inferior to the Bridger group, and . . . supposed to be Lower Eocene." This new genus, *Bathmodon,* was another animal the size of a rhinoceros. Once in the field that year, Cope soon also began to find wonderful material of other large mammals and to send his papers home. While Leidy's paper on *Uintatherium* had been, like many of his short papers, almost telegraphic in its brevity, Cope took the concept "telegraphic" literally. He sent a telegram to the American Philosophical Society to try to establish priority for the discovery and naming of three species of a new genus of these great mammals. Unfortunately the telegraph operator misspelled the names (or misread Cope's writing), so the Minutes of the Philosophical Society meeting of September 20 read: "The Secretary announced that he had received a telegram from Prof. Cope, dated Black Buttes, Wyoming Territory, August 17, announcing the discovery of Lefalophodon dicornutus, birfurcatus, and expressicornis, Cope."[2]

To correct this, at the December 20, 1872, meeting Cope had his original telegram read out: "I have discovered in Southern Wyoming the following species: LOXOLOPHODON, Cope. Incisor one, one canine tooth; premolars four, with one crescent and inner tubercle; molars two; size gigantic. L. cornutus: horns tripedal, cyclindric; nasal with short convex lobes. L. furcatus, nasals with long spatulate lobes. L. pressicornis, horns compressed sub-acuminate. Edward D. Cope, U.S. Geological Survey." The society's secretary, Professor J. P. Lesley, added a note in the minutes: "The above telegram was so badly transmitted by the operators as to be read with difficulty, and the precise forms of the specific names could not be certified until the return of Prof. Cope from the field." Lesley's note presumably was intended to provide his friend Cope full credit for the names and descriptions as of August 17 and no doubt had been specifically requested by Cope.[3]

This all seems rather contrived, and what Cope had done was really

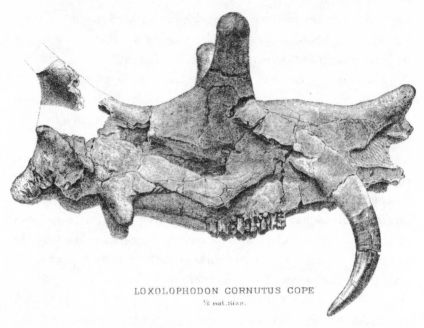

LOXOLOPHODON CORNUTUS COPE
⅕ nat.size.

Cope's drawing of the skull of *Loxolophodon* (now *Uintatherium*) (from Edward
Drinker Cope, *The Vertebrata of the Tertiary Formations of the West*, 1883)

not different tactically from what Marsh had done with his fossil bird
from Kansas when he sent a twelve-line letter (addressed to Professor
Dana) to the *American Journal of Science* announcing, "On my return,
I shall fully describe this unique fossil under the name Hesperornis regalis." Of course a telegram was a rather dramatic way of staking a
claim, and the whole affair might seem rather amusing. But it was not
the slightest bit comical at the time. Not only had the telegram read at
the September 20 meeting given wrong names, it had been printed
without any descriptive information. Did it, then, constitute a bona fide
description? Did it correctly establish a date and Cope's priority?

Part of the naming problem was that there was not yet any agreement
as to what would constitute a formal description of a new species. Today
it must include a full description of a selected "type" (the unique specimen that becomes the name-bearer of the species), a figure, and a clear
statement of the features that conclusively distinguish the new species
from all others. Cope's 1872 Lefalophodon telegram could hardly have

been less like such a description. But much the same could be said for most of Leidy's early papers, including, for example, the announcement of his Judith River dinosaurs. Leidy, Cope, and Marsh all had a tendency to put out brief notes first and then follow up much later with a fuller account, often in a synthesizing monograph. Leidy's Judith River dinosaurs of 1856, for example, were not fully described and illustrated until 1859.

In addition to the rather desperate move of sending a telegram, Cope had hastily written eight papers describing the fossils discovered on his western trip of 1872. Like Leidy's paper on *Uintatherium* and *Uintamastrix*, these were dashed off in the field and dispatched to Philadelphia for publication (by the Philosophical Society). One of these papers described a large "horned" mammal—the one he referred to in his letters home as *Eobasileus*. Confusingly, he created *Eobasileus* by reassigning one of the species—*cornutus*—that he thought had previously been assigned (in the telegram) to *Loxolophodon*.

All might have been well if *Uintatherium, Loxolophodon, Eobasileus, Dinoceras,* and *Tinoceras*—not to mention *Bathmodon* and *Titanotherium*—had been uniquely different one from each other. But they were not. Worse, Cope and Marsh had each known *at the time* that some of the fossils they were collecting, and possibly all of them, were being found by the others. Judge Carter and Dr. Carter no doubt took pride in spreading information about what had been found around Fort Bridger. That is probably how Cope knew to write to his father: "Eobasileus is the most extraordinary fossil mammal found in North America, and I have good material for illustrating it. Marsh and Leidy have obtained it near the same time and I have no idea whether they have fathered it in advance of me or not."[4] Another source of information was B. D. Smith, who had already stirred up trouble between Cope and Marsh by collecting for both of them. He wrote to Marsh: "We got one tusk and part of the jaw nearly one foot long. . . . I think it the same kind that Prof. Lidy got part of the tusk hear that he is blowing about."[5]

Back home for the winter, each man had the opportunity to see what the others had written. Marsh quickly realized that, despite their agreement,

Cope had not sent him copies of his papers as soon as they were published: neither the final versions nor the publication day pamphlets. Reading though Cope's works when they did arrive in New Haven, he saw that he had been beaten by Cope for priority in naming the new mammals from Wyoming—unless, that is, Cope had been cheating. So he made what he no doubt considered a preemptive strike. He traveled down to Philadelphia and, at the December 20, 1872, meeting of the American Philosophical Society, with Cope (and probably Leidy) in attendance, he presented what the minutes recorded as a "short account of the more remarkable results of his exploration in the Rocky Mountains since 1870, viz: His discovery of the first American fossil pterodactyles, bats, marsupials, birds with biconcave vertebrae, monkey (Eocene) of a low type, and *dinocerea, a new order of horned proboscidians with canine teeth*" (emphasis added).[6]

In this paper Marsh pulled no punches either in boasting of his own discoveries or in savaging Cope across the board. "[Marsh] had described three species [of 'pterodactyles'] from the Cretaceous of Kansas," it read. "Prof. Cope had subsequently re-described two of the species in the Proceedings of this Society . . . but [Marsh's] names . . . had priority. . . . [Marsh] recently assigned . . . [gigantic Eocene mammals] to the new order Dinocerata. . . . Prof. Cope has given the name Loxolophodon semicinctus, to a single tooth, which may possibly belong to this group." He even took a swipe at Leidy on the way to another attack on Cope. "Dr. Leidy has described a characteristic specimen as Uintatherium robustum, and a canine tooth, apparently of the same animal, under another name. The remarkable feature of the skull in this group was first indicated in the name Tinoceras, which the speaker had proposed for one of the genera. Prof. Cope subsequently proposed the name Eobasileus, but was mistaken in regard to the main characters of the skull. What he called incisors were canines; and the large horns were not on the frontals, but on maxillaries." And so on. Marsh published similar papers attempting to demolish Cope's taxonomies in articles in the *American Journal of Science* and the *American Naturalist*. At this time he also described a new genus in this group—*Dinoceras*. His final message was that his own names of

Dinoceras and *Tinoceras* were valid; none of Cope's were.[7] It must have been a difficult meeting for the members to sit through.

Marsh's major achievement with this paper was to steal a march on Cope by placing the new species (however many there were, and whatever their correct names) in a new order of mammals: Dinocerea (which he soon after changed to Dinocerata). At the meeting, Cope was powerless except that he immediately "dissented from the propriety of at present erecting the proboscideans so discovered into a separate order, merely on the ground of their possessing horns and canines, and gave his reasons."[8]

So now Cope went to work and produced a long analytical study of even broader scope (read to the Philosophical Society on February 21, 1873), in which he laid out a new plan of classification of all "the short footed ungulates of the Eocene of Wyoming," recognizing the existence of three major groups: Proboscidea (elephant relatives), Perissodactyla (odd-toed hoofed animals related to living horses and rhinos), and Artiodactyla (even-toed ungulates related to camels and cattle). In the Proboscidea he included true proboscidians, Marsh's Dinocerata, and the Pantodonta (another group already known from Europe but lacking the horns). He stated somewhat disingenuously that "whether all the animals to be included in the Proboscidia possessed a proboscis or not, is of secondary importance." In this paper Cope grudgingly adopted Marsh's term Dinocerata, although he "would have preferred using [a name] already employed to coining a new one."

Naturally, Cope's view of the Dinocerata was that there were (so far) four genera: *Loxolophodon, Eobasileus, Uintatherium,* and *Megaceratops.* In other words, none of Marsh's names were valid, and all of his own were.[9] At the April 4 meeting, Marsh fired back with a paper titled "On the Gigantic Mammals of the American Eocene," in which, unsurprisingly, he once again claimed that all his names for the Dinocerata were right and all of Cope's wrong.[10]

Throughout these presentations the two men hurled accusations at each other over real and supposed errors in their work, down to the most trivial details, and up to the question of whether these animals were, or were not, related to elephants. Both reversed positions on this latter point at one time or another. In general Cope adopted a lofty literary tone: "The absence of foundation for Professor Marsh's recent animadversions, and though these latter present internal evidence of idiosyncrasy which almost disarms reply, yet . . ." Marsh's style was more matter-of-fact: "Unfortunately he still misinterprets the structure of this group . . . on nearly every page of the paper, moreover, new errors may be detected. Prof. Cope's defence . . . lacks both candor and accuracy."

And so it went while the rest of the academic community watched with fascination and horror. The matter might have rested there, awaiting the discovery of further material that might (and eventually did) sort things out. But then Marsh significantly raised the stakes by opening a second front in the war over names. He complained both to the Academy of Natural Sciences (of which Cope was corresponding secretary) and to the American Philosophical Society that there had been cheating over the crucial issue of the dates of Cope's publications. Cope's names, therefore, were not only scientifically unjustified, they were invalid by reason of lacking priority of date.

The Minutes of the Philosophical Society record that at the April 4 meeting (with Cope present): "Prof. Marsh read a paper on Prof. Cope's determinations of the dates of papers read before this society, which he afterwards withdrew by permission of the meeting." The precipitating issue was the famous telegram. If, indeed, that constituted a publication, what was its date? August 15, when it was sent, August 17, when received, September 20, when announced to a meeting of the society, or December 20, when read in full form? And what name did it establish, *Lefalophodon* or *Loxolophodon?*

Part of Marsh's complaint was that Cope had failed to hold up his end of their agreement on exchanging publications on the very first day. Marsh stated that he had not received "a single copy up to October 8th; when the last paper of my series was published, and I started for the West."[11] But—the reader will sigh at this news—the situation was more complicated even than that. As was the case with the *Lefalophodon* telegraph, there were several potential publication dates for Cope's papers (and for Marsh's own). For example, Cope's *Eobasileus* paper was one of five mentioned in the minutes of the American Philosophical Society's meeting of September 20 as "communications . . . received from Prof. E. D. Cope under the following titles." It might be possible to take that as having meant that the papers had been formally "read" (read out) to the society, in the way his telegram was (at the same meeting), but there is no evidence that such a thing occurred. In the index of volume 12 of the Society Proceedings, the *Eobasileus* paper ("Second Notice of Extinct Vertebrates from Bitter Creek, Wyoming") is actually listed under the "stated Meeting of August 15th." But the minutes of that meeting list only the first three of Cope's Wyoming papers as having been "received from Prof. Cope." Finally, volume 12 of the proceedings was not actually distributed until February 1873. Preprints (what Marsh called "pamphlets") of that paper had been printed and were available for distribution on August 22. Printed copies of the very first of Cope's series of eight Wyoming papers for 1872 ("received" August 15) were available as early as July 29, which is truly remarkable since Cope had then been in the field for only two weeks. No wonder Marsh was suspicious.[12]

For Cope to have claimed priority on the basis of the early preprint dates might have been acceptable if he had sent copies to Marsh and Leidy immediately. Even though Cope was in Wyoming, some of his colleagues later attested to having received early copies. But it seemed peculiar that, as recorded by notations in academy librarian Edward Nolan's handwriting, Cope did not send copies of his Bridger papers to his own Academy of Natural Sciences until October 29. Marsh did not receive copies of Cope's works until November, when they were forwarded to him in Wyoming. Marsh claimed that some of these were merely uncorrected proofs, and in fact that also was the case for two of those received by Nolan at the academy on October 29.

Marsh's complaint was a serious one, a criticism not only of Cope but of the two institutions. The minutes of the Philosophical Society meeting of May 16, 1873, reported "A discussion respecting the time and manner of publishing the Proceedings of the Society, in which Genl. Stokes, Dr. LeConte, Prof. Cope, Mr Whitman, Prof. Barker, Mr Lesley, and Prof. Cresson took part." As usually happens in such matters,

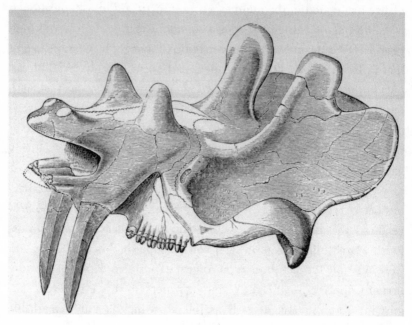

Drawing of the skull that Marsh named *Dinoceras;* now *Uintatherium* (from the *American Journal of Science,* 1891)

this one petered out without real resolution. The only men who really cared about it were probably never going to be satisfied; everyone else was rapidly tiring of their histrionics.

The *American Naturalist* announced that it would not publish any more of these attacks, although the editors allowed the antagonists final salvos in an appendix, which they had to pay for themselves. Cope never again used a telegraphic announcement to establish priority over a name. One effect on Marsh, however, was a steady resolve to make this group of mammals his own, and in 1886 he published a massive, superbly illustrated memoir on the Dinocerata.[13] It established these mammals as truly an American phenomenon and was a personal triumph. But then, as we shall see, even this superbly successful work, published at government expense, would come back to haunt him.

In the end (or, more accurately, at the present time) it turns out that the suspicions first voiced out in Wyoming in July 1872 were correct: these rivals did all have the same materials. Marsh's *Dinoceras* and

Tinoceras were really Leidy's *Uintatherium*. Leidy's *Uintamastrix* was his own *Uintatherium*. Cope's *Loxolophodon* was also the same as Leidy's *Uintatherium*. These uintatheres form the basis of Marsh's Dinocerata. Cope's *Eobasileus* was really Leidy's *Titanotherium* and therefore belonged with Leidy's *Palaeosyops* in the different group of giant, hornless mammals called titanotheres. Cope's *Megaceratops* was really the animal that Leidy in 1871 had described as *Megacerops* and it, too, was a titanothere. Once again Leidy, steady and calm, turned out to have had things right. But Marsh might have claimed a small victory over Cope's *Bathmodon*, which is now considered part of Marsh's *Coryphodon* (and a member of yet another group of horned mammals, the Amblypoda).

The accusations of sharp practice and downright cheating—to say nothing of the embarrassment over the confused names—totally poisoned what, if anything, was left of a relationship between Cope and Marsh; their mutual attacks became ever more personal and their disagreements eventually became public.

TWENTY-TWO

Going Separate Ways

Hayden urged Leidy to return to the West in 1873, stating: "The coming year I will most gladly aid you to visit two localities neither of which will be visited by any one but you, if you go . . . then make a trip to Judith River which you will have all to yourself. Should Marsh or Cope desire to go to these localities I cannot hinder it, though I will not aid them in any way." He ended his short letter on a rather chilling and perhaps tactless note: "I am writing this letter within ten miles of the spot where the Sioux Indians are making raids."[1] In April 1873 he offered more positive blandishments: a position as naturalist to a government expedition. "I think there will be a good chance for you to visit the Upper Missouri this summer. About 2,000 troops are going up on the U.P.R.R. this summer." But the last sentence of this second letter also carries a sting: "Keep the matter still and away from Cope." Cope, of course, was then working for Hayden![2]

The expedition Hayden referred to was probably that led by General David S. Stanley, a show of force that "was sent up the line of the N. Pacific R.R." to the Yellowstone River.[3] Here was a case where Leidy's caution paid off, as the expedition suffered many losses. Leidy did return privately to the West in 1873, and he had to pay his own railroad fare as the Union Pacific was now reluctant to provide free passes, or even half fares, for scientists. For this second trip (perhaps wondering if it would be his last) he took his wife and a group of friends, including two ardent mineral collectors (Joseph Wilcox and Clarence

Bement), the botanist T. C. Porter, and Henry Chapman as zoologist. Passing through Chicago, all were amazed at the devastation caused by the recent great fire. Charles Dolley, a favorite student of Leidy's and a cousin of Dr. Joseph Corson, was the youngest member of the party. In a personal memoir he later wrote that "squares and squares of demolished buildings still lay practically undisturbed. . . . It seemed impossible to me, that the city should ever rise again from that vast area of desolation. I then little understood the spirit that animated the people of the West."[4]

Leidy's party marveled at the richness of the farmlands between Chicago and St. Louis; while on the way to Omaha they were reminded of the brutality of the West when they saw "more than once the horse thieves who had been hung to the cross arms of telegraph poles, capital punishment being then the fate for stock stealing." They found Omaha to have only two hotels, one of which was closed. At the one that was open they had to share a room with other guests and also with bedbugs. But the next day Omaha impressed them with its range of shops, and "among these frontier merchants I found a Swiss watch maker who was able to repair the gold watch which I had inherited from my Father," Dolley wrote.

As we approached the Green River region in Wyoming the land became more arid until at Carter Station the vegetation consisted chiefly of Sage brush and Grease-wood. At many of the stations numbers of Indians had gathered to see the train which was still a great wonder to them. . . . Carter was the terminus of our journey. It consisted of a small frame station and nearby the house of the station agent, where an acre or two of the adjoining land had been cleared, planted to which was irrigated by means of a windmill drawing water from a driven well. All around as far as the eye could see was the desolate sage brush country. Army wagons were awaiting our arrival and we were soon speeding over the rough trail to Fort Bridger, which is some nine miles to the Southwest of Carter Station. . . . [At Fort Bridger] it was no unusual sight to see fifty or more Indian ponies standing at the tie-rail in front of the trading post, while the bucks and squaws were selecting goods inside.[5]

This time Leidy was better equipped for freshwater researches and evidently spent even more of his time during this trip collecting in lakes and streams than prospecting for fossils. James Van Allen Carter took him north this time, to "country on Ham's Fork." He had an excellent eye for spotting Indian artifacts, but evidently his heart was no longer in the fossil business. This was, in fact, not Leidy's last trip to the West, however. In 1878 he returned, but exclusively to collect living creatures from the freshwater lakes and streams and from Great Salt Lake. He then published, as part of the U.S. Geological Survey reports, a massive monograph reviewing the microscopic rhizopods (amoebas and their relatives) of North America.

The year 1873 also saw the last of Marsh's Yale trips, and this time it was with a huge group—thirteen students. They started from Fort McPherson in mid-June along with a large party of soldiers, who were sent not only to support Marsh's group but also to make a show of force for the Lakota Sioux to the north. The students were introduced to the West in a tough trek north past the 1870 Loup Fork site to the Niobrara River. Marsh employed a colorful local man named Hank Clifford to guide them through the Niobrara country, although Clifford stated that he was "in doubt in reference to the Indians North they have bin stealing horses from the Whites And killing Now and then a white man for luck and I think that it would require an escort to travel threw that Country this Summer."[6]

A man of almost unparalleled profanity who, in addition to great skill at finding fossils, was a good hand at conning money and gifts out of Marsh, Clifford was yet another of the many so-called squaw men— frontiersmen who, like George Croghan, Owen McKenzie, Alexander Culbertson, and many others, had married Indian wives. Typically the squaw men married into the highest levels of Indian society and, as a result, had unusual access to the whole region. Clifford would later help Marsh get access to the Black Hills of South Dakota.

In late July the party moved on to Fort Bridger, continuing work in the Black's Fork and Henry's Fork regions and exploring the Eocene badlands in the Washakie Basin, north of the Uinta Mountains along the Wyoming-Utah border. Cope had been the first into this region the year before. The local Shoshone were relatively peaceful, and in ten days Marsh's students challenged Carter's poor assessment of students as "helpers of science" by collecting another five tons of material, including an almost perfect skull of the "sixhorned" mammal that Marsh had called Dinoceras. Then the party headed for Salt Lake City, where they divided up, some going on to Oregon to collect in the John Day country again, then by sea to San Francisco, as before. They broke the return trip by train to make yet another visit to Kansas. Altogether they were out for about five months.

The 1873 Yale trip is interesting because there exists a summary account of its expenses. Oscar Harger, who had gone on the first trip as a student and stayed on to work in Marsh's laboratory, was the treasurer for 1873. The total expenses were $1,857.50, which included $300 for Harger's salary. This is a far cry from the $15,000 (more than $200,000 in today's dollars) that Marsh said the second trip had cost him. This time, the students were paying their own way and also covering some of the expedition expenses.

Material collected by the four western expeditions with students would be more than enough to keep Marsh occupied in the laboratory at New Haven for years. Much of it actually remained unstudied for decades, and Marsh never made another such expedition himself, preferring to use the services of local collectors. Nonetheless, he always wanted more. He purchased extensively from European dealers, such as Bryce M. Wright on Great Russell Street, London. For American material, during his four years of western expeditions, Marsh had very carefully cultivated local people who would continue to collect for him or point the way to promising sites. First there were the professional men like Judge Carter, Drs. Carter and Corson, and army personnel like George Sternberg and Theophilus Turner. Then there were more or less authentic frontiersmen like Hank Clifford, John Chew, and Sam Smith (and a host of equivalents all across the country), who were essentially put on retainer to prospect systematically for new finds. The result was that he created the foundation of a network of men right across the country who would, over the next twenty years, continue to supply him with the cream of fossil discoveries. All he had to do was pay them—usually about fifty dollars a month, or less—and send them instructions whenever a promising new place was discovered. Clifford collected for Marsh in this way until 1891. His letters no doubt made frustrating reading for Marsh. In October 1875, for example, he wrote: "[I] started for the Bone field got about ten miles from the [Red Cloud] agency and the Indians Stole my horses."[7]

All these men also reported back to Marsh about Cope's activities. Sam Smith announced in summer 1873: "Cope came to see me after you left he whined around and tried to get me to work for him but it was no go. And then he had the cheek to tell folks that he had Employed me for next Summer he got a Cold Shoulder at Bridger from Everybody."[8] At the same time, however, it seems that someone, probably Sam Smith, was reporting Marsh's movements to Cope. Cope's field notebook for July 1873 shows a detailed itinerary for Marsh as he traveled from Fort McPherson to the Green River.

At Fort Bridger, Marsh so assiduously cultivated James Van Allen Carter that he acted for years as Marsh's agent, supervising the comings and goings of B. D. Smith, Sam Smith, and John Chew and making sure they got paid. This turned out to be a problem, as Marsh, for all his vaunted skill in organization and his evident affluence, was very often late in forwarding money, leaving Carter to fund things out of his own pocket. Managing Sam Smith was not easy anyway because, like John Chew, he needed to follow the main chance to make a dollar or two and could not always wait around to hear from Marsh. Typical is a letter Carter sent Marsh in July 1877: "About May 1st Sam Smith was on hand and awaited instructions and a settlement of old scores. I could not give him either, but proposed a trip for him, which he agreed to make, but was hindered by finding his horses gone. . . . He has been at work for the Judge all spring and is now, but liable at any day to be out of employment."[9] Nonetheless, such was Marsh's charm that Carter kept organizing things for him until the fort was closed in 1878 and he left to open a pharmacy in nearby Evanston, Wyoming. Judge Carter died three years later.

In return, Marsh had quickly seen what would warm these lonely frontier dwellers to him. To those who merely collected for him he might send a revolver (as he did Mudge) or even a rifle. He sent a Sharp's rifle to John Chew, and then Sam Smith begged for one: "pleas send me one of Sharp's Riffels like John has got and 200 rouns of Catteriges."[10] (All of Marsh's men collected elk, bear, and other skeletons—even robbed Shoshone graves—for him.) To the more professional of his advisers he sent books—books on geology and other sciences, and novels for their wives. He always sent copies of any publications in which he had mentioned their names. Cope never developed such social skills.

Benjamin Mudge in Kansas became the dean of Marsh's local collectors, especially after 1873, when the Kansas legislature directed the board of regents of the State Agricultural College to modernize its curriculum, spending less time on the classics and languages and more on practical subjects suitable to its mission. Courses were added in dressmaking, printing, carpentry, blacksmithing, wagon making, and telegraphy. Mudge, who had variously taught mathematics, physics, chemistry, and the biological sciences, was reduced to teaching geology, astronomy, preparatory geography, and college elocution.[11] He and two other colleagues fought the regents bitterly over these issues and in January 1874 they were summarily fired.

Mudge quickly wrote to Marsh: "When you were here, you stated that you would like to employ one or two young men to collect fossils in Western Kansas. As perhaps you may have learned;—I have been summarily discharged (with two other professors) from this College. This has been done by an incompetent, conceited, clergyman [Joseph Anderson; Mudge might have added that he was a journalist and politician], who is acting as president. This places me at present out of employment; and perhaps I can collect for you as well as a younger man, either alone, or with a young man."[12]

Marsh took him on, paying him one thousand dollars in total for that year, an investment that yielded an incalculable return of thirty-three boxes of fossils. Mudge worked for him until his death, all too soon, in 1879.

Cope's field collecting in 1873 was, if anything, more prolific than in 1872. Again Annie, no doubt concerned about Cope's tendency to overdo things physically as well as emotionally, accompanied him as far as Denver. Rather than returning to either Kansas or Wyoming, Cope started his travels at nearby Greeley to collect in the country recently explored by Hayden. It was also where Marsh in 1870 had discovered a major area of Oligocene badlands. They traveled along the Chalk Bluffs between the North and South Platte Rivers, a landmark that had figured prominently in Judge Carter's diary of his journey west in 1857. Conditions for collecting were good, although the fastidious Cope complained that "the disagreeable part of this business is the necessity of associating with such men as one has to employ. It is almost enough to prevent me from undertaking it."

"I have at least 70 species of Vertebrata, all mammals except five. I found some hundreds of jaws of rodents with a good many perfect crania. I have explored two horizons, the lower and richer containing 50 of the species. It is largely a new fauna also, and quite distinct from those of Nebraska or Wyoming. I have some 15 odd toed hoofed [Perissodactyla], 10 Carnivora, 13 rodents, 3 Insectivora but no Proboscidea as yet. . . . [T]he most remarkable . . . are the species of huge Perissodactyla with horns . . . which corresponding largely with the horned proboscidians I found in Wyoming."[13]

Cope took a break in the middle of the field season and went to visit Annie in Denver. On his return trip, which meant going to Cheyenne by train, he saw Marsh on the platform, evidently having come up from Denver on the same train. "[He was] running about in some excitement." It would be interesting to know whether Cope had suffered a bout of illness at this time, because he suddenly wrote out a will; in his field notebook the entry dated July 12, 1873, makes fascinating reading, if only because he did not propose to leave his collections to the Academy of Natural Sciences.[14] The reason for this was that he was angry that the academy refused to hire professional curators.

> Knowing the uncertainty of life, I write the following as my will & testament.
>
> I leave all my personal real estate to my wife, the latter to be equally divided with my daughter Julia at her coming of age & a liberal amount of income to be spent in giving her a first class education (with religious grounds). The property to be sold or not for division according to the necessity of the case, arising from the deficiency or not of income for this purpose.
>
> I leave all my scientific books, papers & collections of all kinds to the Wagner Free Institution of Science provided that restitution of all in that collection wrongfully taken & held be made as far as is practicable and provided that a chair of palaeontology & zoology be established there with a salary of not less than $2,500 p. annum. [The Wagner Free Institute of Science was also based in Philadelphia.]

Cope's party then went all the way south and west up the valley of Bijou Creek, across the Colorado Divide to collect fishes at the Late Eocene lake beds at Florissant. Here, in yet another spectacular fossil site that had recently been discovered by members of the Hayden survey, huge eruptions of volcanic ash settling over the countryside had preserved a petrified forest and, in the lake beds, extraordinarily detailed fossils of every sort from insects and fishes to leaves.

Cope could not leave the West without a quick trip to Fort Bridger. However, he discovered that during his visit late in the previous year Marsh had done his homework well with the Carters and other locals. Cope was now persona non grata. Sam Smith reported gleefully to Marsh (in a letter quoted previously): "He got a Cold Sholder at Bridger from Everybody he sleep in the Government hay yard at night took his meals at Manley's that was hitoned for a Bone Sharp."[15]

James Van Allen Carter, ever the diligent conduit of information to Marsh, confirmed that Cope had made an attempt to hire Sam Smith again: "I know nothing of Cope. A letter in his hand writing recd here four days ago addressed to Sam Smith is postmarked 'Greeley Colo.' I scarcely think he intends coming here—at least to work, there being some country called the 'bitter creek region' to which his instructions etc as far as I'm able to learn have been addressed."[16]

Cope's field notebook reveals a different story. It shows that despite the duplicitous Sam Smith's protestations of loyalty to Marsh, he had in fact worked for Cope again. "Received 10/11 1873 of Ed. D. Cope the sum of Fifty one 000/000 dollars on acc. Samuel Smith for collection of fossils in May 1873 to be retained by me Subject to Smith's orders. J. M. Carroll."

By the end of the 1873 field season Cope had become extremely unhappy with his arrangement with Hayden. Money was the main issue; not only was he an unsalaried member of the survey, Hayden was often very slow and conservative about reimbursing Cope's field expenses. Cope also felt the pressure that Marsh and his Washington cronies in the National Academy of Sciences were putting on Hayden. Among other things, they were mounting a whispering campaign against Cope. Hayden did his best to balance all sides, but in the end Cope decided instead to take up an offer to work with Lieutenant George M. Wheeler of the U.S. Army Corps of Engineers, providing geological and paleontological expertise for his Geographical Surveys West of the 100th Meridian. For this, Cope reported optimistically to his father, he would actually be paid $2,500 per year, with "$30 per month additional for provisions when out in the field and all expenses of expeditions paid."[17]

Now Cope experienced another limitation of being formally associated with the federal surveys. Marsh, with his superior funding, could go wherever he pleased (even if it did not please others who were working in the same area); Cope had to tag along wherever the survey teams were working. Hayden had not seriously controlled where Cope went to collect in the Bridger region during the previous year. For this year's work Wheeler sent him to New Mexico, a place as yet unexploited by either Marsh or Leidy. Wheeler's mission was topographical mapping, and for that he wanted Cope principally as a geologist, certainly not a paleontologist, and that was bound to create problems. And it was to prove as difficult to get reimbursement for expenses from Wheeler as it had from Hayden, which meant that Cope had to take more loans from his father.

Annie and Julia went west with Cope as far as Colorado Springs, spending the summer there while he first went on to Pueblo, Colorado, to get his instructions from Wheeler. Wheeler had organized a group of seven made up of a zoologist named Dr. H. C. Yarrow, a topographer named Ainsworth, Cope and his assistant W. G. Shedd, together with a cook, a teamster, and a laborer. At first things seemed to go fairly well. The party traveled south through the Sangre de Cristo Mountains, where Cope reported to Annie, "This pass is beautiful and a perfect flower garden."[18]

To his father, however, he revealed how much he was already chafing under the new rules. "I have been hard at work on the stratigraphical geology, a business which I do not object to, but which there are others who can work out. . . . All this comes from the system of orders and regulations . . . which are useless for explorers. . . . It is absurd to order stops here where there are no fossils, and marches there where fossils abound!"

By the time they reached Taos, Cope's temper was beginning to boil over and he began to make side expeditions away from the main party. This put his assistant Shedd in a difficult position—he was being asked to collude in mutiny—something for which he was less constitutionally fitted than Cope. A local priest had told Cope about some Pliocene-Pleistocene badlands near San Ildefonso, which Cope promptly explored, finding fossil deer, mastodon, rhinoceros, camel, antelope, weasels, mice, and a condor. Ominously, he warned his wife: "Thee need not mention these at Colorado Springs till I write further."[19]

The reason for this caution was that Cope was planning a formal revolt. Yarrow would not agree to Cope's insistence on following up leads for new fossil beds, rather than sticking to the mission. The problem seemed to be not so much that Yarrow disapproved of such diversions;

rather, he felt that he had to stand by the letter of his original orders. Cope called this, rather unfairly, a "lack of courage" and went over his head, going with Yarrow to Santa Fe to appeal to General Greg, the district commander. "To my delight the Gen. at once took my view of the case and set the Dr at liberty to violate and disregard the points which I had found so objectionable. He will now have some authority to fall back on in case Lieut. Wheeler complains. . . . [E]verything will I hope go on swimmingly."[20]

Soon Cope headed for the San Juan River region, where "our principal summer's work will lie." It was a bold, in fact desperate, move, but it paid off. Eventually they found "Eocene Bad Lands in great amount on South heads of the San Juan R. . . . 40 odd miles S.W. of Sierra Amarilla."[21] Cope had previously predicted that the Eocene fauna of Wyoming had derived from the south. Now he found himself exploring a huge basin of early Eocene age with a whole new fauna, different from and more primitive than that of the Bridger. It was brilliant coup. And beneath that was a huge Cretaceous basin, of freshwater origin (William Webb would have been delighted to learn), as opposed to the marine Cretaceous of Kansas.

Interestingly, Cope's letters now began to show an appreciation of the local people, their culture, and the landscapes that had been quite strikingly missing from his Kansas and Wyoming reports. He taught himself some Spanish and wrote to Annie in his typically condescending way, as was often the case, commenting particularly on the women: "I rather like these Spanish Americans. They are of medium, some above medium, size and all well and stoutly built. They are often very dark and of straight black hair. They are lively and pleasant. The chief fault in their expression is the absence of intelligence. . . . [T]he signoras and signoritas are often handsome, and only need intelligence to bring out real beauty." He became curious about the evidence everywhere that the land had once been much more heavily populated, the ruins of the circular forts often being ten to twenty miles from any present water.

In letters to his father and to his wife he describes the climate as delightful. Camp life here agreed with him: "Days warm and nights cold. I usually have ice water to wash in the morning, and after breakfast a ten mile ride to the bluffs to work. I have a grand appetite and am getting fat, fatter than I have been since the fever days at Bridger. . . . The mountains are covered with pine and oak to their feet, or where marked are mostly deep red, with white mixed. The creek runs in a cañon with banks of soul and contains excellent water. Coyotes and owls enliven the now moonlight nights, but wolves and panthers we have not heard. Deer are plentiful and bears particularly so."

At last he was really doing something new and of his own making. For once he was happy, although a bad moment arrived when a letter came from Yarrow, who had been ordered back to Washington. Ainsworth had earlier accidentally shot himself, and Cope was afraid that the whole venture would be terminated. Quite to the contrary, Yarrow "fitted me out with men, mules, and provisions and I am now commander-in-chief of the party and W. G. Shedd is my assistant."[22] Eventually even Shedd cheered up, Cope wrote: "[He] carries himself much more pleasantly. He seems to see at last that there can not be more than one director, and is not so contradictory about everything as he was."[23]

All in all, the year's work was a huge triumph for Cope and a vindication of the daring and risk taking that others felt was merely self-serving arrogance. A hundred years later, one of the great twentieth-century students of fossil mammals, George Gaylord Simpson, concluded that the work he had done in the lowest Eocene of New Mexico "was definitely Cope's greatest find in field geology, and the grand paleontological promises are still being followed up by his successors."[24] Cope and Wheeler did not repeat the experiment of working together again, however.

TWENTY-THREE

Two into Four Won't Go

For all the lighthearted accounts of bumbling scientists and jolly field-work depicted in *Buffalo Land,* the romanticized accounts of the Yale expeditions, and the frenetic pace at which Cope and Marsh found and described new species, the period between 1869 and 1874 might really be termed the paleontological fall from grace. In 1869, four strong characters had been sharing the field of fossil vertebrates; if they were not exactly working side by side, at least they were cooperating in a general sense of tolerance. By 1874, there were only two. In 1869, Leidy had been at the top of his game as professor of anatomy at the University of Pennsylvania, and in full flow writing up the vertebrate fossils that Hayden continued to bring in from the West. Five years later, however, at the peak of the excitement caused by the avalanche of new materials, especially those from the Eocene formations of Wyoming and Utah, Leidy abruptly gave up the study of fossil vertebrates.

Hayden was still spending a huge amount of his time out in the field collecting, and keeping busy during the winters either in Washington sorting and parceling out his collections for study or in Philadelphia as professor of geology at the University of Pennsylvania. Worried about his long-term prospects in Washington, he wrote to Leidy: "I am anxious that as much of my scientific work as possible be done in Phila."[1] In 1869 neither Cope nor Marsh had ventured personally into the West. Cope was busily engaged with his role as corresponding secretary at the Academy of Natural Sciences, with his studies of living fishes, amphibians, and reptiles, and also with collecting and studying fossils from the Cretaceous marls of New Jersey, where he now lived. While his *Elasmosaurus* had been something of a poisoned chalice, his New Jersey dinosaur *Laelaps,* the mosasaurs, and his herpetological writings were all great successes. Western collectors were beginning to send him specimens. In New Haven, Marsh was busy establishing the Peabody Museum at Yale and had begun work on the first horse material from Antelope Station.

Hayden initially had amicable relations with all three, Leidy, Cope, and Marsh. In addition to his scientific partnership with Leidy, the Academy of Natural Sciences was buying duplicate specimens from him, both of fossils and of preserved recent animals. He was sending fish specimens to Cope, and he had also been selling duplicate sets of the western specimens to Marsh, to whom he wrote in May 1867: "You stand second in the list for duplicates of everything I collect in the U.S. Surveys."[2] Hayden assured Marsh that "there is no suite of turtles in the world superior to yours. It would be fully equal to Phila. But your large turtle is not quite so large as our largest, but your suite is better than the S.I. (Smithsonian Institution) suite and (inter nos) my good friend Prof Baird threatened not to aid me again if I did not make theirs as good. Your entire collection is good and I doubt whether you would be willing to take $1,000 for it—However there is nothing too good for my good friends at Yale."[3]

Hayden had good reasons for staying close to Marsh, because the Yale professor had political influence in Washington and had written to the Interior Department supporting Hayden's work. He also had funds. At that time Hayden did not feel secure about his federal funding, and the academy in Philadelphia was a rather parsimonious sponsor. When he sent Marsh some plant fossils in 1869 he also wrote with a pitch for funding just like the one he had proposed to Leidy a decade or more earlier, and in the same wording (rather imperious for someone holding his hand out): "If I should fail to get an appropriation this Spring I wish you to place a $10000 at my disposal. I do not know that I shall need it all but I desire it subject to my draft in order that I may meet expenses on the Plains. Newberry will give a $1,000 and several others will do something. I am going to get up a subscription to make an enormous collection, making use of the railroads to do all I can and then go across into Montana and descend the Missouri River in a Mackinaw gathering up everything in my way. See what you can do in case I should need it. There is a beautiful lot of fossil fishes along the road that ought to be preserved."[4] In the event, however, he wrote a month later saying that the government had given him $10,000—for the Nebraska survey.

When Cope and Marsh started making western expeditions of their own, there might possibly—just possibly, given a great deal of cooperation and good will—have been room for all to work side by side. Instead, Cope and Marsh not only pushed their way into direct competition by horning in on Hayden's field localities, they fought bitterly with each other for the next thirty years. And Leidy yielded the field to them. The familiar explanation for Leidy's retreat from what was increasingly becoming a field of battle is that he had no taste for mixing it up with the impulsive, bad-mannered, and aggressive twosome of Cope and Marsh or for competing with them, either for Hayden's materials or directly in his own field expeditions. He is quoted as saying: "I can't stand this fighting. It disgusts me and I am going to drop Paleontology and have nothing more to do with it, because of the way Marsh and Cope are in each other's wool all the time." He told the British geologist Archibald Geikie: "I have got to get out because when anybody found a fossil they used to send it to me and I got it for nothing. Now today Cope and Marsh pay money for such things and I can't compete with their long purses."[5]

The situation was even more complicated than that, however. Leidy's self-esteem, and perhaps also his confidence as a paleontologist, had taken a serious knock with the "H" review. The hurt from this continued to sting a quiet but proud man like Leidy for years. Apart from the injustice of the criticisms, he had lost face with Cope, his raw junior colleague in Philadelphia. Relations with Cope had been rocky since the business of the *Elasmosaurus* head-tail. But one of the main keys to the disintegration of the foursome turns out to have been the behavior of Hayden.

Hayden had started to feel political pressure to allow Cope access to his collections as early as 1868. This is evident in a letter that he wrote to Leidy from Washington on December 11 of that year: "There is a splendid lot of vertebrate fossils here [describing his recent collections from Kansas] which Prof Baird would like you to have and which I most sincerely wish you had. There are enough new things to keep you at work a good while, you will be utterly surprised at them. Now do you wish them? If so, you will prove that fact by coming on here next Saturday night. If Cope comes, he will make a demand for them at once and Prof Henry cannot refuse him and will not do so. I write so that you may have the first chance at them, and partly at Prof. Baird's request."

Marsh had been the first to try to chip away at Hayden's effective monopoly on actually collecting western fossils—by the simple expedient of making his own expeditions west. There were two ways he could have gone about this: he could have explored for new localities by gathering information from informants other than Hayden (George Sternberg and Mudge in Kansas, for example), or he could have trespassed directly into Hayden's field territories. He chose the latter route.

The situation created by Marsh's expeditions was extremely difficult for Hayden. He was trying to do his own work, both in surveying new territory and collecting in localities that had already been discovered. He was trying to get material to Leidy because he needed Leidy to write contributions to the survey reports. All along, though, Leidy had made things more difficult by refusing to travel out West himself. He would have had a far greater moral and territorial claim, on for example the mammals from the Bridger Formation, if he had followed up the first gifts from Doctors Corson and Carter, and made his own collections. Keenly aware of this, Hayden continually urged Leidy to join him in the field.

Marsh's dilemma was that, while taking the easy route of visiting previously discovered localities, he risked using up his financial and personal resources only to find that he was duplicating Hayden's collections and Leidy's descriptions. As Hayden wrote to Leidy: "He says his expedition has cost him $20,000 and now I have taken the cream off of it. I have only followed my office and my instructions."[6] In fact, as

Hayden's increasingly anxious (and numerous) letters to Leidy show, once Marsh started out on his first expedition with the Yale students, Hayden tried to be generous and to accommodate Marsh where he could—by not going to Kansas at the end of that year, for example. But he drew the line at accommodating Marsh with respect to the exciting new exposures of the Bridger Formation.

Marsh's western expeditions thus put double pressure on Hayden. Hayden felt strongly that his first obligation was to Leidy. On September 1, 1870, he reported to Leidy from the field at Black Forks Station: "Marsh has written me a note which I enclose to you as my best and oldest friend. I told him last spring, as I did Cope, that all fossil vertebrates remains that I have ever collected in the world would be sent to you. You could do as you choose about letting any one have them. So I shall send all to you. I shall send a lot more from Bridger, perhaps by Freight. If Marsh is offended he has no right to be in anyway. The field is mine by right of occupation and the vertebrate field is yours by right of occupation 18 years ago." It is worth noting that Hayden left it up to Leidy to choose whether Cope or Marsh should share in the material.

The second pressure on Hayden was that, if he were to give Marsh access to prized sites, both known ones and those yet to be explored, there would be less material available to be written up for his own reports, and therefore not only to fulfill his mandate but also to advance his own career. Marsh continued to press, however, and he had powerful allies in Washington, who, for example, had produced the essential army escorts for his expeditions. Therefore, in all the vastness of the West, Hayden and Marsh constantly fell over each other.

Marsh's own attitude inflamed the situation. Hayden complained to Leidy: "He is more ambitious than Cope ever was. . . . He speaks in the highest terms of you, but feels cut to the heart that I did not give him all my collections or leave them on the ground. You see his style of reasoning. . . . I write you that you may have the facts and be on your guard when Marsh comes to see you. Listen to all he has to say and give him such consideration as you choose. . . . Marsh claims to have made a great haul, about a dozen new species from the Loup Fork group. 9 boxes from White River way down on Green River, lots of things from

California. Say as little about me to him as you can and make things as harmonious as possible."[7]

The situation became even more complicated when Cope made his first trip west in 1871. Cope furthermore became a really serious competitor to both Leidy and Marsh with his formal association with the government survey the following year. From Hayden's point of view, this association was positive because Cope's studies would considerably augment the all-important survey reports. In truth, the volume and range of material being collected meant that Cope was urgently needed for the description of reptile and fish material, thereby allowing Leidy to concentrate on the mammals. As for Cope, he was quite happy to edge out Leidy, whom he saw as worthy but dull, and he resented how Leidy had been so open in his correcting of the infamous *Elasmosaurus* head position. He had been elated when he read H's scathing review of Leidy's *Cretaceous Reptiles of the United States* in 1865, and he felt himself very little in sympathy with his "Philadelphian brother Professor."

Once Cope went out on his second western trip he quickly began to act no better than Marsh. The issue, once again, was the rich collecting fields of the Fort Bridger region. Dr. Carter had proposed that Marsh and Leidy conspire to exclude Cope; when Marsh did not respond, Carter wrote again, the querulous tone showing the tension that was building: "Presumably [Leidy] will visit our locality in this month sometime. I do sincerely hope you will come and that you two will be here together. Prof. Cope has been here some time, but has done nothing. . . . [Hayden] wrote to the Judge asking him to apprise Profs. Leidy and Cope who would come out as members of his party. . . . He first learned that Dr Leidy thought of coming so he hastened to take his visit unto himself. Now this is false. Dr Leidy comes at my personal invitation."[8]

With Cope undeterred, Hayden tried to explain to Leidy how he had "asked [Cope] not to go into that field," as we have seen. No doubt Leidy, who knew Cope well, understood the difficult position Hayden was in, but Hayden scarcely helped matters by repeatedly begging him to sort things out: "Write a little letter in the case of Marsh and others. I did not arrange for Cope to go any where. I merely offered to aid him after he had made all his plans and would not depart from them. I trust my explanation will be satisfactory to you."⁹

During the troubled year of 1873, Marsh complained to Hayden that Cope was making derogatory remarks about him in his reports for the Hayden survey. He also tried to get Hayden involved in the dispute over dates of publication. Hayden replied to Marsh rather stiffly: "I have consulted with Prof Baird & Mr S. C. Scudder who happened to be here and they both agree with my own decision that I can do nothing in the case except when Prof Cope uses personalities. I insisted on that and obtained a promise to that effect, and I do not now know of a single personal allusion to you that could be offensive in the report for 1873. As to dates, claims for species or discoveries, those matters should be settled by other parties and they will be undoubtedly in due time. I do not consider myself competent to decide disputed claims. . . . Prof Cope is one of the Collaborators of the Survey, you are not, and have refused to become such though requested by me to become so many times within the last three years. You call upon me to decide against Cope in a matter which Cope claims to be as much in the right as yourself and which must be settled by experts. Gill and Leidy should take the matter up and investigate all the circumstances and their opinions placed on paper would forever settle the difficulty."¹⁰

The result of all this was that, in these beginning years of the Cope-Marsh wars, Hayden was the man in the middle. Unfortunately, it was a role he couldn't cope with, so to speak. Being something of a paranoid, he was always the first to find signs of conspiracies and to stir up trouble. His response to pressure was always to try to put Leidy—more senior, more statesmanlike—in the middle instead. As his relations with Marsh deteriorated, he constantly wrote to Leidy in terms like those in his letter of October 13, 1870, from Green River Station, which sums up the whole problem: "If he [Marsh] calls upon you . . . you must make all smooth with him and not commit me. I do not know what he has but I suspect not much. If you must[,] assume the ground that you had the right to report on my fossils . . . but I leave the whole matter to you. . . . Do not forget the essay and Report. Make it as full as possible, I hope to get out a handsome volume this winter." And, as the Cope-Marsh relationship soured, particularly when accusations of personal attacks on each other's reputation started to fly, the risks to Hayden himself increased because of Marsh's political connections.

For years, Leidy had tried to guide Hayden and to patch up his quarrels. As the political situation heated up, both in the field and back East, he wrote to Hayden advising him to change his behavior: "Permit me as a friend to say a word or two intended for your eye only. I wish you to be respected and liked as you have always been, but I find some of your friends and acquaintances begin to speak of you coldly. They say in general that you were once amiable and kind, but fear that prosperity is making you indifferent and arrogant. It costs nothing to remain as you always were, but on the contrary pays well!!"[11]

Hayden replied hotly: "Those who know me best would be surprised at such a statement and I ask you, do you see any thing or have you seen anything that pointed in that direction[?] . . . There is not a man on earth who has been a true friend to me all the way who can say that I have ever deserted him[.] Whenever I have seemed indifferent or arrogant there has been a persistent reason. . . . [T]he consequence is that I get a 'sore head' every little while. I cannot comply with all the demands that are made on me. . . . There is a clique in Phila which meets one evening a week. . . . I was treated coldly as much for the warmth with which I defended your character and that of Prof Baird as for anything else. I have no doubt that my character was discussed in much the same way. Marsh did me great harm at Bridger and Salt Lake and of the strength of his talk I was treated and so was my party with great coldness. He was most active against me. But they will accomplish nothing."[12]

After years of fruitless urging by Hayden, the direct incursions by Cope and Marsh into the Bridger region finally drove Leidy to make his own expeditions to the field in the summers of 1872 and 1873. In his Bridger Formation monograph of 1873, Leidy described thirty-one new species, but at the very end he mentions Marsh and Cope, who had by then described (however validly) more than ninety species between them. Clearly, he recognized that they had overtaken him. "The Green River Basin has been sedulously explored from Professor O. C. Marsh with the most important and fruitful results. In the abundance of fossils and the number of extinct genera and species of vertebrates they represent, his collections are perhaps not exceeded by any obtained from any one locality elsewhere in the world. . . . I may further remark that during the last summer Professor Cope made an extended exploration of the Green River basin, and obtained large collections of fossils, to a full account of which we look forward with much interest."[13] This was nothing less than a farewell address.

One can only sympathize with Leidy in opting out, as Cope and Marsh had sufficiently fouled the nest. But Hayden had made Leidy's position ultimately untenable by constantly trying to force him to adjudicate and smooth over his difficulties with the only other men in the field, both of whom Leidy considered colleagues. In fact, Leidy's own feelings for Cope, his junior Philadelphia colleague, were always less warm than for Marsh, with whom Leidy continued a collegial correspondence for years. And Leidy—"the last man who knew everything"—had other strings to his bow. He eagerly returned to his first love—parasitology and microscopy—although he continued to publish several papers per year about vertebrate fossils, returning time and again to his fossil horses.

One last twist in this disreputable saga is that Hayden, who could not avoid getting embroiled in intrigues, was the next to be driven out of the field, as Marsh and his allies succeeded in turning the political balance in Washington.

TWENTY-FOUR

To the Black Hills

To explorers in the Dakota Bad Lands, the densely forested Black Hills stood as a dark foreboding presence to the north, a place of ancient contorted rocks thrust up from deep in the earth. It was a place where no paleontologist needed to go, and where the Indians, to whom it was sacred ground, would have made him most unwelcome.

Under President Andrew Johnson yet another well-meaning attempt had been made in 1866 to secure peace in the plains; it meant persuading the Indians to retreat to designated reservations and to accept (in effect to become dependent on) government aid. This scheme had turned sour; it was a bad scheme anyway, and corruption in the Bureau of Indian Affairs and poor management at all levels meant that the Indians had given up a lot—including much of their hunting grounds—in exchange for very little. Red Cloud fought a two-year war that resulted in the Fort Laramie Treaty of 1868, which committed the government to removing its forts from the Bozeman Trail—cutting through prime hunting lands on the way to the Montana gold fields—and assigned a huge swath of land in South Dakota to the Lakota and Dakota Sioux.[1]

The Black Hills were literally and figuratively central to this land and this treaty. But two events soon changed everything yet again. In 1871 it was revealed that the Northern Pacific Railroad was going to drive west from Bismarck, right across the northern, most disputed, part of the Sioux lands; in 1874 the presence of gold was confirmed in the Black Hills.[2] All this occurred in the "progressive" climate of a

South Dakota legislature that was in no mood to allow Indian treaties to interfere with western development. Within a few years it would all boil over and create the conditions for the pyrrhic victory of the tribes at the Battle of the Little Bighorn and then their inevitable defeats.

In 1873 the Lakota were bitterly unhappy with the level of provision of supplies from the Bureau of Indian Affairs and generally distrustful of both the federal government and its local agents. When, that summer, General David Stanley took an expedition that included Custer and the Seventh Cavalry along the Yellowstone River as a show of force to support the railroad surveyors, Crazy Horse inflicted major losses on them. The government's response was inevitable: more force and the hunting down of any bands that left reservation lands. Leidy might have been offered a position with the Stanley expedition, but it was at this point that Marsh became involved in Army-Indian politics. Even before the Stanley expedition had gone out, General William T. Sherman instructed General E. C. Ord (commanding the Department of the Platte) to "send out a scout from Ft. McPherson on the 15th of June, to proceed directly north until it reaches the Niobrara River, thence . . . across the country back to Fort McPherson. The principle object of this scout is to make it a little embarrassing to the Sioux [by a show of force south of their reservation,] who seem to be preparing to annoy the Stanley Expedition. The next object is to gratify the professors of Yale College by letting Prof. Marsh gather the bones of dead elephants and other animals. . . . I should think two companies would be enough. . . . The Commanding Officer will be required to show all the kindness possible to Prof. Marsh and boys with him, and especially to give them every protection."[3] This, then, was the first leg of Marsh's 1873 expedition.

The next year, through his assiduous and successful cultivation of many senior army officers in order to get escorts for his expeditions, Marsh made himself part of this volatile mix. It is clear from a letter that General T. H. Stanton wrote him that Marsh had been playing an astute political game to keep the army brass supportive of his fieldwork. "My dear Professor, Accept my thanks for a copy of the Tribune containing the article in reference to your discoveries in the west. I was much interested in it, and knew that the views of the correspondent, must be, in good part your own. I wish you would arrange it to come out here by 1st July and go with me on my official tour to Red Cloud and Spotted Tail Agencies. It is north of the Niobrara, on the White River, and that region is very rich in fossils. The remains of a mastodon are reported near Red Cloud, and I saw many astonishing teeth and bones there on my last trip, from which I only returned yesterday. Near Spotted Tail 25 miles north eastward between the White & Big Cheyenne rivers the ground is reported to be 'heaped up' in many places with fossil remains. I have no doubt you could find much to interest you. Gen. Smith would furnish you all the escort you would need."4

After the fiasco of the Stanley expedition, Custer was sent to investigate the reports of gold being found in the Black Hills, commanding a force of a thousand men that was obviously intended to strike fear in the natives. As the expedition included geologists and gold-mining experts, as well as members of the press, there was little doubt about its purpose, and little surprise about the consequences. The party discovered that the rumors of gold were true, and General Sheridan decided that it would be necessary to build a fort near the Black Hills to keep the Indians under control.

Sheridan wrote from his headquarters in Omaha: "My dear Marsh, I will start Genl. Custer on an expedition from Fort Sinclair directly across the country to Bear Butte at the foot of the Black Hills. He will then examine the Black Hills Country & the Belle Fourche . . . [where] the greatest accumulation of fossils is sure to be, is to be passed over. I do not intend to let Genl Custer be embarrassed by any out side purpose except yourself if you should desire to go."5 Marsh, however, was not only occupied during the summer of 1874 with the fossils from his previous four years of expeditions and with his continuing spats with Cope, but construction of the new Peabody Museum was beginning. It was not a time to leave New Haven. Aware also of the inflammatory nature of this expedition, he refused the invitation, carefully keeping his options open by sending his assistants George Grinnell and L. H. North. But he itched to get his hands on more fossils.

Sheridan's paymaster, General Stanton, increased the pressure by sending a telegram from Fort Laramie: "A vast deposit of fossil remains of extinct marine and other animals had been discovered ten miles north of Red Cloud, covering an area of about six miles square."[6] This location meant that the discovery was probably in the White River Oligocene. Marsh sent back a telegram, asking for specimens to be sent to him. Stanton replied: "Mr J. W. Dear, Trader at Red Cloud Agency, Neb. has a very nice collection, and he promised me he would hold it, until I could inform you. I go into that country again 1st of Nov. Cannot you come with me . . . [?] [There are] one three horned head (rhinoceros, or elephant), . . . immense serpents heads, turtles, and bones of all sorts, jaws, vertebrae, teeth are scattered over the country, making it look like a vast bone yard."[7]

Demonstrating the extent of Marsh's connections in the Department of the Army, Ord also wrote with the news of the discoveries from near the Red Cloud agency. "Thinking this would interest you I send this memo and as we have a garrison at Red Cloud of 5 comps [companies,] . . . anything that I can do to promote the success of such a visit . . . [T]his time you had better take Sioux guides from Red Clouds band—instead of Cody & co."[8]

The fact that Stanton's site was reported to be near the Red Cloud and Spotted Tail Indian agencies would have given pause to any man who read the newspapers. Perhaps the situation intrigued Marsh. When Stanton wrote for the second time, Marsh's thirst to acquire more specimens and to reinforce his stamp on western science, together perhaps with his having an eye to solidifying his place at the Washington political table, took over. Public concern over the continued bloodshed, and disagreements over the treatment of the Indian tribes, had already surged with the exposure of similar scandals in 1871 (the "Indian Ring"). Soon Marsh would see for himself how little had changed. He headed to the Black Hills, assembling with a party of soldiers at the Red Cloud agency on November 4. Taking Ord's advice, he hired as a scout Hank Clifford, his old guide from the Niobrara. The squaw man Clifford was a perfect choice for such a mission: his wife was the daughter of Red Cloud.

At the Red Cloud agency, where he found the chiefs Red Cloud and Sitting Bull and some twelve thousand people camped along the banks of the White River, Marsh tried to persuade Red Cloud and his people to allow him to collect in the Bad Lands south of the Black Hills. Indeed, he needed Red Cloud's men to help escort him there. But the Sioux were deep into a dispute with the local Indian agent over rations and the need for a census, and little inclined to cooperate. The Sioux clearly did not trust Marsh, did not trust the soldiers with him, and claimed to be fearful of possible confrontations with their relatives the Miniconjous to the north. The latter had been making raids into the Black Hills region. In any case, now that gold had been discovered, it was hard for any of the Sioux to believe that Marsh was only interested in digging for old bones.

Marsh countered by promising to pay for specimens and to take back to Washington their complaints about their treatment, including the issue of the quality and quantity of rations and supplies provided by the government. He tried giving the Indians a big feast but never felt that they would abide by any agreement he made with them. In the end, Marsh reportedly slipped out of camp and the group made the fifteen-mile trek with horses and wagons to the fossil beds where, in fierce cold, they collected madly for a few days before retreating. They amassed a huge collection of fossils and left just ahead—or so they were warned—of a Miniconjou raiding party.

Most of the authority for these tales comes from Marsh himself, who sent a long story about his trip to the *New York Tribune* in December. His account is heavily romanticized and shows Marsh in a heroic light. The fact is that some of the story has to be taken with a grain of salt. A party of soldiers and collectors slipping undetected out of the Red Cloud agency—with its thousands of people, dogs, and horses—seems unlikely. If there was a moonlight flit, it must have been watched carefully by the Sioux, who would have tracked their every move south. The party was certainly watched constantly while collecting. The raiding Miniconjous may have been real, or they may have been an excuse for the Sioux not leaving camp. Ord wrote to Marsh in December congratulating him "on the success of your trip. . . . I was glad to hear that you didn't allow the bluff game of the Indians to deter you from your examinations."[9]

Nonetheless, Marsh returned east genuinely concerned by the plight of the Indians and the extent of local and governmental corruption. He had seen for himself that what newspapers like the *New York Tribune* had been saying for months was true. He had seen emaciated cattle brought in and claimed as being top weight. He had seen the short weights and terrible quality of the foodstuffs. For someone so cautious and usually so politically conservative, however, it is surprising that Marsh actually did take Red Cloud's complaints, enthusiastically, to Washington, and to the president himself. The political moment was well timed. Grant's policy for calming down the Indian situation, which had worked fairly well up to then, was in danger of falling apart as the flow of emigrants to the West was growing daily. More and more Indians were refusing to stay cooped up on the reservations, and in their resentment they were becoming an ever more serious threat to the streams of emigrants along the trails and to the construction of the railroads.

Peace with the Indians was essential. So Grant was inclined to agree with Marsh. The problems, however, were not in his office but in the Department of the Interior and its intransigent secretary, Columbus Delano. To make any headway, Marsh had to enlist not only the support of political friends but also the press—and the *Tribune* was only too happy to oblige. A commission of inquiry was formed. Marsh did his homework well, writing to Stanton among others for information about the flour and beef contractors at Red Cloud. Stanton later wrote to say that "Capt. Egan, 2nd cav at Red Cloud 16th says, of your committee 'look out for a first-class white-wash.' Other information is to the effect that there are only two members of the Com. who desire to get at the bottom of things. These are Senator Howe and Profes. Atherton. But your persistence has had the effect of calling out Mr Welsh and will generally wake people up on the subject, even if the Com. find everything lovely."[10]

Over the next months, Marsh managed to make a sworn enemy of Secretary of the Interior Delano but, in the end, Delano resigned and a combination of forces produced an improvement in the Indians' lot and a cleanup, if only temporary, in the Bureau of Indian Affairs. The exact weight of Marsh's influence in the resolution of all of this is unclear. Equally shadowy is what Chief Red Cloud made of it. Red Cloud and Spotted Tail did go to Washington in 1875 to meet with Interior Department officials but acted in a strange and confused manner. Red Cloud later presented Marsh with a peace pipe and the magnificent Indian regalia and artifacts that today are the glowing center of the Peabody Museum's western anthropological collections. It would be nice to report that he and Marsh became blood brothers. But if Red Cloud, proud and enigmatic, was reluctant to play the role of a tame Indian, and kept his thoughts to himself, who could blame him? It would also be nice if Marsh had been the only hero of the affair, but to get anything changed in Washington (especially with an election approaching) was impossible without a massive campaign by the press and a whole number of politicos, all with their own agendas.

Inevitably the upshot of the whole Black Hills issue was that the government decided to move the Sioux to a reservation farther north, allowing the Northern Pacific through and opening the Black Hills to miners and settlers. In 1876 a force of Lakota (without Red Cloud), Ogallala, Arapaho, and Cheyenne gathered in southern Montana under Sitting Bull, Crazy Horse, and Gall, where they massacred Custer and three hundred of his men at the Battle of the Little Bighorn (Battle of the Greasy Grass). Thereafter the Sioux were progressively herded and harassed northward. Sitting Bull was eventually killed at Standing Rock in 1890, and the massacre at Wounded Knee happened the same year.

For someone as committed as Marsh to carefully cultivating government sources and to ensuring his own access to the West, taking on the Interior Department was a brave and risky thing to do. The Red Cloud affair was so untypical of Marsh that one has to ask whether there might have been more cynical motives as well as humanitarian concerns behind it. Marsh was then a relative newcomer on the Washington scene but with a number of approving and influential friends. President Grant and his cabinet were unlikely to be reelected. It may well be that Marsh saw an opportunity to make a name for himself both in the public eye and in the political arena. He certainly managed to show himself to be tough and principled, a powerful friend and a dangerous man to cross. Two years later he was elected vice president of the National Academy of Sciences, and then almost immediately became president when Joseph Henry suddenly died. At that point he was well positioned politically to take on a different target, the Hayden survey, and the everpresent Professor Cope, which he did to great effect.

To the Judith River

Both Cope and Marsh paused in their fieldwork in 1875. Indian troubles may have been part of the reason, although not in the Bridger region where Sam Smith and John Chew continued to collect for Marsh, nor in New Mexico. Principally, both Cope and (even) Marsh were feeling the financial pressures of constantly working out West and they had inordinate backlogs of material to describe. They were also exhausted by the constant bickering, each feeling he had no option but to keep up the pressure on the other. It was a time also for Cope to start to exert himself more as an intellectual in the field of paleontology rather than just a collector and describer of fossils by adding to the series of essays that he had been writing on fossils and evolution. These works show that he was seeking not only to explicate Darwinian theory, but also improve on it.[1] This led him progressively into Lamarckian ideas about the possibility of characteristics acquired during an organism's lifetime being inherited.

Marsh still had his collectors in the field; one group was dispatched to Santa Fe to collect in the New Mexico Eocene—this time Marsh was following in Cope's wake. Cope had not been able to leave the army of collectors out in the field that Marsh had, but a wonderful opportunity presented itself at the end of the year when a letter arrived from a young student of Mudge's in Kansas. Charles H. Sternberg was the younger brother of Captain George M. Sternberg. He grew up in New York State and Iowa (his father was a Lutheran minister), and then went with another brother to work George Sternberg's farm near Fort Harker,

Kansas, which at the time was the western terminus of the Union Pacific. A longtime amateur fossil collector, he had sent material to the Smithsonian; like so many others, his first collections were of fossil leaves. Now he wanted to work full-time as a fossil collector, but Marsh had no funds to employ him. With a nice even-handedness, Mudge suggested that he write to Cope. His letter was so effective that Cope, instantly liking Sternberg's style, replied with a three-hundred-dollar advance for expenses. Sternberg started by collecting mosasaurs for Cope in western Kansas, and he went on to become one of the great fossil collectors of the century and the founder of a whole family dynasty of collectors. Late in his career he helped open up the great dinosaur beds of western Canada, and his four sons continued in the business. Sternberg's autobiography, *The Life of a Fossil Hunter,* makes wonderfully exciting reading, as Sternberg was not afraid to dramatize what would in any case have been a fascinating career as a new kind of western pioneer.

At some point in mid-1876, Cope had decided to make a trip to the Judith River badlands where the young Hayden, twenty years earlier, had discovered the first American dinosaur teeth. Although Hayden had been back to the Judith River region, he had not found any more dinosaur fossils there, and he had no reason to go again. Given the brewing troubles with the Indians, much of which had been triggered by Custer's Black Hills expedition and the discovery of gold, and remembering Marsh's difficult foray into the Black Hills in November 1874, it really would not have been sensible to venture west in 1875. The situation was no more peaceful in the spring of 1876 and was shortly to descend into outright warfare. That being the case, if Cope was to have headed west at all, it might have made more sense for him to go back to Fort Bridger. Instead, Cope summoned Sternberg to join him and set off for Montana.

This was an occasion when Cope could have benefited from an exchange of intelligence with Marsh. Just as Cope was pondering his possible trip to the Judith, Marsh's collector Hank Clifford wrote to New Haven as follows: "I received yours of April 9th today with the map of maj Stanton trip last fall which will help me a great deal when I start North which I cant say when that will be as the Indians is raiding[.] [W]ith all small Parties that start in the direction of the Black Hills I have mad three attempts in the last six weeks to go after Bones and the Indians stoped me every time so I shant Try it again until Crook gets out after Them he is going to start the sixteenth of This month & thinks in about a month They will be a little more quite [quiet] as I could not go after Bones I am building a mail station. . . . You speke of that big head on the Niobrio I will try and get that as soon as Posibal. The Indians is steeling horses and killing so much now that I am afraid to go any place at present They are so bad on the Cheyenne and Black Hill Road the Stage Company had to take there Stage and Stock off and quit running."[2]

Whatever intelligence Cope was receiving, he decided to go anyway. His party included Sternberg and a local man named J. C. Isaac, who had worked for Cope before. When they met in Omaha they did not make a very convincing group of explorers. Cope arrived by train with his wife Annie (who would spend the summer in Ogden, Utah) and was surprised to see that Sternberg had a crippled left leg. Cope himself looked alarmingly pale and weak to Sternberg. After taking the narrow-gauge train to Franklin, Idaho, they could travel farther only by road, and it took another eight days of debilitating, bone-rattling travel in a Concord coach to reach Fort Benton. (They could also have gone up by steamer, but the schedule of steamboats was limited.)

It was now less than eight weeks since the defeat of Custer's forces at the Little Bighorn (on June 25, 1876), two hundred miles off to the southeast. Somewhere out on the plains Sitting Bull and several thousand Sioux were on the move, presumably heading north toward them, and somewhere also there were another thousand or more Crows and Blackfeet. Most locals advised the group not to go out, and few men were willing to risk accompanying them. Still, Cope was undeterred, paying one thousand dollars a month each to hire a guide and a cook. He wrote to Annie in Omaha to reassure her that the rumors about Sitting Bull were "cock and bull stories," saying that his party would be left alone because Sitting Bull and every able-bodied Sioux would be preoccupied elsewhere.

Cope's group set off downstream from Fort Benton, following the north bank of the Missouri. This was particularly difficult and inhospitable country once one left the river. Even today, if you take a ruler across the map of Montana from east of Fort Benton to the Yellowstone River at, say, Glendive, the line cuts across few paved roads. In his journal of the Lewis and Clark expedition, Meriwether Lewis wrote on May 31, 1805, of this region and the river itself: "The hills and river Clifts which we passed today exhibit a most romantic appearance. The bluffs of the river rise to the hight of from 2 to 300 feet and in most places nearly perpendicular; they are formed of remarkable white sandstone which is sufficiently soft to give way readily to the impression of water; two or three thin horizontal stratas of white free-stone, on which the rains or water make no impression, lie imbeded in these clifts of soft stone near the upper part of them; the earth on the top of these Clifts is a dark rich loam, which forming a gradually ascending plain extends back from 1/2 a mile to a mile where the hills commence and rise abruptly to a hight of about 300 feet more."[3]

When Hayden explored the region south of the river twenty years before Cope, he wrote, "Near the mouth of the Judith River . . . is a wild, desolate and rugged region which I have called the 'Bad Lands of the Judith.' No portion of the Upper Missouri country exhibits the effects of erosion and denudation on so large a scale, and to add to the picturesque effect of the scenery, the variegated strata are distorted and folded in a wonderful manner. . . . [T]he surface of the country is cut up into ravines and cañons, with nearly vertical sides, rising to a height of 400 to 600 feet above the bed of the river, with scarcely a tree or a shrub to greet the eye of the observer."[4]

Closer to Fort Benton Cope found that the country around the river—"the continuation of the plains"—was green and fertile. This was good country for cattle grazing. "It abounds in buffalo, antelope, deer, wolves, Indians, etc. . . . [I]n crossing from Benton we had very little wood, fair water and splendid grass so far as the eye could see. Scenery very fine; the Bear Paw Mts. on the North and the Little Range on the South covered with snow." The farther downstream they went, the wilder the scenery became. Cope wanted to head some forty miles beyond the mouth of the Judith to the region of Cow Island, a ferry boat stop, from where he planned to strike out southward into the "interior."

Near the mouth of the Judith River was a small trading post with the ambitious name of Fort Claggett. On the other bank of the river was a large Indian encampment, the sight of which must have caused some hearts to race. This was not Sitting Bull and the Sioux, however, but a party of some one thousand "River Crows, with a few Piegans and Mountain Crows. I was introduced to Bear Wolf war chief of the Mt Crows who has taken 26 scalps and stolen 90 horses from the Sioux! . . . Last night a chief of the River Crows, Beaver Head and his squaw, slept in our camp under the wagon, and took breakfast." Cope found the Indians to be "all in good humor" and he found that he could greatly amuse them by taking out his dental plate to wash it: "One man (Mountain Jack) rode several miles to see it."5 (A very similar story is told of Mudge.)

The party collected for two weeks along the Judith before heading downstream to Cow Island to explore "a bad land country of soft black earth cut up with terrible canyons, with the worst water I ever saw." By this time Cope had amazed Sternberg with his physical resilience, and despite the conditions they got a good haul of fossils. Out of these collections came the first decent material of the horned dinosaurs, or Ceratopsia—the group that includes Cope's earlier *Agathaumus* and later became famous for *Triceratops*, which was discovered in 1887 by John Bell Hatcher, one of Marsh's collectors. Cope named his new genus *Monoclonius*. He also collected more material of Leidy's *Palaeoscincus*. But the problem was: how would they get their seventeen hundred pounds of fossils out of these deep canyons and back to the river?

The situation was really quite desperate. In 1805, William Clark had written: "The Stone on the edge of the river continue to form verry Considerable rapids, we [which] are troublesom & dificuelt to pass, our toe rope which we are obliged to make use of altogether broke & we were in Some danger of turning over in the perogue in which I was, we landed at 12 and refreshed the men with a dram, our men are obliged to under go great labour and fatigue in assending this part of the Missouri, as they are compelled from the rapidity of the Current in many places to walk in the water & on Slippery hill Sides or the Sides of rocks, on Gravel & thro' a

Stiff mud bear footed, as they Cannot keep on Mockersons from the Stiffness of the mud & decline of the Slipy. hills Sides."[6]

Cope's report to Annie concerning his own journey along the same section rather understated things: "We had a difficult task to get down to the Missouri through the canyons and precipices. Had to let the wagon down with ropes." Having accomplished that, Cope and Sternberg left Isaac at the camp and tried to find their way back to Cow Island, following the river and then climbing up to cut across country in the places where cliffs to the water's edge made a passage along the river impossible. "At one point I made three attempts before I could get down to the high bank and my horse came down in some perilous places, where he could only slide." They reached Cow Island long after dark, but their difficulties were still not over. The last boat of the year was due to arrive the following day, and the fossils were back at the camp, three miles downstream of the Cow Island steamboat landing. When Cope told his wife that "I had a lively time getting to the boat with my fossils," he was again understating matters.

Cope's earlier fossil collections from the Judith had been brought down to Cow Island for him on one of the flat-bottomed scows (Mackinaws) used by the river men. Cope had no other option but to buy the boat and set off with Sternberg downstream to the camp. Isaac should have been there but had gone off to look for them, much to Cope's disgust. After he turned up, the three men used their horses to drag the boat back upstream. "After sundry adventures in which we all got very wet, and the horses rolled down the bank into a mud hole," they got to Cow Island just in time to meet the steamboat.[7]

PART FOUR

Toward the
Twentieth Century

TWENTY-SIX

The Rise of Dinosaurs

Even with Hayden's original specimens from the Judith River in 1856 and Cope's new material from 1876, in twenty years not many new dinosaurs had been found in North America. From New Jersey there were Leidy's *Hadrosaurus* and Cope's *Laelaps;* in 1859 J. S. Newberry collected some material in Utah, but it was not described until 1877, when Cope gave it the name *Dystropheus*. Marsh had found *Claosaurus* in the Cretaceous of Kansas in 1872. In his expedition to Wyoming in the same year, and again in 1874, Cope had discovered fragmentary materials of dinosaurs that he assigned to no fewer than three species of *Agathaumas*.

A year later everything changed: following a few months of intensive collecting, American paleontology acquired further direction and eventually added a whole new element to popular culture. Whatever the mastodon had offered in the way of a totemic American giant creature, whatever surprises Jefferson had thought the West would hold, whatever sensation the Leidy-Hawkins *Hadrosaurus* had caused—and no matter that European dinosaurs like *Megalosaurus* and *Iguanodon* were billed as "monsters" of a highly respectable forty feet or more—Colorado and Wyoming were about to yield dinosaurs on a different scale altogether—creatures far beyond any previous imaginings.

The news came without warning and, as is usually the case with momentous discoveries, took some time to be digested and appreciated. For Marsh, it came in the form of a letter from Professor Arthur Lakes, of Jarvis Hall College in Golden City, Colorado. As all good letters should, it went straight to the point: "Dear Sir, A few days ago whilst taking a geological section and measurements and examining the rocks of the banks of Bear Creek near the little town of Morrison about 15 miles West of Denver, I observed . . . some enormous bones apparently a vertebra and a humerus of some gigantic saurian."[1]

Lakes was an Englishman who had studied at Oxford before emigrating to the United States around 1868. He ended up teaching writing at Jarvis Hall, which eventually became part of the Colorado School of Mines. A tall, handsome man, he was a part-time preacher, a gifted writer, and, fortunately for us, a diarist. Not only did he write prolifically about his collecting expeditions, he also greatly enjoyed painting watercolors in what seems now—to put it kindly—a primitive or natural style. He left us a colorful firsthand record of the sort that is sorely lacking for so many other paleontologists of the day. Lakes's field journals make fascinating reading; better than any others, they capture the magnificent landscape and the *feeling* of the work.

Once he settled in Colorado, Lakes's interests in natural history quickly became centered on geology and fossil collecting. He and his friends (especially the retired navy captain Henry Beckwith) spent a lot of their spare time exploring the region around Golden City, collecting insects as well as prospecting for fossil remains such as leaves, which they found in good numbers. Whenever possible, they extended their travels into the region of easily accessible valleys and hogback hills just east of the Front Range of the Rockies. In his journal, Lakes wrote: "At Morrison is a remarkably fine development of the stratified rocks of the Triassic, Jurassic, and Cretaceous periods. The sandstone lying uplifted along the base of the mountains forming long parallel ridges . . . with a low valley between them. Whilst . . . Bear Creek Canyon cuts through the whole series of rocks and gives an admirable section."[2]

His timing was perfect. Very probably, if Lakes had not made the discovery, someone else would have, for soon it was observed that dinosaur bones littered the ground like brushwood along the Jurassic exposures near Morrison, Colorado. In fact, Lakes had written to Marsh and sent him some reptilian bones "from Morrison Bear Creek" the previous year. "I have been hoping to hear from you of the safe arrival of the saurian remains," he wrote, and he had even sent "a small box containing two teeth . . . & a bone supposed to be pterodactyls?"[3] Marsh had apparently not replied.

The bones Lakes and Beckwith found in March 1877 were enormous. Lakes described the discovery of two of them as follows: "As I jumped on top of the ledge there at my feet lay a monstrous vertebra carved, as it were, in relief on a flat slab of sandstone. It was so monstrous, however, thirty three inches circumference so utterly beyond anything I had ever read or conceived possible, . . . we stood for a moment without speaking gazing in astonishment as this prodigy and threw our hats in the air and hurrahed: and then began to look for more. Presently Cap B cried out why this beats all!! At his feet lay another huge bone resembling a Herculean warclub ten inches in diameter and about two feet long."[4] Lakes and Beckwith were so excited that as soon as they could they began camping at the site so as to spend all their waking hours collecting.

Marsh's response to Lakes's letter of April 2, 1877, about the "enormous bones" is unknown. In fact, it may not have been completely obvious to him what Lakes was offering or that there was any urgency. Lakes had written that he was planning to collect more and to study the bones better when he got time, but thought he would "meanwhile acquaint you with the fact & if of sufficient interest to you shall be glad to communicate with you & receive any instructions or directions from you in regard to the bones." He also continued with the less welcome news that, "As soon as I had discovered them I wrote a letter to Dr Hayden."

On one hand, blessed with hindsight, we would imagine that Marsh would have leaped at the chance to get first crack at Lakes's finds. On the other hand, except for the reported size of the bones, Marsh had no way of knowing that Lakes had anything particularly interesting— interesting enough to compete with all the other bones being shipped to New Haven by other western collectors eager to make a dollar or two. Another factor may have been that, from the size of the specimens, these had to be reptiles, so probably they were related to mosasaurs, unless they were more modern in age and came from a mastodon, of which there was now a surfeit. Lakes himself thought they were from a mosasaur. So far Marsh had not worked on reptiles; those had been

Cope's province. Marsh had been preoccupied with his toothed birds and giant mammals.

Some authors even think that Marsh ignored this first letter. But he didn't. He was obviously intrigued enough to write back, and sufficiently acquisitive to offer to look at the bones—if Lakes would ship them to New Haven. That he must have replied immediately is clear from the fact that Lakes wrote again to Marsh only eighteen days later, thanking him for sending copies of some publications and his offer "to identify said bones." He also sent "a tolerably accurate drawing . . . of both bones which will I think enable you to identify them almost as well as if I forwarded them to you." But Lakes seemed to be playing his cards close to his chest: "I am not yet quite decided what to do with the bones, and I hope to add somewhat to their number before I decide."

Six days later, Lakes wrote again describing remains "that I have been disposed to attribute to no less than 6 different animals if not separate species." Marsh must have said that he wanted the bones because Lakes ended the letter: "Hoping you are still desirous of receiving the bones. You will not be disappointed when they arrive." And there was a P.S.: "The boxes are in a building close to the Railway station at Morrison awaiting your letter to ship them immediately."

Almost as if the continuing tragedy of the Marsh-Cope relationship had a life of its own, it was at this point that Marsh seems to have hesitated. Lakes kept collecting and putting things in the store; soon he had a great deal of excellent material. In May he sent Marsh some bones, but then he also sent some vertebrae to Cope in Philadelphia, mentioning that he had "two skulls and some teeth." Naturally Cope asked for those. Then, at last, Marsh made his decision and typically took the whole matter over. He sent Lakes a check for one hundred dollars on June 9, although it was late in reaching Lakes, who was on the point of agreeing to work with (and for) Cope: "Despairing of hearing from you, I was on the look out for anyone who would help me or make some sort of an offer to purchase the specimens."

Swinging into action now, Marsh telegraphed Mudge in Kansas, telling him to go to Colorado to supervise the dig and report back. However, in a repeat of the situation in 1874, Cope had already received the new material from Lakes and, with his usual madcap speed, had begun to write a paper on it. This paper was destined never to be completed, but he did read a note on the bones to the American Philosophical Society that summer. Perhaps it was the news of this that spurred Marsh into action. Then poor Lakes had the unhappy duty of writing to Cope asking him to forward all the material to Marsh.

Mudge arrived at the Morrison dig on June 29. Lakes recorded the scene in his journal: "As we were eating our dinner under the trees an attractive looking little old gentleman rode up on horseback and asked if Prof Lakes was there. He introduced himself as Prof Mudge and we were soon deep in the matter of bones and saurians and relations with Prof Marsh." (There is something telling about the fact that relations with Marsh should rank high on the agenda.) "Professor Mudge we found quite an acquisition to our party. He appeared to be about sixty years of age but was lithe and active as a boy and full of interesting information."[5] Mudge had plenty of anecdotes about collecting in Kansas and recalled the time he was challenged by a group of Sioux and "made himself agreeable to them . . . by grinning with his false teeth and throwing them out beyond his gums, a feat which . . . they begged him to do again and again." Evidently Cope did not have a monopoly on this device for disarming Indians.

Mudge found that Lakes and his friends had already boxed up more than a ton of specimens in addition to those that had been sent to Cope. He immediately wrote to Marsh, who authorized him to hire Lakes at $125 per month, and the work started in earnest. The party excavated a large quarry—the first of many—more than thirty feet long. It may have taken a while to get things going, but by mid-July they had already sent enough material to New Haven for Marsh to publish a notice of his first dinosaur.[6] "We soon had a letter from [Marsh] from Yale informing us that our discoveries were Dinosaurs of a new and gigantic species. The vertebrae thirty inches diameter was a portion of the sacrum i.e. the vertebrae to which the tail is attached and the 'Hercules' war club was probably a humerus."[7] Marsh named the animal *Titanosaurus montanus*. The name seemed appropriate, as this was by far the biggest dinosaur yet discovered. Within a month Mudge had found a second dinosaur, which Marsh later named *Apatosaurus* (deceiving reptile).

At this point it may be useful to note that the class Dinosauria, founded by Richard Owen in 1842, would turn out to encompass a huge diversity of different kinds of animals, by no means all of them giants. There were tiny dinosaurs as well as hundred-foot behemoths. Today some 550 different kinds are known, and their classification is complex. The dinosaurs that Buckland and Mantell described in the 1820s (*Megalosaurus* and *Iguanodon,* respectively), for example, were both bipedal forms but they were not at all closely related. *Megalosaurus,* and even more so *Tyrannosaurus* (discovered in 1900), gave us one of the most characteristic images of a carnivorous dinosaur: upright, small forelimbs, massive head and jaws; these are called theropods. In 1841 Owen described a new kind of reptile that he thought was rather whalelike and so gave it the name *Cetiosaurus.* This was the first discovery of a dinosaur of the second, even more totemic shape—the huge lumbering beast on four short legs, with a long neck and tail. These were the sauropods, and they now include the characteristic *Brontosaurus* (properly, *Apatosaurus*) and *Diplodocus.* Sauropoda and Theropoda together constitute the Saurischia (reptile pelvis). *Iguanodon* and Joseph Leidy's *Hadrosaurus* belong to the group Ornithiscia (bird pelvis), in which the structure of the hip region had suggested to Leidy and Huxley a close relationship between dinosaurs and birds. The saurischians also had their quadrupedal representatives, among which are such well-known forms as the spiky-backed *Stegosaurus* and the horned dinosaurs like *Triceratops.*

Remarkably enough, although Marsh had gained control of the Morrison beds, all was not lost for Cope. While Lakes and Mudge were beginning to recognize the scale of both the beasts they were uncovering and the extent of the exposures in which they could be found, Cope got news of an entirely different site, of the same Jurassic age, eighty or so miles to the south. Just as Lakes was first writing to Marsh about the Morrison finds, a schoolmaster named Oramel Lucas was collecting fossils in Colorado at Garden Park, near Cañon City. He was an agreeable man and a strict Presbyterian, as Cope later reported to his wife. Like Lakes, his immediate aim had been to collect fossil leaves; like Lakes, he stumbled on a treasure trove of dinosaur remains. He wrote to Cope, who immediately pounced on them, agreeing to buy everything that Lucas could supply at ten cents per pound.

Therefore, in yet another remarkable parallel, by August, just as Marsh was announcing his *Titanosaurus,* Cope had received enough material from Lucas to publish a preliminary description of a huge dinosaur of his own. It was a sauropod from Cañon City that he named *Camarasaurus supremus,* and inevitably he too claimed that it was "the largest or most bulky animal capable of progressing on land, on which we have any account." Equally inevitably, Cope could not resist a comparison with Marsh's just described *Titanosaurus,* claiming that his own animal "exceeds in its proportions any other land animal hitherto discovered, including the one found near Golden City by Professor Lakes."[8] The battle for the first and the biggest that had started with the horned mammals of the Eocene was now carried into the field of dinosaurs. In one last jab, Cope also pointed out that the name of Marsh's animal—*Titanosaurus*—was preoccupied (meaning it had been used already); this forced Marsh to change it to *Atlantosaurus.*

For Marsh all this was particularly galling, because he had a collector of his own in that very region. David Baldwin lived in Cañon City and had even told Marsh about the existence of the large bones. In yet another case of history repeating itself, it turned out that Baldwin had done nothing about getting these kinds of fossils for Marsh because Marsh had never replied, although the following year he had an extensive correspondence with Baldwin when sending him off to collect in New Mexico. Baldwin and Marsh later fell out over the matter of payment for the New Mexico work, apparently in part because Marsh was dissatisfied with Baldwin's haul of fossils. However, in their biography of Marsh, Charles Schuchert and Clara Mae LeVene note that Marsh never opened many of Baldwin's boxes, which, when his assistant Samuel Williston came to examine them decades later, turned out to contain a veritable treasure of exciting new Permian fossil reptiles.

Meanwhile, Baldwin gave up on Marsh and started collecting for Cope.

Once he heard about Cope's good fortune, Marsh fired off a telegram to Lakes, ordering him to go to Cañon City to see what could be done. As it happened, Lakes was away, so Mudge went instead. Mudge may have been a kindly looking old gentleman, but he could be quite ruthless. The very day he arrived in Cañon City he managed—without consulting Lucas—to get into the storehouse where the fossils were being boxed up for shipment to Cope. "I arrived here about sunset yesterday and lost no time in looking at the bones Cope secured here several weeks ago at cost—a low sum. They are not Titanosaurus, but the most anomalous in structure of anything I ever saw or have seen described. I exceedingly regret that Baldwin did not secure them for you, as he might when they were first discovered. . . . I only saw the bones by lamplight—poor at that—and the man in charge did not appear willing to visit his store today—Sunday—nor did I wish to appear to have too strong an interest in the matter."[9]

Three days later Mudge had managed to corrupt Lucas: "He feels that he has sold his big bones too cheap to Cope and we can secure his good will in the future by a little kindness and good treatment. He is a young man trying to obtain an education." Lucas was not sufficiently strict a Presbyterian to resist Mudge, who, with promises of cash, persuaded Lucas that his agreement with Cope only covered the "big bones." To seal the deal Lucas gave Mudge some bones "which look to me like birds. . . . Whatever it is he will sell it when you have decided what it is; and whether it is bird or not. At any rate it looks to me to be new, and valuable."[10] The specimen turned out to be the first very small dinosaur, a tiny (possibly juvenile) creature that Marsh named *Nanosaurus.*

Following this invasion into the Lucas-Cope territory, Mudge opened up a separate quarry to the southeast of Lucas. When Williston came up to him from where he was collecting in Kansas, the two collected a small amount of material of two kinds of dinosaurs, forms that would later be named *Allosaurus* (it was a carnivore like *Megalosaurus* and *Laelaps*) and the huge *Diplodocus,* the totemic sauropod dinosaur, with its long neck and tail, short stocky legs, and tiny head. By the end of the season, however, Marsh had to concede that not enough material was forthcoming and left Cañon City to Cope.

One reason for abandoning Cañon City was that the wheel of fortune turned yet again, and in the summer of 1877 a letter arrived in New Haven from Laramie, Wyoming, about a site that was to promise even more than Morrison and Cañon City—perhaps even more than all previously known sites combined.

"Laramie, Wyoming. July 19th, 1877. Prof. C Marsh, Geologist Yale College. Dear Sir: I wish to announce to you the discovery not far from this place of a large number of fossils, supposed to be those of the Megatherium, although there is no one here sufficient of a geologist to state for certainty. We have excavated one . . . there is several others. . . . We are desirous of disposing of what fossils we have, and also, the secret of others. We are working men and not able to present them as a gift, and if we can sell the secret of the fossil bed and procure work in excavating others we would like to do so. . . . We remain Very respectfully Your Obedient Servants, Harlow and Edwards."[11]

Messrs. Harlow and Edwards, whoever they were, evidently knew their fossils would be of value—after all, Wyoming had been producing fossils and attracting Marsh, Cope, and their workers for the past five or six years. They were also determined to be very stealthy about it all. Their letter used the word "secret" twice and included the sentence, "We have said nothing to anyone as yet," holding both the potential for exclusivity and a threat to sell to the highest bidder. Their letter mentioned a shoulder blade "four feet eight inches" in length—something that could not have failed to attract Marsh's attention at a time when he and Cope were vying for the largest dinosaur.

When Harlow and Edwards offered to send Marsh "a few fossils, at what they cost us in time and money in unearthing," he wrote back requesting them to do so, although the bones did not arrive until mid-October. Immediately Marsh saw that they were from yet another huge sauropod (which he would name *Apatosaurus grandis*). He asked for more and sent a check for seventy-five dollars to seal the deal. Harlow and Edwards replied, reinforcing their need for secrecy: "We are keeping our shipments of fossils to you as secret as possible as there are plenty of men looking for such things and if they could trace us they would find discoveries which we have already made." So Marsh dispatched his top assistant, Samuel Williston, to view matters for himself. But then a new difficulty emerged. Harlow and Edwards could not cash Marsh's check—for the simple reason that Harlow and Edwards did not exist.

When he arrived, Williston sorted things out and was able to report to Marsh that the fossil bed was not at Laramie but near Como Station on the Union Pacific line. Both Marsh and Cope must have passed through that station, and been within less than a mile of Como Bluffs where the fossils were, a dozen times. As for "Harlow" and "Edwards," they were pseudonyms. Each used their middle names: "Harlow" was William Harlow Reed, the section foreman; "Edwards" was William Edward Carlin, the station agent at Como Station. It was all perfectly fitting with the atmosphere of suspicion and paranoia that had overtaken the rival camps in Philadelphia and New Haven. (Marsh had already instructed his field-workers to use code words in their telegraphs. When they needed money, for example, they were to ask for "ammunition"; Cope was referred to as "Jones.")

When Marsh sent him to Como two years later, Arthur Lakes left a description of Como Station: "a lonely spot on the Union Pacific. The station consisted of a red building, a tank like a huge coffee pot and a small section house for boarding the men at work on the track. . . . Behind the station to the south was a high bluff rising about 600 feet above the prairie, the top crested with sandstones of the Dakotah group of the cretaceous. The Face of the bluff was composed of ashen grey and variegated red and purple clays and shale's with some layer of sandstone. This soft material is eroded here and there . . . into many little channels and . . . ravines. . . . The line of excavations in search of dinosaurs was a few feet above the variegated belt between it and the cap or Dakotah sandstones. . . . [T]his bluff was very rich in saurian remains, far more so than those of Morrison and Cañon City in Colorado." Less than a mile to the north was Lake Como, where Marsh had collected specimens of the non-metamorphosing salamander *Sirenodon* back in 1868. Little had he known that he would have found the first specimen of *Allosaurus* on that occasion if only he had kept on walking.[12] Lakes did trek around the lake and noted in his journal, "near one of the sandstone ridges forming the North Bank of the lake, we came upon a quarry of saurian bones but slightly opened: A medium sized sacrum of a dinosaur lay exposed on the surface."[13]

Even without a buyer for their fossils, Carlin and Reed had worked hard the whole summer of 1877, accumulating several tons of giant dinosaur bones by the time Williston arrived. The exposures at Como Bluff were the same age and same formation, the Morrison Formation, as those that Lakes and Lucas had discovered in Colorado. Williston reported to Marsh that the productive exposures extended over seven miles, and this time Marsh acted promptly. Within days of receiving Williston's first letter from Como, he sent off a draft agreement for Reed and Carlin to work for him exclusively at ninety dollars a month. But Carlin and Reed turned out to be tough negotiators. Perhaps the wording of Marsh's draft raised their hackles. Certainly there was nothing deferential in Carlin's letters to Marsh as the negotiations got bogged down: "I have nothing farther to say. All that I agreed to I will do, provided that you also perform your part. I am now waiting Mr Reed's decision . . . his decision settles it. . . . I have agreed with you and done everything in my power to come to an understanding and am tired of the whole business and if you withdraw Mr Williston or Prof. Mudge I want to know of it at once, as I shall conclude then that no further understanding is possible and shall see what I can do elsewhere."[14]

Reed continued to work at the site through the winter of 1877, but the contract was not finally agreed upon until January 1878, when Carlin visited Marsh in New Haven. "I shall expect the agreement to be made out in proper form by some lawyer, exactly according to our understanding and as given by me to Mr Reed at the time and I will come to New Haven again about Friday or Saturday and sign for the same."[15]

None of this augured well, and it would turn out that work at Como Bluff would not be easy. Carlin was evidently the businessman of the partnership, while Reed was the man in charge and a skilled collector. Both were constantly suspicious of Marsh, feeling that they should have had more money. They also bitterly resented the fact that Marsh had sent Williston to supervise their work. When Williston fell ill and returned to Kansas, Reed, essentially worked alone until Williston's brother Frank arrived. By the end of the summer four more quarries had been opened, and out of them poured an almost overwhelming volume of fossils. That year Marsh was able to describe the giant herbivores *Apatosaurus, Diplodocus, Dryosaurus, Barosaurus, Sauranodon, Stegosaurus, Antrodemus,* and the smaller carnivores *Allosaurus, Laosaurus, Labrosaurus,* and *Nanosaurus.*

It was impossible to keep a site like that, so close to the railroad, a secret. The *Laramie City Daily Sentinel* published a long article naming Carlin and Reed and even stating how much Marsh had paid them.[16] Once again Marsh sent Williston to Como, just in time to report that a suspicious looking character named Haines had turned up, asking about the fossils. Williston was sure that he was one of Cope's men; Carlin thought it was Cope himself. It was not, but it seems likely that sometime during the summer of 1878 Carlin, who had worked less and less at the site, had started to excavate either for Cope directly or with men in Cope's employ. In February he had to write to Marsh about what had become a consistent problem in Marsh's dealings with his workers: "I wrote to you a few days ago requesting remittance for work done up to Jan'ry 17th 1878." And again in March: "I have been looking anxiously for a letter containing a remittance of wages due. . . . [T]he man that I have had working for me quit work this morning demanded his wages, and has left the country. . . . I had to *borrow* the money to pay him."[17]

Reed had quit his job with the railroad to collect full-time, but Carlin had kept his. In May he was asking Marsh for payment yet again: "Am compelled again by numerous necessities to ask you to remit my wages to date. I had hoped that this would not be necessary. As I explained to you that my expenses were considerable and they have recently been increased. In order to make our business fair I divide my wages from the RailRoad with Mr Reed and consequently am short of money most of the time. . . . [Y]ou will obliged me very much by remitting promptly at the end of each month."[18]

In September Carlin wrote again: "As a just and honourable man I can hardly doubt but that you intend to pay me the balance due to me . . . which I have been in constant need of for some months. . . . I am very sorry to say that if you do not attend to the matter that I shall be compelled to take legal measures."[19] We do not know what Marsh replied, but Carlin quit the project that month and Reed went home for the winter.

One of Reed's more unsavory tasks had been to destroy, on Marsh's direct instructions, any remaining bones in Quarry 4 so that no one else (that is, Cope and Carlin) could get them. Even so, two strangers invaded Quarry 1, and Reed had to face them down. He obviously needed assistance; Marsh sent Lakes to him, and another eight quarries were opened. Collecting at Como Bluffs continued intensely under Marsh for another twelve years.

That year (1879) Cope had two collectors working nearby, sometimes in the quarries that Marsh's people had abandoned. Reed was hostile to them, of course, but Lakes was more accommodating, and the two parties sometimes got along reasonably well.

Naturally enough, Cope actually did make a sortie to Como himself to see what pickings there might be (just as Marsh had sent Mudge to Cañon City). To his surprise, Lakes discovered that Cope was not the monster he had expected. "I·went down to the lake to bathe . . . [and] on returning . . . the train arrived and a tall, rather interesting looking young man stepped out of the coach and introduced himself as Professor Cope. He brought his blankets and a rubber bed for camping excursions. . . . He entertained his party by singing comic songs with a refrain at the end like the howl of a coyote." Cope left the next day, after pleasant chats about geological matters and England.[20] Lakes wrote later about Cope: "I must say that what I saw of him I liked very much his manner is so affable and his conversation very agreeable. I only wish I could feel sure he had a sound reputation for honesty."[21]

For his part, when Cope arrived at Como, he found that "the men who have been working for me, were not the ones with whom I had been corresponding, which quite surprised me. I found them good young fellows from Michigan named Hubbell. . . . I saw the bones of a Camarasaurus sticking out of the grounds, and the boys have dug up a huge flesh-eating saurian which they send off in the Morning." (The

Hubbell brothers were the ones who had coopted Carlin to collect for Cope.)[22]

As time went by Reed continued to be difficult and could not get along with Lakes (the English preacher), so he left Lakes to supervise at Como while he worked alone at the far end of the bluff, where he opened Quarry 13 and discovered yet another cache of *Camarasaurus, Coelurus, Camptosaurus, Diracodon,* and *Stegosaurus.* Then at Quarry 10 he found the almost complete skeleton of a *Brontosaurus* (now known to be the same as *Apatosaurus*) *excelsus,* which today is the centerpiece of the Peabody Museum's Great Hall at Yale.

When Lakes went back to teaching at Jarvis Hall College, the staffing difficulties at Como continued. The following year Frank Williston deserted to work with Carlin for Cope, and Cope's people became more open about prospecting in and around Marsh's quarries. Eventually, in April 1883, Reed gave up working for Marsh altogether and went back, unsuccessfully, to sheep farming. On his departure he left everything in the charge of his assistant Edward Kennedy, and never again would the work continue at the level of productivity that Reed, through all the difficulties, had so brilliantly delivered for Marsh. Later Reed became a curator and assistant professor of geology at the University of Wyoming.

In addition to the dinosaurs themselves, fifteen years' work in Colorado and Wyoming introduced a whole new element to fossil collecting in the West. While the work was hugely significant for the hundreds of dinosaur and early mammal remains that were excavated by field crews working for Cope at Cañon City and Marsh at Morrison and Como Bluff—and also by each other's crews working as nearby as they dared—the management of field paleontology had taken a new turn. One of the lessons of these three great sites was that, no matter how superb the fossils, a great deal more depended on the men who did the collecting than either Marsh or Cope had at first realized. Instead of single prospectors walking for miles looking on the surface for bones or digging isolated specimens out of the face of a bluff, collecting now involved groups of men quarrying large sites. Often a number of quarries would be opened in close succession along a particular geological exposure. Collecting could be conducted on a large scale. At Morrison, Mudge even resorted to blasting to open up the site. Considerable technical skill was called for.

As a result, field collecting now also involved a serious element of staff management. In addition to the professorial types like Mudge and Lakes and the hired hands like Chew and the two Smiths, whole new categories of worker had arisen: skilled, proud, and ambitious collectors like Carlin and Reed, and loyal staffers like Sternberg and the Williston brothers who had their own professional aspirations. Cope, having the lesser funds, got mired in fewer of the inevitable personnel problems; Marsh, with his curious inability to pay his people on time, had more trouble and his workers at Como seem to have been particularly fractious, with Reed and another of his men, at one point, even getting to the point of pistols drawn.

Security was now much more of an issue. Quarrying on the scale being practiced at Morrison or Como could not be kept hidden. Having men on site permanently made it easier to defend the locations against rival collectors, so it became even more necessary than ever for collectors to work straight through the winter months. Lakes wrote in his journal: "February 12, 1880. Very cold, snow blowing into quarry. R[eed] froze his foot, had to make a fire to thaw him."[23] At Como, Williston had Carlin and Reed put up a tent over the dig (at Quarry 1) so that work could proceed through the bad weather. But nothing could prevent one's rival from opening a new quarry in the same formation a mile or two down the ridge.

Another new factor was demographic. States like Wyoming and Colorado were becoming more populous. Wyoming has always had a small population; in 1870 it was 9,118, but it more than doubled to 20,789 in 1880. Colorado had 39,864 people in 1870 and had more than quadrupled to 194,327 in 1880 (early census figures did not include the Indians). More and more commonly fossil collecting was being done in public view, especially as it was still the case that most new localities were being found near railroad settlements. People heard about fossil collecting from their friends who were being employed by the eastern professors. They could read about the latest discoveries in the local newspapers. Collecting started to seem like a good way to make a few extra dollars, and curio shops started to spring up to sell fossils and—even more lucratively—minerals to the growing population of residents and travelers. No longer was fossil and mineral collecting an eccentric occupation conducted out of the sight of an indifferent public. Not surprisingly, therefore, Reed complained to Marsh in September 1881: "This country is run over with bone hunters and have bin trying to hire my men. They offered Phelps more than you are paying me but I told him they would not give him steady work and you would not hire him again. I think McDermott the section foreman is going to work for them there has bin six men here looking for fossils. . . . I have stuck up a notice that the land is taken."[24] Equally unsurprisingly, the men invading Marsh's sites were not just connected to Philadelphia. Now there were also collectors working for Agassiz at Harvard and, just like Cope, Agassiz tried to hire Reed's workers away from him.

And the dinosaurs all these men were collecting were not the only new kinds of fossils. In Jurassic times, the Como area, like the Bridger basin 150 million years later, was subtropical, swampy and environmentally rich— ideal for the assembly of a diverse array of every kind of fossil from plants to vertebrates. When the protagonists could step back from the search for the biggest and "baddest" dinosaurs, it turned out that Como was also the source of tiny fossils of, arguably, even greater significance. They were the oldest known remains of mammals in America and first turned up at Quarry 1 in the spring of 1878. In 1879 larger numbers were found at Quarry 9, which became known as the "mammal quarry." These were not mammals as we know them, or those of the Eocene, but small primitive shrewlike creatures close in structure to the point of origin of all mammals. Most of the remains consisted of jaws and isolated teeth. Eventually both Cope and Marsh found significant remains of these Jurassic mammals. And in another odd coincidence, the only other mammals close to these shrewlike animals from Wyoming, both in terms of age and primitiveness, were from Stonesfield in Oxfordshire, England—the location of William Buckland's very first dinosaur, *Megalosaurus*. The Mesozoic was indisputably the Age of Reptiles, and every discovery of new and more formidable looking dinosaurs confirmed that. But the mammals that eventually replaced reptiles as the dominant animals of land and sea had been literally and figuratively scurrying in the underbrush all along.

TWENTY-SEVEN

The Good, the Bad, and the Ugly

In the mid-1800s there were many more interesting and pressing issues in the West—the Indian situation, emigration, gold, the politics of territory- and statehood—than fossil bones. Edward Drinker Cope and Othniel Charles Marsh pursued their work and conducted their feuds within a rather small range of intellectual arenas, principally the American Philosophical Society and the National Academy of Sciences. But because public funding of the surveys was involved, paleontology eventually became embroiled in politics and then, inevitably with exposure in the national press, a semiprivate fight became a public scandal.

The story of the constantly changing, interweaving careers of Cope and Marsh was referred to earlier as a tragedy. The essence of classical tragedy is that the central figure—Hamlet or Othello, for example—is, at least in part, a hero. In the recent past, tragedy has sometimes been reinterpreted to portray the central figure as a villain or, worse, a weakling for whom everything he touches is poisoned. That is failure, not tragedy, and it is not how we should see Cope and Marsh. Full of human frailties as they were, they were colossi on the stage of science, heroic in their achievements and in the magnitude of their loves, as well as the ferocity of their hates. Even so, as in all good tragedies, they finally fell from grace.[1]

On the good side, there is the range and scale of their collections and publications. Their work had major significance in Europe as well as at home. Both had tremendous physical and mental energy. Using the considerable financial resources at their disposal, Cope and Marsh opened a Pandora's box of paleontological wonders in a period of less than twenty years. The contributions of the two men to the science of fossil vertebrates were in many ways quite different and were built upon different foundations. Marsh, having been a keen mineralogist previously, and having taught geology at Harvard, came to focus his attention principally on the zoology of his fossils and their classification. Cope, having trained (largely trained himself) in zoology and herpetology, and with no formal geological training, made many important contributions to the use of fossils in western stratigraphy. Both were preoccupied with the precision of identification and classification that was a shared common ground with mineralogy and ultimately had grown from the work of Linnaeus a century before. Interestingly, both had a success in their very early careers with descriptions of important Pennsylvanian-age fossil reptiles from coals: *Eosaurus* and *Amphibamus*. Cope was by far a more prolific author in the number of his publications and wrote many books and papers aimed at a broad audience in which he tackled a wide range of issues. Marsh stuck to paleontology. Cope continued his primarily small-scale field operations right through the decade of the 1880s. Marsh, with enormous financial resources at his disposal, stayed at home in New Haven and presided over the treasures that his army of collectors sent in.

Both men were intensely ambitious and were heavily influenced by European schools of biological science (and particularly by Thomas Henry Huxley). They studied with many of the same people in Europe in the early 1860s. But Cope was far broader and more versatile as a scientist. He would have been famous today on the basis of his studies of living fishes, amphibians, and reptiles alone.

The different ways in which Cope and Marsh ran their enterprises for collecting and describing fossil vertebrates tell us a lot about the personalities of the two men. Marsh, built short and square, was more shy and introverted; he never married. His mode of operation was that of a corporate executive. Like any giant of industry, he strove for exclusivity. He worked ruthlessly to eliminate the competition and to buy up the talents of other workers, especially those who might set up in business as rivals. He worked to monopolize the supply of raw materials and means of production (the supply of fossils). He released a steady stream of new products (monographs) and tried to control the market in two ways, through the *American Journal of Science,* which was (and still is) published at Yale University, and by controlling the culture of science in the United States from his position as president of the National Academy of Sciences. In Washington he showed himself to be a skillful politician, something that Cope could never master.

Cope was a tall, handsome extrovert who loved the ladies (so much so that his wandering eye may in the end have caused his wife to leave him). In another age he would have been described as dashing or even swashbuckling. He operated as a maverick intellectual—on the outside rather than the inside—constantly fighting the system, always trying new ideas. He was daring and even foolhardy in the field and fast and careless in the laboratory. He had greater access than Marsh to the American Philosophical Society's publications and eventually bought a journal, *The American Naturalist,* to help get his prolific flow of scientific papers out. Continuing the business metaphor, Cope was more the freewheeling entrepreneur, investing venture capital in new projects and depending too heavily on his own energies and enthusiasms. Sometimes his decisions were brave *and* right (as when he chose to ignore army orders and explore the Eocene of New Mexico in 1874); sometimes he was reckless; often he was simply spread too thinly across the scientific landscape.

When it came to money, Marsh was a careful man, Cope a spendthrift. In the late 1870s, seeking to enlarge the small fortune that his father had left him, Cope invested unwisely in Mexican silver mines and lost almost everything. That is why his superb collections today are in the American Museum of Natural History in New York. Cope had to sell them, and the Academy of Natural Sciences in Philadelphia, as usual, hesitated just too long over the price.[2]

Both developed excellent field methods. For Marsh, Sternberg invented the technique of making jackets (first in papier-mâché and later plaster-of-Paris) to support specimens like so many broken limbs as they were taken up in the field. Both spent their financial resources freely in the pursuit of fossils that could be said to have become, not just their life's work, but their whole life—aside from their feuding, which became a work of its own. Both illustrated their publications with excellent artwork (Marsh could afford the better help in this department).

Outside of science itself, Marsh is recognized for the stance he took on behalf of the Lakota under Red Cloud. And he accomplished a great deal for science in general as the president of the National Academy of Sciences. He could have done more had he not been so combative, earning a good many enemies as well as friends. Cope was an outsider when it came to politics and scientific organizations, becoming involved mostly with the American Philosophical Society and the Academy of Natural Sciences, where no doubt he often wore out his welcome with his own combativeness and pushiness. He was never invited to be an officer of the society. At the Academy of Natural Sciences, where for years he acted as the unpaid corresponding secretary, he was a thorn in the side. Cope, like Marsh, knew that the future of institutions like the academy, and science in general, lay in a full and proper professionalism. Curators could no longer be unpaid amateurs who did their work as a part-time hobby. In Philadelphia it was a fight that he was destined to lose—and the academy lost thereby also.[3]

Marsh and Cope, together with Leidy, made enormous contributions to the database of evolutionary paleontology and were early converts to Charles Darwin's theory of evolution by natural selection, first published in 1859. They understood the intellectual underpinning it provided for their science, and both sought to present their fossils as a physical demonstration of evolution in action. In 1876, Thomas Henry Huxley, the foremost evolutionary zoologist of his age, visited the United States in order to deliver an inaugural address at the opening of the Johns Hopkins University in Baltimore. When he visited Marsh in New Haven, the two exchanged information about the fossil record of horses. The European record, as worked out by Huxley, showed the later history of the three- and one-toed horses. Huxley lectured on this analysis in New York in September 1876, predicting that "in still older forms, the series of digits will be more and more complete, until we come to the five toed animals, in which, if the doctrine of evolution is well founded, the whole series must have taken its origin."[4] When he visited Marsh in New Haven he was shown the fossils collected by Marsh's western teams that documented the early history. Sitting together in the Peabody Museum, the two men put together an entire story tracing back the ancestry of the modern one-toed horses to the three-toed forms of the Miocene and to "Eohippus" from the Eocene, with its four front toes, each with tiny hooves. Over the years, the genealogical register of horse evolution has become vastly more complex, with many side branches and dead ends, but the essentials of the story worked out by Huxley and Marsh have remained one of the best empirical demonstrations of evolution in action over some 50 million years.

While Marsh (like Leidy) stayed close to the paleontological facts, Cope was the most interested of the three in matters of theory. He had deep philosophical interests, especially in metaphysics and the origins of the mind and free will. He wrote a large number of papers on evolution from 1870 onward that show his grasp of the subject and his original approach to contemporary arguments. In common with other paleontologists, Cope saw that there were great difficulties in accounting for the kinds of massive modifications of animal structure and function that the fossil record demonstrated solely in terms of a very slow, gradual accumulation of chance mutations, as Darwin proposed.[5] He found an alternative, which he sought to meld with Darwinian theory, in Lamarck's ideas about the inheritance of acquired characters. Acquired characters are those that are developed during an organism's life time by use and disuse. In the most hackneyed example, the giraffe might have acquired its long neck through generations of animals stretching to reach the tops of trees. For Cope, one of the most important elements in such a theory was the role of the will, and he saw this as a driving force in human evolution. These theories are now discredited, but in the last third of the nineteenth century, when the science of genetics was still unknown, they were very attractive to many scientists. In 1887, Cope published *The Origin of the Fittest*, a collection of his evolutionary essays, and ten years later he wrote an influential textbook called *The Primary Factors of Organic Evolution* (described further in Appendix C).

Cope was also much more interested than Marsh in the technical foundations of systematic paleontology (the naming and classification of species, for example, and the definition of higher categories). Cope's Law specifies that the ancestor of any group is to be found in generalized rather than highly specialized taxa. Cope's Rule states that in evolution groups tend to proceed from small to large size (an example would be Marsh and Huxley's horses, progressing from the dog-sized, five-toed horses of the Eocene). He also worked out a theory that still largely stands to account for the origin of the complex teeth of mammals from the simple cone-shaped teeth of reptiles.

Cope wrote other papers on the meaning of contemporary developments in cell theory, the evolution of structural adaptations, and animal coloration. In all this he was distinctly current in his thinking. Cope's field notebook for 1874 shows that he was (temporarily at least) intrigued by a form of eugenics called Stirpiculture that was being promoted by John Humphrey Noyes at the Oneida Community in New York as a way of producing a superior race of humans. It was essentially a plan for selective breeding, and in a closed human community like Oneida that meant spouse sharing and preventing some women from bearing children. "Sunday Aug 25th," he wrote in his journal: "Thoughts on Stirpiculture. 1. Property basis. 2. Right of strongest. 3. Preservation of moral vs intellectual. 4. Sexual selection is always w. ref to moral qualities; the good being always preferred by either."[6]

The greatest and most lasting scientific achievements of Cope and Marsh were the fossils themselves. If we put aside for a moment the personal squabbles and discount the mistakes that they made, the duplication of names, and the horrendous duplication of effort, there remains a heroic record of fossils discovered and described. And to that list we would have to add the actual fossil localities that they either discovered or developed right across the country, from New Jersey to Oregon. We can be glad, indeed, that they did not have teaching obligations, for that would have cut badly into their time.

But if all that is the good side of Cope and Marsh, we cannot ignore the bad. They accomplished so much, and principally in a very short period from 1870 to 1890, that one has to wonder how they found time to fight each other. Except, of course, it was the fighting that helped spur them on. Without the one to challenge the other to greater efforts, either alone would probably have accomplished less. On the bad side of the ledger we cannot ignore their constant, often almost petty squabbles over priority, and their dubious practices aimed at establishing priority in naming a new fossil. Both were aggressive and cavalier in the naming of new species. As a result, they made a nightmare out of what should be the simple foundation of biological science. All too frequently they both described the same species under different names. Often neither of them got it right, having tried to put a new name on something that Leidy had described long before them.

Many of the tales told about the behavior of these two are probably now overly embroidered, although there is every reason to believe the story that Cope once stole the skeleton of a beached whale destined for Harvard by simply going to the freight office and changing the instructions for its delivery. Neither Cope nor Marsh would have won many points in a popularity or congeniality contest. Marsh was evidently a hard-driving, secretive, and demanding employer who had no compunction about putting his own name on the work of his assistants. This dependence on, and exploitation of, his employees was to prove an Achilles' heel.

For a careful man, reliant on other people's money, Marsh was surprisingly casual, and even arrogant, in his repeated failure to meet his obligations to field collectors. Not only did Sam Smith have trouble getting Marsh to pay him on time (and Dr. Carter had trouble getting Marsh to forward funds for either Smith or Chew), David Baldwin (who collected for him in New Mexico in 1877) had to launch a concerted effort to get paid what he considered he was due. Carlin had to threaten legal action before he received recompense. Sometimes Marsh simply seemed both mean about money and hard to please. Difficulties often arose because Marsh had not defined carefully enough what he was willing to pay for in terms of his collectors' time and fossils. There seems to be a great deal of truth in the statement made by Lieutenant Carpenter (associated with Wheeler's Survey), who had to adjudicate the disagreement between Marsh and Baldwin: both men showed an "absence of proper business terms."[7] In the case of Thomas Condon's fossils from Oregon, Marsh persuaded himself that he had purchased the specimens that Condon had loaned him.

That sort of misunderstanding finally cost Marsh Sam Smith's sometimes shaky loyalty. After having collected for Marsh for more than a decade, Smith felt badly used. The following letter probably sums up (although with more colorful spelling) how many people felt about Marsh: "I rote to you saying that I and [unclear] would work for $80/00 per month each and you rote to me a very short letter which I would think from the reading of it that you was hot. You said that you would give me my price but that I would have to stand all the Shipping Expence such as Twine, Boxes & paper etc and you also said that you was disappointed in our summer work as much as to say that we did not do our duty. You said that I could take another man at $60/00 per month but you could not put him on vouchers but would to put his wages with mine and then we could divide it to suit our selves, that was hinted that we intended to do the work and draw the Extrymans pay our selves, so I think it best for me to do something Else. I can do better at other work. . . . I have always tried to do the best I could for you but since the work has been Dissatisfactory I shall quit and do something Else."[8] Perhaps most surprising is that even his assistants in New Haven, on whom he relied so much, had difficulty in getting paid on time, and Marsh lost their loyalty too.

Then there was the ugly. Like the little girl in the nursery rhyme, when Cope and Marsh were good, they were very, very good, but when they were bad they were horrid. Among the least attractive practices they and their collectors engaged in were: leaving false clues to distract the other from the real fossil sites; destroying fossils in the ground so that the other should not find them, spying on each other at new sites; and hiring away each other's workers. There is little doubt that Marsh started this sort of chicanery. As early as 1868, when Marsh and Cope were still congenial colleagues, Marsh hired away people to collect in

New Jersey for him, instead of Cope. In turn, Cope cheerfully continued the practice.

Once they ceased to trust each other, everything went downhill.
The style of letters sent from Cope to Marsh give an indication of how
the split progressed. For the period from May 1868 to January 1873
there are some thirteen extant letters. All begin with some version of
"My dear Marsh" or "My dear Professor Marsh." He did not, however,
use the even more punctilious "Esteemed friend" with which he began
his letters to William Sansom Vaux. The real differences show up in the
way Cope signed off. Until 1870 he ended his letters with "from thy
friend," "truly thy friend," or "I remain with much regards thy friend."
Then in February 1870 it became "with kind regards." In 1872 it was
"with regard" and in 1873 this had become a curt "yours truly." Little of
this slide into coolness was due to changing fashions in the formalities
of letter writing.

The lack of trust showed up first, formally at least, in the rancorous
exchanges in 1872 and 1873 over priority for naming the horned mammals. But it seemed that all events conspired against them, although
Cope tried to remain civil. With tragically awful timing, it was just at
this point that Cope received from Kansas and Wyoming the specimens
that had been intended for Marsh. (Marsh's collectors were having a
hard time serving two masters.) Marsh did not hesitate to spread the
word that Cope was behaving dishonestly, and Cope responded with
some heat, demanding: "Now, as to a man of honor I request of you:
1st. To correct all statements & innuendoes you have made to others
here & elsewhere, as to my ?dishonorable conduct. 2nd. To inform me
at once if others make such charges to you, about me."

A draft of Marsh's reply has an icy politeness to it:

Feb 3, 1873 Dear Prof. Cope, In reply to your letter of the 30th
ult. I have only to say that 1st I desire most sincerely to be on
friendly terms with you. 2nd If I have, by word or deed, done
you the slightest injustice, or should in future unintentionally
do so, I will promptly make due amends. . . . If you can truly
say the same, we have at once a basis of appreciation that will

prevent any serious misunderstanding in future. . . . As to the past, I will say frankly that I feel I have been deeply wronged by you in numerous instances. These wrongs I have usually borne in silence, and have even defended you for it in strong terms. I was likewise informed of your efforts to examine the specimens collected for me, and it was believed then that you obtained some of them. Now for all this you have only yourself to blame. I had nothing whatever to do with it.

In regard to the Kansas fossils, let me say with equal frankness that I had lost some valuable specimens and had reason to think they were in your possession. Could I have done less than to give you a chance to explain the matter. You have said distinctly that you have neither Wyoming or Kansas fossils of mine, and I have, therefore, nothing more to say. . . . When others have spoken against you, after the Smith affair last summer I made up my mind that forbearance was no longer a virtue.[9]

Differences that had once been papered over now were out in the open. Marsh did not easily back away from a fight, and one of his weapons was his new political base at the National Academy of Sciences in Washington. Soon he was quietly spreading gossip and encouraging dissatisfaction with the state of the government's western surveys, and particularly Hayden's, as the following exchange of letters shows. First Hayden wrote in his typically insistent and tactless way: "April 2nd, 1874. My dear Marsh, Your name is being used extensively here at this time by certain parties to sanction a statement that the Survey of which I have charge is a fraud. It is working to your disadvantage. Is this use of your name in such a connection authorised by you? Please write a telegraph to me on receipt of this at my expense. I wish to make use of your reply for your own good. Yours sincerely, F. V. Hayden."

As usual, Marsh replied in measured, superior fashion.[10] "My dear Hayden, Your letter of the 2nd came duly, and I regretted extremely to scan it. Your language could admit of only one interpretation & that was an implied threat that if I did not at once endorse your survey I should suffer for it at the Academy. As no personal considerations whatever could induce me to yield in such a case, I made no reply, leaving it for you to act as you saw fit. As the Academy will adjourn before this reaches you, I now answer your letter, with the same candour, but hardly with kind feelings, that I should have done had you written me a straightforward letter about the rumours you allude to."

The exchange with Hayden provides a background for the series of political moves that led to Hayden losing his beloved survey. By the mid-1870s, in a rather unwieldy arrangement, the government was supporting multiple separate surveys: The Interior Department ran the efforts by Hayden and John Wesley Powell in the Rocky Mountain regions. The Army Department was in charge of Wheeler's survey of the 100th meridian and Clarence King's U.S. Geological Exploration of the 40th parallel. Then there was also the U.S. Coast and Geodetic Survey, and the General Land Office, in charge of government lands, had its own surveyors. It made sense to have a single organization, but that left the question of whether it would be under the aegis of the War Department or the Interior Department. Naturally Marsh wanted to have a controlling voice in the result.

One of Marsh's first actions as the new president of the National Academy of Sciences was to write a report, requested by Congress, on the possible rationalization of the surveys. There was a fierce political fight because, among other things, if a single survey was to be created, as Marsh's committee recommended, either the Department of the Interior or the army would house it; the other would lose.

The academy's report advised the formation of a single U.S. Geological and Geographic Survey that, with the U.S. Coast and Geodetic Survey, would be housed in the Interior Department, where the General Land Office was also in charge of disbursement of government lands. When Congress had finished with the issue, a plan to merge just the Hayden, King, and Powell surveys was approved (the rest remained separate). Now another battle was joined: who should head this new entity?

Powell was widely considered too controversial a character; that left

King and Hayden. Hayden was by now Marsh's enemy and also a colleague of Cope. King was Marsh's friend. After a great deal of nasty infighting, the job went to King. Hayden was left out in the cold, relegated to the status of just another government geologist. He left Washington and settled in Philadelphia, writing reports and also books on the West. Cope, of course, lost both his support from the government and the accompanying access to sites. With his ally King at the head of the survey in 1879 (he was later replaced by Powell), it was inevitable that Marsh would supplant Cope as the official government paleontologist. This gave Marsh sole access to the collections from the surveys and also government funds to defray collecting costs. He also continued to use his personal resources to hire collectors to work all across the West. Cope, especially after the disastrous failure of his investments, had to scrape and fight for every fossil. One interesting result of this was that Cope spent a great deal of time in the field, while Marsh spent almost none.

Disenfranchising Hayden (and Cope) was a brilliant coup for Marsh to have stage-managed from his newfound position as president of the National Academy of Sciences. It is the more impressive because Marsh had been a member of the academy for only four years. Cope, naturally, had opposed everything concerning Marsh in the academy (which he had himself been elected to very early), so he had suffered a political loss on top of everything else.

A certain poetic justice prevailed when Marsh himself later fell out of favor. In his new capacity as government paleontologist, Marsh planned to produce a series of lavish monographs on his researches. The first of these was his book on the toothed birds from Kansas. This superbly produced volume, *Odontornithes,* especially the limited edition prepared at Marsh's own expense, is today a collector's item. A second work, *Dinocerata,* was, if possible, even more lavishly produced. However, those who live by the sword die by the sword. Cope was not one to take things lying down, and in the mid-1880s he campaigned vociferously against the King-Powell survey and Marsh. He had hit on the right political note and had a powerful ally—Alexander Agassiz, son of Louis and now director of Harvard's museum. The theme of their attack, which for Cope was mere convenience but for Agassiz was a matter of a philosophy, was that government funds were now being wasted in the surveys. They were not really necessary scientifically, and—here was the political point—any science would be better done by the private sector.

Substantial government appropriations were at stake, and there were plenty of men in Congress keen to press the charge that government funds were being used to create vast private collections at Yale (and in Philadelphia) and that, even if the surveys themselves were necessary, expansive research on fossils was not a good use of government money. Cope was asked to return all his specimens to the Smithsonian, but he could easily show that he had collected the vast majority of them using his own funds. It had turned out to be an advantage not to be paid properly! Marsh was particularly vulnerable because of his lavish government-funded monograph on toothed birds. One can easily imagine how this could be made to seem a ridiculous expense.

It was both a classic war of philosophies and a case of what practically was best for the government to fund—pure science or practical science. It seemed to many legislators that the government should not be spending large sums of money on something as impractical as fossil collecting. The end result was that the survey budget was slashed by about a quarter, many positions were lost, and all funding for Marsh was gone. Powell telegraphed Marsh demanding his resignation.

Marsh had lost his position with the survey, his access to specimens, and his salary. With the financial crash of 1890, now even he was short of funds. And another fallout of the debacle was that Marsh got a letter requiring him to return to the Smithsonian all of the materials at Yale that had been collected with government funding (some $150,000).

One of the fascinating features of the episode is that geological survey and paleontology, which might otherwise seem a rather benign and harmless sort of activity, should become the focus of such intense ill feeling. Everyone involved seemed to belong to a different quasi-political group and each was vilified by the others. The popular press loved this sort of thing and made sure to stir the pot whenever the opportunity presented itself. Cope, Marsh, and Powell obliged them. The real ugliness started when, while all the political maneuverings were going on, a rather desperate Cope finally lashed out at Marsh in public. Perhaps noting Marsh's success in promoting himself during the Red Cloud affair by working with the *New York Tribune,* Cope had fallen into, or rather, it has to be admitted, had launched himself into, the clutches of a fairly ruthless newspaperman, William Hosea Ballou.

Cope had for years been collecting instances of Marsh's real or supposed villainy and that of his henchman Powell. He fed Ballou the dirt and Ballou, who was an independent journalist, hawked the story around the papers until the *New York Herald* bit and plastered the row between Cope and Marsh all over its pages. SCIENTISTS WAGE BITTER WARFARE evidently sold newspapers. Cope's charges were a reckless mixture: Marsh was guilty of plagiarism, false reports, bribery, running the Geological Survey as his personal research organization, corruption in awarding jobs, and so on. It was thoroughly disgraceful and there was enough mud for some to stick.

The most telling part of Cope's charges against Marsh came from Marsh's own staff, as he reaped the consequences of his unbendingly magisterial style of management at the laboratory that he set up at Yale. For years his assistants had felt used and abused by Marsh, who had allowed them to draft his papers but never acknowledged them, let alone shared authorship. Cope had saved a letter from Samuel Williston (Marsh's assistant for eleven years) for just this occasion: "I wait with patience the light that will surely be shed over Professor Marsh and his work. Is it possible for a man whom all his colleagues call a liar to retain a general reputation for veracity[?] . . . The assertion of Professor Marsh that he devotes his entire time to the preparation of his reports is so supremely absurd, or rather so supremely untrue, that it can only produce an audible smile. . . . I never knew him to do two consecutive honest day's work in science. . . . The larger part of the papers published since my connection with him in 1878 have been either the work or the actual language of his assistants. At least I can positively assert that papers have been published on Dinosaurs which were chiefly written by me."[11]

The affair might easily have fizzled out, as the public's attention span for this sort of sensationalism is short and newspaper editors are always looking for the next cause célèbre to boost sales. Marsh might have decided to ignore Cope. But he chose to reply, damning Cope personally as a liar and accusing him of having stolen specimens from collections all over America and Europe. As for the charges of his assistants, they were all simply minions who had done no important work. He even dredged up the old six-horned mammals row from 1872–73, which the reading public evidently found a terrible bore. Cope and Marsh's scientific colleagues were embarrassed and appalled at the sight of these two going at each other in such an undignified way. When the dust had settled, clearly no one had won and both had lost a great deal.

The final verdict on these two squabbling men should perhaps be a Shakespearean "a plague on both their houses." But whatever their defects as men, as scientists at least, they were worth far more than that. In the end Cope has to be seen as the more brilliant scientist, but Marsh was the better scientific strategist. While they might possibly have accomplished even more if they had worked in tandem, as rivals they drove each other to greater and greater successes. Toward the end they wore each other out and, in that exhaustion, bad temper got the better of them.

Ironically, the spectacular flameout of a situation that had been smoldering for years had one positive effect. The feuding that had once been an embarrassment to the whole scientific community eventually gave the field of bone hunting an attention-grabbing story line, and even an aura of glamour, that even Hollywood could not have invented. Never again would there be difficulty in attracting recruits to the subject. Pictures of Marsh's crew of Yale students, armed to the teeth and looking like extras from a spaghetti western, and mental images of Cope's and Marsh's men scouting each other's diggings, rifles in hand, and with one eye for the fossils and another for the Indians, suggested a line of activity—conducted by scholars, no less—that would be hard to beat for excitement.

TWENTY-EIGHT

Going Public

Although they are now familiar to us, it was only slowly that the great discoveries of vertebrates in the American West made by Hayden, Leidy, Cope, Marsh, and their many associates reached a broad public audience. This may seem surprising given the fast early start for Charles Willson Peale's mastodon exhibitions in Philadelphia and on tour in Europe, and then the success (and continuing fame) of Owen's stunt in hosting a dinner inside the incomplete *Iguanodon* reconstruction for the Crystal Palace Exhibition in England (a direct copy of the dinner that Rembrandt Peale held under the skeleton of the great mastodon in 1802). Even the popularity of Leidy's *Hadrosaurus* mounted by Benjamin Waterhouse Hawkins in 1868 at the Academy of Natural Sciences, with duplicates sold to Princeton, the Smithsonian, and the Royal Scottish Museum in Edinburgh, did not immediately launch the sort of media rush that we see today. The place that dinosaurs occupy in our popular culture is really only a product of the 1890s and later.

In most popular and semipopular science books of the mid-nineteenth century, the most prominent position was given to ichthyosaurs, plesiosaurs, and above all to the pterosaurs: "The most wonderful animal of this [Jurassic] or any other age, was the Pterodactyle—a creature which was not altogether unlike the fabled dragon of the Middle Ages."[1] The reptile-bird *Archaeopteryx* failed to make an impact for many years and was most often classified simply as a fossil bird. This was because the British Museum specimen lacked the head (and therefore the teeth) and a number of its other reptilian characteristics had not been recognized. Not until well after the new Berlin specimen was described in 1871 did *Archaeopteryx* achieve a prominent place both in science and in the popular imagination.

The relatively low-key reaction to dinosaurs before, say, 1875, was mostly because the available material (with the exception of *Hadrosaurus*) was quite fragmentary and Hawkins's Crystal Palace reconstructions already seemed faintly laughable. Ichthyosaurs and pterosaurs, on the other hand, were already known from superb skeletal materials and seemed eminently realistic as living creatures. Not until the 1880s did dinosaurs and the giant fossil mammals from the American West find their way into textbooks and the popular literature. And that depended on a mixture of popular interest in fossils, the wave of exploration and discovery after 1877, and a great deal of scientific work (and imagination) that brought the animals alive.

When the American Centennial Exposition of 1876 was being prepared in Philadelphia, Cope proposed to Baird at the Smithsonian that it would "give great impetus to scientific study among the people—which is sadly wanted—if a series of reconstructions of great fossil vertebrates were to be exhibited." The Smithsonian was already planning to exhibit its copy of the reconstruction of Leidy's *Hadrosaurus* made by Hawkins. Baird tested the idea with Marsh, as he could scarcely have agreed to something that would give either one of them—Cope or Marsh—a lone starring role. First he asked: "Have you ever thought of exhibiting a choice series of your more interesting fossil remains from the West at the Centennial?"[2]

Cope's idea was to have Hawkins make a number of three-dimensional life reconstructions, not just skeletal mounts of dinosaur bones, for the centennial. It was basically a revival of Hawkins's earlier, but aborted, New York Paleozoic Museum. Baird later tried this idea out on Marsh too: "We have had a great pressure brought to bear upon us to expend a portion of our Centennial appropriation in employing Waterhouse Hawkins to make some restorations of prehistoric animals." But Baird seems really to have been trying to sabotage the idea, continuing: "First, the money is needed for other more serviceable purposes; second . . . I have no confidence in Hawkins' or any other person's ability with the materials at hand."[3]

Marsh rejected the idea that reconstructions had any place in science (or at least in the aspects of science that would be exposed to the public). Undoubtedly also he did not want to be associated with someone like Hawkins, who had worked so closely in the past with Cope and Leidy. He turned down Baird with the rather dismissive observation, "I do not think it possible to make restorations of any of the more important extinct animals of this country that would be of real value to science."[4]

Perhaps Marsh had in mind a fear of continuing the amateurish reconstructions of dinosaurs made in the Crystal Palace exhibitions (beloved now for their very unreality). When, two decades later, proud (or amused) hosts took him to see them during a visit to England, he immediately complained at a meeting of the British Association for the Advancement of Science that "their friends have done them . . . injustice in putting together their scattered remains, and restoring them to supposed lifelike forms. . . . So far as I can judge there is nothing like unto them in the heavens, or on earth, or in the waters under the Earth."[5]

Marsh's opposition to displaying mounted skeletons of his extraordinary array of fossil vertebrates might seem strange to our eyes. But it was entirely consistent with his approach to science as something not to be diluted by attempts to make it accessible to the general public. More than that, however, it was impractical. Fossils are extremely heavy, and Hawkins's *Hadrosaurus* mount showed the extent to which it was necessary to incorporate intrusive ironwork—and no small amount of artistic interpretation—into the display. It was not until Henry Fairfield Osborn pioneered the display of mounted skeletons at the American Museum of Natural History in 1891 that the modern era of fossil display began (to which was added, a century later, the robotic versions that must surely have Marsh spinning in his grave).[6]

Marsh's view (shared by most of his contemporaries) was that only the scientist, not some sculptor-technician, could interpret the skeletons properly. This still meant using an artist, but under close supervision. He employed many really gifted artists to make graphic reconstructions of his material, starting with the lovely images of *Hesperornis* and *Ichthyornis* in his Odontornithes monograph. Through the 1880s and 1890s, Marsh and his artists produced a glorious cavalcade of drawings of restored skeletons, giving us some of our most iconic images of such dinosaurs as *Dinoceras, Brontops, Triceratops, Stegosaurus,* and *Brontosaurus.* His work in this area reached a peak with *The Dinosaurs of North America,* one of his last reports for the U.S. Geological Survey.[7]

Surprisingly, given his dismissal of the value of "reconstructions" he published much of this work in short notes devoted simply to the drawings, without extensive scientific justifications. These drawings quickly found their way into textbooks. Ironically, his reconstructions reified particular interpretations of their posture and life habits that have been hard to change, even as new evidence has become available. In a sense, therefore, Marsh was right about resisting the move to produce elaborate reconstructions, and his own work turns out to demonstrate the dangers he warned against.

Marsh even refused to have his superb fossils exhibited in his own Peabody Museum in any form except for a few isolated bones. He only reluctantly allowed a papier-mâché reconstruction of a *Dinoceras* skeleton to be made in 1885.[8] Fittingly (for a rich man) the paper was recycled dollar bills. Marsh continued his obdurate opposition to making physically mounted reconstructions of his skeletons until late in his career, but in 1891, when he was in a position of prominence in Washington, he had to agree to have his artists create some enlarged drawings of dinosaurs to be displayed in the U.S. Geological Survey's display at the Columbian Exposition of 1893 in Chicago. The affair turned into a nightmare. F. W. Clarke, chief special agent of the survey, really wanted to borrow the skull of *Dinoceras.* But that turned out to be too heavy to mount on the wall. So Clarke had to make do with a papiermâché mockup made by Ward's Natural Science Establishment, an early provider of teaching materials. There was endless bickering about the costs, and finally Marsh sent some huge canvases, so large that Clarke could not display them all. Marsh imperiously insisted that they could be mounted only in the way he required, and the canvases ended up on display at the Smithsonian (which is what he had intended all along). The affair itself demonstrated how Marsh had descended into a kind of egotistical shell.

On the other hand, the art of reconstructing fossils in fleshed-out, putatively lifelike poses seems to have fascinated Cope. There exist several drawings in which Cope recreated his dinosaur *Laelaps* as it might have appeared in life. One of them appears as a doodle in his field notebook for 1874. In 1878 Cope installed a life-sized reconstruction (of which no record seems to remain) of *Atlantosaurus* in the former American Centennial fairgrounds in Philadelphia.

Life-restorations of prehistoric organisms—animal and plant—had considerable educational potential—most readily realized through the skill and imagination of painters. In fact, since at least the 1830s artists had tried to create realistic portrayals of life in ancient times, not only taking individual fossil creatures and restoring them to a supposed life-appearance but placing them in scenes with authentic ecological settings, as determined by knowledge of the geology and the associated remains. The first man to attempt this was Henry de la Beche, an English geologist contemporary with William Buckland and Mary Anning (he later became the first director of the British Geological Survey). In 1830 he painted a famous watercolor reconstruction of a scene of Jurassic life titled *Duria Antiquor* (The Ancient Life of Dorset). An engraving was made of this and copies sold to support his friend Mary Anning. Many artists followed de la Beche's lead.[9] Benjamin Waterhouse Hawkins, in addition to his dynamic reconstruction of *Hadrosaurus,* painted some large murals of Mesozoic life at Princeton University around 1870. But the greatest paleontological artist of them all, at the end of the nineteenth century and for the first half of the twentieth, was Charles Knight.[10]

A New Yorker, trained at the Metropolitan Museum of Art and the Art Students League, Knight was coached in paleontology by Cope. He was a superb wildlife artist as well as gifted in reconstructing fossils and their ecological contexts. He took the practice of creating entire ancient landscapes in mural form to new artistic heights and the best possible levels of scientific accuracy. His huge murals with their great life and energy can still be seen at the American Museum of Natural History, the Field Museum in Chicago, and the Los Angeles County Museum of Natural History. They introduced the Mesozoic and Tertiary world to countless thousands of museum visitors. In perhaps the last of this heroic genre, in the mid-twentieth century Rudolph Zallinger painted two enormous frescoes (*The Age of Reptiles* and *The Age of Mammals*) that are on display in the Peabody Museum at Yale, where they look down upon the magnificent collections of fossil reptiles and mammals made by Marsh, now arrayed in all their glory for the public to enjoy. And today so many gifted artists continue the tradition of reconstructing fossils as they might (or, sometimes, might not) have appeared in life that it would be invidious to try to list all their names.

As far as the public perception of dinosaurs (and by association, other fossil animals) is concerned, everything began to change in 1885, when remains of *Iguanodon,* the dinosaur first described by Mantell in 1825, were discovered in a coal mine at Bernissart, Belgium. And not just a few bones—there were no fewer than twenty-nine more or less complete skeletons, with more left in the flooded mine. The remains were studied by the great Belgian paleontologist Louis Dollo, and a series of mounted skeletons—a veritable herd of iguanodons—was put on display in a hall of the Belgian Royal Museum of Natural History in Brussels. This was indeed sensational stuff. The king of the Belgians was shown the skeletons and pronounced them like "giraffes." But these "giraffes" were bipedal. They demonstrated quite clearly that Leidy and Huxley had been right in reconstructing *Hadrosaurus* as a biped years before. Casts of the *Iguanodon* skeleton in full bipedal pose quickly found their way to the British Museum in London and to the Oxford University Museum, where they remain to this day.

Immediately after the Bernissart discovery, the French popular science writer Camille Flammarion published *Le Monde Avant la Creation de l'Homme,* a comprehensive, profusely illustrated book of geology and prehistorical life for a popular audience.[11] And Flammarion had no hesitation about seizing on the sensational aspects of the discovery of these huge dinosaurs ("quelles masses prodigeuses . . . quatorze metres de longeur [forty-five feet long!]"). A dramatically posed *Megalosaurus* (Buckland's dinosaur) was featured on the title page, and his introductory chapter included a drawing of an *Iguanodon,* standing erect, theatrically represented as munching something from the fifth floor of a Paris apartment building.

Surprisingly, this sort of sensationalism still took a while to catch on in the American press, perhaps because Flammarion's works were never translated into English. Once it got itself aroused, however, the popular press in America had a heyday with dinosaurs. Its fascination with dinosaurs has never waned since 1897, when a sensational article ("Gigantic Saurians of the Reptilian Age"), lavishly illustrated by Charles Knight, appeared in *American Century.* The subject was taken up again the following year in the tabloid *New York Journal and Advertiser* and the *New York World* after new discoveries of the huge *Brontosaurus* were made. "When the Brontosaurus walked, the earth trembled," the *New York Journal* reported. "One man could not lift its smallest bone. Its petrified skeleton weighs 40,000 pounds." The *Journal* included a front-page drawing of a *Brontosaurus,* looking a little embarrassed, standing against a skyscraper, with the caption, "How the Brontosaurus would look if it were alive and should try to peep into the eleventh story of the New York Life Building."[12] This blatant copy of Flammarion finally established dinosaurs as a phenomenon of popular culture.

As for the science behind the fossils, like America itself, that had also changed. Less than fifty years had passed since Hiram Prout described the first fossil mammal from the Dakota Bad Lands, but he would now have found the paleontology of the West unrecognizable.

TWENTY-NINE

1890

The End of the Beginning

The fossil collectors of the old West, with dirt under their fingernails and often sick from poor water and bad food, created a golden age of paleontology. Just like the homesteaders and miners trekking westward, and the residents of the new towns springing up across the West, they were pioneers. The ancient animals whose bones they dragged out on the backs of their mules now stand in the great museums of the world as monuments both to science and to individual perseverance. The even larger numbers of less complete remains that are assembled in museum research collections form the raw data of evolution in action and evidence of a changing, dynamic earth. These objects, the men and their travels and labors, created a small but significant part of western history. They helped define the West in terms of geography and geological history and created a unique place for American fossils in the long record of life on earth. They were part of the story of the western expansion of the nation—having followed the American frontier from Virginia, Pennsylvania, and New York through the Midwest to the Rocky Mountains and beyond.

By 1890, the country had become settled from coast to coast, and the U.S. Census Bureau announced that it was no longer possible to define a western frontier. This fact soon developed great significance through Frederick Jackson Turner's ideas about the defining role of the frontier in American history, although his famous statement that "the

317

existence of an area of free land, its continuous recession, and the advance of American settlement westward explain American development" has long since become so qualified as to be almost meaningless.
The concept of the frontier was surely only one of many defining elements of either the American character or the course of American history. But in the world of bone hunters the same initial premise seems to
hold true: after 1890, the frontier phase of fossil collecting had ended.
By 1890, every state in the Union had been explored for fossil bones
and the exploration of western Canada had begun. By 1899, both Cope
and Marsh were dead.

The frontier period of bone hunting had concentrated on exploration and discovery. Hundreds, even thousands of species were discovered and described. There then followed a new world of synthesis.
Collectors still went out, many of them year-round, as they do today,
but at home scholars had turned from simple documentation of the fossil record to understanding what it meant. When he was not concerned
with Washington politics, Marsh's work during the 1880s largely involved drawing together comprehensive reviews of whole groups,
building up the patterns of evolution of the toothed birds (Odontornithes), horned mammals (Dinocerata), all the Jurassic dinosaurs, the
Cretaceous pterodactyls, and the Jurassic mammals, together with discussions of the classifications of dinosaurs and other groups.

Cope continued to spend much more time in the field than Marsh,
collecting in New Mexico (1881), Colorado and Oregon (1882), Dakota,
Wyoming, New Mexico, and Mexico (1883), and Kansas (1884). Like
Marsh he also applied his restless mind to pulling together major review
papers. Perhaps the best known of these still goes by the name of
"Cope's Bible"; *The Vertebrata of the Tertiary Formations of the West* is
a simply massive tome of over a thousand pages and 134 plates, summa-
rizing a life's work by Cope and everyone else who had collected in the
western basins and badlands.[1] He also found time for major reviews of
the living amphibians of North America, and of the fossil oreodonts,
perissodactyls (horses and related forms), artiodactyls (cattle and their
relatives), and Cetacea and Sirenia (marine mammals). On top of this,
he drew upon his experience for major writings on evolution and a

range of philosophical issues, while still keeping up a flow of numerous descriptions of new species.

The cast of characters changed, too. It would be all too easy to give the impression that Cope and Marsh were the only serious paleontologists working with the amazing fossil vertebrates of the West. But that would be as unfair as it is untrue; Cope and Marsh, being somewhat larger than life, are simply the ones who (as they would have wished) dominate our attention. In fact, as early as that pivotal year of 1877 when the dinosaur beds were discovered, a new cohort of professional bone collectors had begun to build their careers, and they would come to dominate the field just as Cope and Marsh had.

It began one day in 1876, when three undergraduates at Princeton College of New Jersey, William Berryman Scott and Henry Fairfield Osborn, with their friend Frank Speir, were chatting. Scott announced: "Fellows! I have just been reading in an old *Harper's* an account of a Yale expedition to the Far West in search of fossils; why can't we get up something like that?" Speir and Osborn replied, "We can, let's do it."[2] Of course, it turned out to be complicated to organize, but a year later the first Princeton expedition headed west. The result was that Scott, who had been intending to become a physician, and Osborn, who was supposed to follow his father into the railroad business (he was president of the Illinois Central) instead eventually became professional paleontologists, part of a large contingent that would soon replace Leidy, Cope, and Marsh.

As they started to plan their trip, Scott and Osborn naturally got in touch with Marsh, who rather coldly discouraged them—not surprisingly given that he was so bitterly competing with Cope to get every fossil in North America for himself. Next they approached Cope, who, perhaps also predictably, was not much more encouraging, although later, when he had seen how serious they were, he became their ally and their friend. The youngsters had asked Cope for advice about collecting in western Kansas, but he was not willing to share his mosasaurs with untested amateurs. Instead the party decided to head for Colorado and Wyoming (no doubt with Cope mischievously encouraging them to butt in on Marsh's territory there). In the summer of 1877, a rather large expedition headed for Colorado under the supervision of physics professor K. C. Brackett. The college trustees subsidized the trip to the extent of $10,000, in addition to which each of the sixteen student members chipped in $150. The party also took along two students as photographers, and a proctor and a janitor from the college went along as "cook-teamsters." An unusual co-leader was General J. Klarge, whose role was to help train the students to fight Indians. In his memoirs, Scott recalled his attempts to teach the students the discipline of close-order drill.

Free passes on the railroads got them all as far as Colorado Springs, but that was the easy part. Scott recorded later: "No plan of exploration had been made, no localities suitable for collecting had been fixed, in fact, the expedition started to deteriorate into an aimless wandering about."[3] At Florissant they chanced on a productive fossil bed, although they seem not to have known that fossils had already been found there for several years. Soon they were all tired and ill tempered, and Brackett and Klarge, men of exactly opposite temperament, had a bad falling out. So Scott, Osborn, and Speir, with Klarge, desperate to retrieve something of value, set off independently for Fort Bridger, Wyoming.

Evidently, either Cope or Leidy had put the Princeton party in touch with people at Fort Bridger. At first, James Van Allen Carter anxiously reported to Marsh in New Haven: "As far as I'm able to judge from present indications the Princeton party has given us the slip. No word of or from them here for over a month."[4] It also turned out that the "Princetons" might even have written directly to Sam Smith, and Carter wrote: "Latterly he [Smith] has seemed very uneasy and has been subjected to strong influence to join another party. He has always been faithful to you tho', and hence has not consented to give his time to any one else."[5] Three weeks later Carter was sounding defensive: "He anticipated looking for work with the Princeton party, and I wanted it understood that I would employ him for you at once. . . . I have heard nothing from the Princetons, and it seems somewhat doubtful about their coming!"[6] At last, in August, Carter wrote again: "Princeton is here—working in Grizzly Butte & on Cottonwood Creek & Henry's

Fork. They do not intend going beyond Green River in a southerly direction. Talk some of a trip to Bitter Creek but this they cannot make at this season." On the subject of Smith: "Sam Smith has not yet returned from a trip north in interest of the Judge, and I do not think is apt to put in any time this season."[7]

According to Scott, Dr. Carter gave them useful information about where to explore and helped them find local guides. The students collected along Smith's Fork, Cottonwood Creek, and Henry's Fork, putting together some very creditable finds of fossil mammals. Ironically, Speir, the one member of the three who did not became a paleontologist, turned out to be by far the best at spotting fossils in the rock. The following year, Scott and Osborn returned to Wyoming in a smaller party of five students and tried once again to get Sam Smith to lead them. This time they collected farther to the east (around Rock Springs) and south (Church Buttes).

After graduation, Scott set off for Europe (apparently with Osborn's father footing the bill), spending a year studying with Huxley in London before going on to the continent and completing a medical degree. He returned to Princeton in 1882 as an assistant professor of geology and never left. Once back at Princeton, Scott set up an active field collecting program. In 1882, he took a group from the college to the Pine Ridge agency and gathered fossils in the old White River Bad Lands. In 1884 they went to Fort Custer and collected in the Bighorn Basin. The following year they were back at Fort Bridger, where they worked with "one of our old guides." In echoes of his earlier duplicity with respect to Cope, it turns out that this was none other than Sam Smith, who must have worked for the original Princeton group in either 1877 or '78 after all. In 1886 they were back at Fort Bridger again, and then in 1887 in Oregon. In 1890 they returned to the White River Oligocene. Perhaps the most striking fact about this fieldwork is not that the Princeton men continued to find wonderful specimens at the old localities, but that they returned, year after year, to the same old places. Scott took over the large fossil mammals of the West as his special subject, writing one of the great books on the subject, *A History of Land Mammals in the Western Hemisphere.*[8] It is ironic, then, that although Scott detested O. C. Marsh all his life, in 1985 Princeton University gave up its paleontology program and its fossil vertebrates all went to Yale.

Meanwhile, like Scott, Osborn made his European tour after graduating, studying in London with Huxley and in Cambridge. But where Scott was a geologist, Osborn was a zoologist. In 1881 he received a doctorate from Princeton and settled down to teach comparative anatomy there (with his father subsidizing his salary). Ten years later he was recruited to go to Columbia University as a professor and appointed simultaneously as a curator at the American Museum of Natural History. At the museum he began actively to build up superb collections and greatly expand the scientific staff. A huge part of the expansion came through purchases of the bankrupt Cope's mammal collections in 1895 and, after his death, of his South American and European collections, together with his American fishes, amphibians, and reptiles. Among Osborn's favorite topics was the evolution of the elephant group (proboscideans). As a scientist, he was rather like his intellectual mentor Edward Drinker Cope: in his more than a thousand scholarly papers and books he relished ideas and speculation. He became known for proposing what are now seen as wrongheaded ideas about evolution and eugenics. But he brought to the subject an organizing genius and an expansive approach (perhaps easy for a man with huge family wealth behind him) that had previously been lacking. For him science was a great adventure to be shared with everyone. When he became director of the museum, Osborn began the process of making dinosaurs one of its showcase images and attractions.[9] Fully mounted specimens were displayed in the museum halls, presenting fossils to the public as high drama and fossil collectors as heroes. It was a complete reversal of the attitude that Marsh, who had held everything so close to his chest, had brought to the science.

Osborn's arrival at the American Museum of Natural History launched what has been called a second "Jurassic dinosaur rush."[10] Not to be outdone by New York's success, the Carnegie Museum of Natural History in Pittsburgh, with the financial backing of its eponymous founder, launched its own major collecting expeditions, as did all the other major museums. They also bought extensively from the growing cadre of professional collectors in the West. Andrew Carnegie himself wanted only the biggest and the best. His museum sent collectors to the West, and perhaps their smartest move was to hire William Reed away from the University of Wyoming as their local expert. In 1902 Carnegie got what he wanted, in the form of *Diplodocus carnegii,* a sensational, complete sauropod discovered at Sheep Creek, not far from Como Bluff. It was even more impressive than Marsh's *Brontosaurus.*[11] Later that year, King Edward VII of England visited Carnegie at his Scottish castle and saw a drawing of the new *Diplodocus.* He decided that Britain had to have one of these creatures too, so Carnegie had a plaster copy made, bone-by-bone, and in 1905 a version of America's biggest and best dinosaur stood in pride of place at London's Natural History Museum. Eventually a total of nine copies were made and given to museums in Europe and South America.

Another milestone in the popularization of American paleontology came in 1902, when expeditions to eastern Montana from the American Museum of Natural History, led by Barnum Brown, bagged a superb skeleton of *Tyrannosaurus rex*—the tyrant-king reptile. *Tyrannosaurus* dramatically filled all those hopes and dreams that Jefferson had had for his great-claw and the supposedly carnivorous mastodon. It was impossingly big and definitely ferocious. With *Tyrannosaurus* and *Diplodocus,* the archetypal monstrous carnivore and herbivore, paleontology had twin icons whose silhouettes are now instantly recognizable everywhere. (Ironically, Cope probably had acquired the first *T. rex* specimen in 1892, a vertebra from South Dakota that he called *Manospondylous gigas,* a name that simply didn't convey the same aura of all-American size and invincibility.)

Among the new professionals entering the field was Jacob L. Wortman, who for many years had been Cope's principal scientific assistant. He went first to the American Museum of Natural History with Osborn, and then became the first curator of vertebrate fossils at the Carnegie Museum of Natural History in Pittsburgh. He was joined there by O. A. Peterson, one of many of Marsh's assistants to branch out on his own. As was documented in the newspaper exposés of the

Marsh-Cope feud, the men who had worked for Marsh—including Peterson, Samuel Williston, Oscar Harger, and John Bell Hatcher—tired of playing second fiddle to the great man. They were doing a lot of Marsh's research, and even drafting his scientific papers—without acknowledgement. He refused to let them publish anything on their own, even discoveries they themselves had made. Oscar Harger was one of the most brilliant of the students who had taken part in the Yale student expeditions, and he was a key contributor to *Odontornithes*. Given a free hand, he might have become a fine paleontologist, but he died young, in 1887.

Perhaps the most independent (and therefore successful) of Marsh's assistants was Samuel Wendell Williston.[12] He joined Marsh's team in 1876 on the recommendation of his teacher at Kansas, Benjamin Mudge. He had already had a varied life as a railroad surveyor and medical student and learned fossil collecting from Mudge. In 1877 he ably supervised the excavations at Como Bluff, and while simultaneously working for Marsh in New Haven he also earned a medical degree from Yale, in 1880. Marsh would have preferred to keep him as a collector and assistant, but Williston always had greater ambitions. With Marsh blocking his development as a paleontologist, he picked up on an old interest and began a parallel career as a scholar of insects. He became an expert on the fly family, earning a Ph.D. from Yale in 1885. After he had taught anatomy for a while at Yale and served as medical officer in New Haven, in 1890 he was appointed professor of geology at the University of Kansas. He had also been offered a chair at Kansas State University (the former Agricultural College), but he always resented the way the institution had treated his old teacher, and he turned it down. At Kansas, Williston could at last flourish. He built up the university collections in Cretaceous vertebrates and made himself, among other things, the foremost expert on the Cretaceous mosasaurs. At the end of his career he moved to the University of Chicago.

Perhaps the most charismatic of the new breed was John Bell Hatcher. Hatcher was born to a poor family in Illinois, grew up in Iowa, and worked as a coal miner to save enough money to fund his education. Collecting Coal Measures plant fossils led him to study geology at

Yale, graduating in 1884. He then went to work for Marsh as a collector in Kansas, Texas, Nebraska, South Dakota, and Wyoming. Some of the best horned dinosaur (ceratopsian) material at Yale was his. He found excellent new mammal material for the Yale collections in the old Dakota Bad Lands, where it all started.

One of Hatcher's typically novel contributions to science was the discovery of tiny Cretaceous mammal teeth in Wyoming. He saw that ants, when excavating their colonies, would carry out the larger chunks of gravel and, often enough, mammal teeth with them. Collecting the fossils simply meant finding the ant hills. "It is well to be provided with a small flour sifter with which to sift the sand contained in these ant hills," Hatcher wrote, "thus freeing it of the finer materials and subjecting the coarser material remaining in the sieve to a thorough inspection for mammals. By this method the writer has frequently secured from 200 to 300 teeth and jaws from one ant hill. . . . Another way to secure these small teeth is to transport the material to a small stream and there wash it in a large sieve in the water, the finer material being washed away, but this treatment is too harsh to give the best results, what few jaws there are always being broken to bits."[13] Dry and (careful) wet screening are today common techniques in the search for vertebrate microfossils that Hatcher precipitated.

With his rather fiery temperament, Hatcher could not abide working for Marsh and in 1893 he joined Scott at Princeton as curator of the vertebrate collections in the museum. He made more superb collections from the White River Oligocene for Princeton and then embarked with his brother-in-law, Peterson, on the most flamboyant and risky phase of his career: three years of expeditions to Patagonia. He had very little money of his own but stated that he would put up the funds and also contribute fifty dollars a month for expenses. No one is quite sure where he found the money—the most popular story (one of those stories that, if not true, ought to be) is that he earned it playing poker. In 1900 the Carnegie Museum at Pittsburgh appointed him curator and (with Charles Gilmore and Earl Douglass) he was crucial in collecting and mounting the *Diplodocus* skeleton for which the museum is justly famous. But he died of typhoid in 1904, at the age of forty-three. By then, vertebrate paleontology had become an elaborate science with a host of young professionals working in the universities and museums, much as it is now.

A final change in the world of fossil collecting, driven in part by the growing public fascination with dinosaurs, concerned the image of the people who hunted for the bones—the ways in which they viewed their work and themselves. In this they were, once again, part of a larger historical movement at the heart of which, in a curiously divided way, was the land itself.

In addition to the popular myths of the West—its rugged individuals, its harsh but rich and beautiful, God-given lands, its (noble) savages and the abundant ennobling (not to say enriching) opportunities—the period of the mid-to-late-nineteenth century was also marked by two opposing philosophies. On one hand, it was a time of science and rationality drawing upon the earlier traditions of the Enlightenment and driven in no small part by the industrial revolution. Science and technology were clearly the movements of the future. On the other hand, and in great part as a backlash against the mechanistic materialism of the scientific age, this was also the last great phase of the Romantic Movement. It was a time when American artists and writers turned to the majesties of nature for intimations of the "sublime."[14] The mountain landscapes of the West provided a limitless source of inspiration for artists like Jacob Miller, Karl Bodmer, Albert Bierstadt, and Thomas Moran, who had accompanied early expeditions up the Missouri. In the geology of the mountains, in particular, artists found a dramatic manifestation of God's creation and of nature's powers.[15]

Fossil collecting straddled these worlds. Field-workers in the American West, with their eyes hugging the ground looking for signs of fossil bone in the mountains or on the plains, were surrounded by some of the most glorious scenery on earth. Even the barren, semidesert badlands of the high plains of Wyoming and western Kansas had a special, harsh beauty for those who could see it. Arthur Lakes, for example, collecting in Wyoming for O. C. Marsh, never tired of the landscape. He wrote (on June 11, 1878):

Adjourned to a spot three miles East of camp . . . [t]o a quarry
where we had found the remains of a small crocodile. . . .
About noon heavy clouds began to loom up over the south and
quitting our work we hastened to take shelter under the castellated sandstones on the top of the bluff. Under them we
stretched ourselves like modern saurians awaiting the coming
storm; watching the instincts also of a few butterflies and
moths that came in for shelter . . . a few drops of rain and then,
with a deafening crash followed by a blinding flash, the storm
hit full on the back of the cliff behind us; then very heavy hail-
storm, stones as large as bantam eggs. The sight was very fine
with the sun shining brightly on the showers of white grass.
The heavy clouds with repeated flashes moving off to the
northward over the wild rocky outlines, the roar of the thunder
and the rushy sound of the hail and the sense of deep solitude.
Then in a few moments the storm was passed; the sun shone
bright and a few faint notes of joy broke out from the song sparrows in the sagebrush and all nature seemed refreshed.[16]

Perhaps nothing in geology has ever been as lyrical as Clarence
Dutton's description of seeing the Vermillion Cliffs at Grand Canyon
on a late afternoon in 1881.

It seemed to us that all grandeur and beauty thereafter beheld
must be mentally projected against the recollection of those
scenes, and be dwarfed into commonplace by the comparison;
but as we moved onward the walls increased in altitude, in ani-
mation, and in power. At length the towers of Short Creek
burst into view, and, beyond, the great cliff in long perspective
thrusting out into the desert plain its gables and spurs. The day
was a rare one for this region. The mild, subtropical autumn
was over, and just giving place to the first approaches of winter.
A sullen storm had been gathering from the southwest, and the
first rain for many months was falling, mingled with snow.
Heavy clouds rolled up against the battlements, spreading their

fleeces over turret and crest, and sending down curling flecks
of white mist into the nooks and recesses between towers and
buttresses. The next day was rarer still, with sunshine and
storm battling for the mastery. Rolling masses of cumuli rose
up into the blue to incomprehensible heights, their flanks and
summits gleaming with sunlight, their nether surfaces above
the desert as flat as a ceiling, and showing, not the dull neutral
gray of the east, but a rosy tinge caught from the reflected red
of rocks and soil. As they drifted rapidly against the great barrier, the currents from below, flung upward to the summits,
rolled the vaporous masses into vast whorls, wrapping them
around the towers and crest-lines, and scattering torn shreds
of mist along the rock-faces. As the day wore on the sunshine
gained the advantage. From overhead the cloud-masses stub-
bornly withdrew, leaving a few broken ranks to maintain a feeble resistance. But far in the northwest, over the Colob, they
rallied their black forces for a more desperate struggle, and answered with defiant flashes of lightning the incessant pour of
sun-shafts.

The half-tones at length appear, bringing into relief the
component masses, . . . the salients silently advance towards
us; the distorted lines range themselves into true perspective;
the deformed curves come back to their proper sweep; the an-
gles grow clean and sharp; and the whole cliff arouses from
lethargy and erects itself in grandeur and power as if conscious
of its own majesty. Back also come the colors, and as the sun is
about to sink they glow with an intense orange vermilion that
seems to be an intense luster emanating from the rocks themselves.[17]

The student members of the Yale College expeditions had a more
vaguely rosy notion of what they were getting themselves in for. They
did not imagine that they were going west simply to grub around in the
dirt, nor did they have aspirations to make great scientific discoveries.
They were going to traverse the vast plains, climb the mighty Rockies,

fight the Indians, and shoot the buffalo. Thereafter, successive Yale (and Princeton) parties posed for pictures in new western hats and shirts, draped with belts and bandoliers, and carrying their trusty (as yet virgin) rifles and Bowie knives. This was really the stuff of romance, and if the reality usually turned out to be a lot dirtier and less fun, at least the magnificent scenery would periodically awe and enchant them. The Indians were just as unpredictable and fearsome and the frontiersmen just as bearded and gritty as legend foretold.[18]

Charlie Betts's reading of western novels shows up in a passage he wrote describing how, near the North Platte River, "we followed the old California emigrant trail. . . . Here we were found by some soldiers, who had been sent back to guide us through a labyrinth of shale and sandstone known as Scott's Bluff. It was pitch dark when we began to pick our way through these narrow and rugged defiles, where, at every turn, deep cañons yawned at our feet. Fitted by nature for ambush and surprise, this had been the Indians' favorite spot to fall upon the emigrants; and those dim bluffs, that towered so gray and ghostly silent, could tell many a tale of lurking warriors, of desperate fights and massacres."[19]

Sam Smith of the Rocky Mountains clearly saw himself as a successor to the legendary mountain men who had first ventured into the Green River country. As in the western tradition (or at least the myth), he came to a bad end, disappearing in the late 1880s. When his body was found, it was assumed that he had been murdered. He, like Hank Clifford, John Chew, or John Reed, had lived every day in that world of fights and massacres. It would be fascinating to know what these men thought about the land they lived in, or the old ways of life that were slowly dissolving as "civilization" spread west along with the telegraph and Judge Carter's grand piano.

If the Yale students saw themselves as part of a great adventure that was romantic, at least with a small "r," Leidy, Cope, and Marsh were made of sterner stuff, or they put a more sober face on things. The West, to them, was a place to quarry out facts. It was a place where ambition ruled, rather than the soul. For an American scientist, the appropriate response to the magnificent geological landscapes was rational rather than emotional. They were part of the America that had been built on reason and had grown out of the Enlightenment in a spirit of philosophical, scientific, and ordered freedoms.[20] Very little in their diaries, letters, and formal writings suggests that they were particularly moved by the grandeur of the landscapes in which they worked. The scientific way was different, and they had been well indoctrinated by people like Thomas Henry Huxley—an archenemy of anything suggesting the vague sentimentality of the Romantic Movement. They did not see it as their role to marvel at the mysteries of nature (let alone find God there); their mission was to describe, classify, and rationalize nature.

To be sure, Cope and Marsh in their writings show an appreciation of the simple beauties of the western landscapes (although it was evidently difficult for anyone to find western Kansas soul-inspiring). Cope's field notebooks are filled with sketches of the New Mexico mountains that are quite beautiful in themselves. But Cope and Marsh did not love the West—either the land or its peoples—in the way that Hayden or Dutton did. When it came to science they were both intensely serious men, bent on a mission, and with no time for frivolity. Their work was hard manual labor. When they shot game, it was for food rather than sport. They also carried considerable responsibility, first and foremost for getting their parties safely from place to place. They simply were not romantics in any sense.

Similarly, William Berryman Scott was a highly focused (and rather dull) man, writing in his memoirs: "It was no yearning for the 'open spaces,' no desire for the 'simple life,' that sent me to the West year after year. The discomfort, camp life, the pains of rheumatism and sunburn, the difficulty, often the impossibility of cleanliness, to say nothing of the separation from my family, combined to make these expeditions a hardship, but . . . later there was compensation in the evident delight of discovery, in pushing forward the frontiers of knowledge."[21]

This image of a dogged pursuit of the facts has led at least one historian to the harsh conclusion that, although Darwinism gave paleontology an intellectual hard edge, for the most part "in the spirit of American sciences generally, the palaeontologists shied away from fundamental questions, fixed their attention on immediate, visible goals, and excelled as technicians, notable for their skill, daring and ingenuity in preserving, restoring, and mounting specimens."[22] This might have been true of Leidy, but for Cope, Hayden, and Marsh, at least, this simply was not the case. Not only did they have an acute sense of the sciences they were creating, they had contributed significantly by framing and answering "fundamental questions" in both geology and evolution. Their work is undervalued because it was not quantitative or experimental and therefore could be characterized as old-fashioned.

Few, if any, sciences have been as successful as paleontology in remaining intellectually serious and yet broadly accessible at the same time. The major factor that sets paleontology aside from other sciences (and makes it easy to dismiss as a higher version of stamp collecting rather than "real" science) is its transparency and accessibility to the public. A great deal of its popularity also arises from the mysteries inherent in the fossils themselves, especially the dinosaurs, paradoxically both strong, powerful, and dangerous but also small-brained, very safe, and very dead.[23]

Some of the accessibility of the science of fossils as a whole also comes from the image that the new generations of bone hunters helped create for themselves. In the romance of wild places, the paleontologist becomes a player in a world that seems glamorous and exciting. After the discovery of fossils in the old West, important collections were no longer made by European gentlemen in suits and ties (perhaps having removed the jacket) directing a couple of workmen in a small quarry in England or New Jersey. Instead, fossil collecting had become prospecting. A man with a horse and a pick—and of course a rifle—could venture out west and, like his gold-seeker cousins, bring back untold wealth from the rocks.

A new paleontological tradition developed, therefore, after the turn of the twentieth century and involved a dual rubric: the formal laboratory science typified by the growth of the research universities, and fieldwork evoking the dying embers of the Romantic Movement. As a result, the prevailing popular image of the modern paleontologist is almost schizophrenic—both a contemporary scientist and a last Romantic. He (increasingly also she) is a scholar and also a rugged individualist, the noble explorer who pits himself against the wilderness and brings back treasure. As someone whose life in the field is inspired and powerfully determined by the land itself, he is perhaps as much Hank Clifford and Sam Smith as he is Othniel Charles Marsh; he is certainly the dashing Edward Drinker Cope and Roy Chapman Andrews, the twentieth-century explorer of China and finder of the world's first fossil dinosaur eggs. There is perhaps also a touch of Teddy Roosevelt. All in the name of, and pursuit of, science.

No matter that the vast proportion of paleontology has come to be conducted in less than glamorous conditions and concerned with distinctly less dramatic organisms than "sixhorned" mammals and huge dinosaur bones. No matter that it has become largely a laboratory science; paleontology is still associated in the public mind with a romantic aura of rugged individualism and enterprise, richly rewarded. And indeed, every summer, professors from the great scholarly institutions throw off their jackets and ties and exchange them for the casual shirts, jeans, and boots of the prospector. Each autumn they bring back their fossils to the laboratory. And still there are very many questions to be answered: from the origins of life itself to the diversification of our prehominid ancestors.

For the public and for many paleontologists, professional or amateur, fossils continue to represent a happy fusion between nineteenth-century romanticism and the cold hard clarity of contemporary science. Where the steamboat and the railroads opened up the American West for bone hunters in the nineteenth century, in the twentieth the internal combustion engine and four-wheel drive have completed the task. Where a horse could go in days, a jeep can go in hours. A jet plane can get you to that jeep anywhere in the world in less than twenty-four hours. If there are no roads at all, then a helicopter will get you in and out. If the fossil in its plaster jacket is too heavy for humans to shift, that helicopter can hoist it to safety. The Arctic and the Antarctic have been successfully prospected for fossils, as have the interior of China and the tropics of South America and Southeast Asia.

Collecting and studying fossil vertebrates is now a truly international effort, with arguably the most exciting discoveries, such as tiny dinosaurs with protofeathers living side by side with large predatory mammals, currently being made in China. Its intellectual base has vastly extended. A glance at the contents of modern journals like *Nature*, *Science*, and *Paleobiology* would cause someone like Cope or Marsh to rub his head in bewilderment—what do all those mathematical formulae have to do with fossils?

The legacy of the mastodon lives on: we still study fossils to learn what they can tell us about how extinct organisms once lived, and for what they can tell us about the patterns and rhythms of life throughout the history of the earth. Were Jefferson alive now, he would be surprised to learn that we have long since accepted the fact of extinction and of evolutionary change in life over vast eons of geological time; but I believe that when he examined the evidence he would be persuaded. Most of all, Jefferson would recognize how the lure of new treasures lurking in unexplored lands still draws us on. We still are fascinated by the discovery of large and ferocious-looking fossil creatures; finding the biggest and the best is still a matter of constant media attention. And while paleontology may have long since transcended national and international boundaries, at its heart there still remains the image of the lonely explorer heading west with a mule and a pick, opening up new lands, and finding wonderful things.

Appendix A
The Geological Column

The familiar geological table and associated geological time scale of textbooks has a long history. If we had a perfect sequence of rocks in one place we could drill down and progressively find modern soils at the surface and, at the very bottom, rocks from the Precambrian that are nearly 2 billion (2,000 million) years old. In between we could trace the sequence of strata to which familiar names have been applied over the years: (from the top down), Quaternary (Holocene and Pleistocene), Cenozoic (Pliocene, Miocene, Oligocene, Eocene, Paleocene), Mesozoic (Cretaceous, Jurassic, and Triassic), and Paleozoic (Permian, Pennsylvanian, Mississippian, Devonian, Silurian, Ordovician, Cambrian), with the Precambrian underlying all. In the sides of the Grand Canyon one can find most of those from the Triassic downward.

These names obviously have been coined from a variety of sources: some refer to relative ages, or to classic places where they outcrop or were first recognized, and others are named for ancient European tribes (for example the Silures and Ordovices). The story of how this sequence of structures came to be recognized and named could occupy many a book in itself, and it continues as more and more details emerge about very ancient life in the Precambrian rocks.

When William Maclure produced his first geological map of America in 1808 he used a terminology that was still actively being revised. The simple early classification—Primary, Secondary, Diluvial, Alluvial—that Maclure used was the product of a number of other European geologists, notably Abraham Gottlob Werner (1750–1817) at the School of Mines,

Freiberg. This was at first a classification based solely on rock types (defined by their mineral makeup). Very quickly, the system was also defined in terms of origins. Primitive rocks (our Precambrian) were those unstratified granites and crystalline rocks that were thought to have been set in place by God at Creation. Secondary rocks had been much altered by later natural processes in the earth's surface. They contain fossils. Diluvial rocks were a set of sediments laid down by the Flood of Noah and therefore contain elements of both primary and secondary rocks that have been disturbed and redeposited. The superficial Alluvial layers on the surface of the earth were basically soils and sediment created very recently by erosion and deposited by rivers, streams, and seas. Werner later added an intermediary Transitional category, lying between the Primary and Secondary. The Transition Series was thought to lack fossils but, because the rocks were stratified, they must have been modified from the Primary rocks and were in every sense (time and composition) transitional between the two.

In the early part of the nineteenth century, three British geologists, Henry de la Beche, Adam Sedgwick, and Roderick Impey Murchison, teased the Transitional Series apart. They found that all of the Transition involved fossils, although they were fossils of very primitive kinds compared with living organisms or even the rich fossil beds of the Coal Measures. Investigating very ancient-seeming rocks in England and Wales, they recognized three distinct elements (Cambrian, Silurian, and Devonian) within the Transition; a fourth (Ordovician) was added later.

The Carboniferous (Pennsylvanian and Mississippian in North America), which includes the Coal Measures, was formally defined by William D. Conybeare and William Phillips in 1822, and the Permian by Murchison in 1841. The Triassic was defined on the basis of German deposits by Friedrich August von Alberti in 1834; the Tertiary was named by Charles Lyell in 1833, and the Quaternary by Jules Desnoyers in 1829. The last major unit to be formally named was the Cretaceous, by Jean-Baptiste d'Omalius d'Halloy in 1822, on the basis of the chalk and greensands of northern France. With this, the basic components of the fossil-bearing parts of the geological column as we know it were largely in place.

Soon, all rocks and fossil occurrences could be placed within this sequence of deposition. The study of geology, however, was hampered for a long time by lengthy and bitter arguments about the age of the earth and the manner of formation of these structures. For example, had basalts been precipitated underwater, as Werner thought, or had they been formed under the influence of heat and volcanoes, as promoted by the Scottish geologist James Hutton? Several American authors in the first part of the nineteenth century noted wryly that more scholars in northern Europe (where there are no active or recently active volcanoes) opted for the watery Neptunist model, while those of southern Europe were Vulcanists. Interestingly, most American geologists tended to the Neptunist view, partly out of respect for Werner's immense authority in the field of practical mineralogy, and partly because it fit better with the biblical version of early earth history when, before Creation, all was a watery void. The only real answer could, of course, be that both water and heat have been (and are) involved in the formation and evolution of the earth's crust.

Appendix B
Leidy on Evolution

In the introduction to his 1853 monograph "A Flora and Fauna Within Living Animals" (written in 1851), Joseph Leidy laid out his thoughts on the origin of life on earth. This was intended not as an exposition of evolutionary theory but as an essential introduction to his treatment of some of the simplest forms of life then known—microscopic organisms living parasitically and symbiotically in animal tissues. His writing, however, gives us a nice view of the state of thinking about the earth and evolution at the very midpoint of the century.

Leidy's prime target was the theory of spontaneous generation, a concept that had once seemed self-evident. If one puts dry straw into water and leaves it, soon the water will abound in tiny animals and plants; they were first given the obvious name of "Infusoria." Similarly, maggots would apparently generate themselves in old meat. The notion was that these creatures had somehow come into being out of nothing. The problem with this explanation was that "spontaneous generation" provided an exception to the common observation that all life (at least all macroscopic life) proceeds from preexisting life. But various early experimenters such as Francesco Redi (in the seventeenth century) and Lazaro Spallanzani (in the eighteenth) had already shown that boiling water prevented any such generation. Meat that remained covered (keeping out flies) did not produce maggots. Finally, Louis Pasteur—the most familiar name among these—in the nineteenth century (after Leidy's monograph) showed that there was no such thing as spontaneous generation, only contamination by the spores or eggs of existing creatures.

Leidy did not believe in spontaneous generation, but there remained the philosophical difficulty that, unless something like evolution has occurred in the history of life, the appearance of new forms in the fossil record must have come about through some kind of spontaneous generation—with God as the obvious prime mover of such creation. At the very least, if (as in the deist view) God had merely started nature going and then left it to its own devices according to laws that he had set in place, there must have been an initial case of non-natural (spontaneous) generation. And, if that one case is to be granted, why not more? In denying spontaneous generation, therefore, Leidy had to tread carefully around the subject of religion and evolution.

Leidy first set out firmly the postulate that there had been a time when there was no life on earth. "The oblate spheroid form of the earth, and the physical constitution of its periphery, indicate that it was once in a molten state." There had been an "incalculably great" period "before the earth-crust . . . had sufficiently cooled by the radiation of its heat for living beings to become capable of existing on its surface." Leidy showed that the temperature must have fallen to below 65 degrees centigrade, the point at which proteins like egg albumen coagulate, before this could happen.

Leidy saw life as originating "in a formless liquid state. The first step in organization is the appearance of a solid particle. An aggregation of organic particles constitutes the spherical, vesicular, nucleolated, nucleated body, the organic cell, the type of the physical structure or organization of living beings." Leidy further argued that the basic elements of life are "susceptible of a great variety of modifications, within a definite range, without its destruction." Each species then is a function of particular environmental conditions. However, each species is also "an immutable organic form. The study of the earth's crust teaches us that very many species of plants and animals became extinct at successive periods, while other races originated to occupy their places. This was probably the result, in many cases, of a change in exterior conditions incompatible with the life of certain species and favorable to the primitive production of others. . . . Probably every species has a definite course to run in consequence of a general law: an origin, an increase, a point of culmination, a decline, and an extinction."

Leidy preceded his statement that "there appear to be but trifling steps from the oscillating particle of inorganic matter . . . gradually up to the higher orders of life" with the following argument concerning the origin and proliferation of life on earth. "Of the life, present everywhere with its indispensable conditions, and coeval in its origin with them, what was the immediate cause? It could not have existed upon earth prior to these essential conditions; and is it, therefore, the result of these?" The possible causes were: (a) "pre-existing natural conditions," (b) a natural beginning followed by "transmutation under the influence of varying exterior conditions," or (3) "all species in all times originated through supra-natural agency." Leidy's conclusion was that, as a supranatural agency was "only, of course, an inference, in absence of all facts; and if living beings did not originate in this way, it follows they are the result of natural conditions."

In all this, the combination of Leidy's positing "transmutation" and his dismissing of supranatural agency as an "inference," in favor of an entirely natural origin of life or of non-life, was very daring for 1850s Philadelphia, and it was entirely out of character for the man when compared with the reticence of his later, paleontological writings. Once bitten, twice shy. The loss, undoubtedly, is ours.

Appendix C
Cope on Evolution

Whatever criticisms one may make of Edward Drinker Cope's style as a scientist and a person—hasty, aggressive, careless—one always comes back to his essential brilliance, the vast range of topics that he mastered, and the depth of his knowledge in all of them. Nowhere is this more evident than in his two books on evolution, *The Origin of the Fittest* and *The Primary Factors of Organic Evolution*. The former, a series of essays written over a twenty-year period, allows us to see the development of his ideas. In both books he is not afraid to venture where the angels fear to tread—the relation between evolution and metaphysics. He also focused intently on a primary difficulty in Darwin's theory of natural selection.

In Cope's view, Darwinian natural selection and sexual selection presented no difficulties. The key issue in evolution was the origin of the variation on which selection acted. As there was then no understanding of the nature of genetics, the laws of inheritance, or the mechanisms of morphogenesis (by which the fertilized egg turns into an adult), many authors like Cope probed into the available facts to articulate principles (always given fancy Greek names, after the fashion of the day) that putatively underlay morphological adaptation. One was kinetogenesis, which involved the relationship between mechanical function and structure—the shape of teeth for chewing, limb joints for moving, and so on. Physiogenesis posited a causal relation between size and shape, and between the senses and animal coloration (Cope had a lifelong fascination for blind cave-fish, for example). In development, change could be effected by relative acceleration or retardation of the maturation of different organ systems (length of limbs in horses, for example). Any increase in one area, however, was accompanied by a reduction elsewhere. Another way of dividing up processes in evolution was to distinguish the anagenetic (exclusively vital) from the catagenetic (physical and chemical). Another was to postulate the effects of a "growth force" (bathmism, or bathymogenesis) whose effects "directed" evolution.

In the introduction to *The Primary Factors of Organic Evolution,* Cope listed some opposed viewpoints in contemporary theory. We can rephrase these as questions: is variation *directed* or *random;* are variations caused by the interaction of the organism and its environment; are characteristics acquired during the life of an organism inheritable; are movements of the organism caused or *directed by sensation* and other conscious states or *instinctual* (selected); is the rational mind developed by experience, through memory and classification, or is mind a product of "natural selection from multifarious mental activities"?

A key issue for Cope was the evolution of the mind and free will, where he saw a continuum between man and the lower animals. In many of the issues he pinpointed, what once seemed a dichotomy was identified as a more complex network of causes and effects. What stands out is how late-nineteenth-century scholars like Cope who grappled with these problems turned, inevitably it seems, to the works of Jean-Baptiste de Lamarck. One of the difficulties with Darwin's theory was that it depended on the chance origin of (it was presumed) minute variations within populations—the sort of differences that one can see any day by comparing the people sitting in a bus. Was that a strong enough phenomenon to drive the accumulation of the differences between, say, a reptile and a bird, or a fish and an amphibian? The Lamarckian view included the theory that variation was directed and that characters *acquired* during an individual's lifetime of experience, and even through volition, somehow became incorporated in the genetic foundation of the next generation.

If this were true, for example, a blacksmith's children would have large muscles. This theory had its attractions, and Darwin himself, in the six subsequent editions of *On the Origin of Species,* slipped in a number of Lamarckian references, if only as a sop to his critics. Cope well knew that there was strong evidence against the inheritance of characteristics acquired during life by use, disuse, or the power of the will. He lists, for example, the persistence of the foreskin even after generations of circumcision. On the other hand, he always circled back to the simple principle that all characters were at one time acquired (epigenesis).

The point of these comments is not to expose Cope to criticism for his views, but to record his mastery of a subject and indeed the originality of his thought. Many modern biologists have spent an entire career thinking about evolution without producing books of such authority, but for Cope this was merely one of many intellectual interests. And out of his thinking have come two general propositions that have stood the test of time remarkably robustly. Both derive from his detailed knowledge of the fossil record. Cope's Law states that in evolution, a new group will arise from an ancestor that is more generalized than specialized. A corollary of this is that extreme specialization in evolution is always a dead end. While this law is open to semantic quibbles (what is advanced, anyway?) and even the charge of circularity, Cope's Rule has stood the test of time even better. This rule states that, during the lifetime of a lineage, in general, there will be size increase. Ironically, prime examples of the law and the rule would be the evolution of huge one-toed modern horses from their dog-sized, five-toed Eocene ancestors, as documented by his rival Marsh.

Notes

Abbreviations

AJS   *American Journal of Science*
AMNAT   *American Naturalist*
ANS   Academy of Natural Sciences, Philadelphia, Archives
             of the Library
APS   American Philosophical Society, Philadelphia
JANS   *Journal of the Academy of Natural Sciences*
PANS   *Proceedings of the Academy of Natural Sciences*
PAPS   *Proceedings of the American Philosophical Society*
Phil. Trans. *Philosophical Transactions of the Royal Society of London*
TAPS   *Transactions of the American Philosophical Society*
USGS   United States Geological Survey
Yale   Yale University Library, Manuscripts and Archives Division

The collected letters and papers of Othniel Charles Marsh are held in the Manuscripts and Archives Division, Yale University Library, and are cited here with the single word Yale. Some citations are from Charles Schuchert and Clara Mae LeVene, *O. C. Marsh, Pioneer in Paleontology* (Yale University Press, 1940).

The scientific correspondence of Edward Drinker Cope is largely in the Central Archives, Library of the American Museum of Natural History. Some citations are from Henry Fairfield Osborn, *Cope: Master Naturalist* (Princeton University Press, 1931). The letters of Ferdinand Vandiveer Hayden to Joseph Leidy are in the Archives of the Library of the Academy of Natural Sciences, Philadelphia, here designated ANS.

Quotations from the letters of Benjamin Franklin are from *The Papers of Benjamin Franklin,* ed. Leonard Labarree and others (American Philosophical Society and Yale University Press, 1959–2003), cited here as *Franklin Papers.*

Quotations from the field journals of Arthur Lakes are from Michael F. Kohl and John S. McIntosh, eds., *Discovering Dinosaurs in the Old West: The Field Journals of Arthur Lakes* (Smithsonian Institution Press, 1997), designated as Lakes Journals.

In all quotations from original documents, I have retained the authors' spelling and punctuation, no matter how eccentric.

ONE

Fossil Hunters on the Frontier

Epigraph: George Catlin, *Letters and Notes on the Manners, Customs, and Condition of the North American Indians,* Letter no. 9, Yellow Stone, Missouri, 1844.

1. Texas and New Mexico were added between 1845 and 1848; Oregon, Idaho, and Washington in 1846; Nevada, (most of) Arizona, and California in 1848.

2. For background reading see, among many other works, Bernard De Voto, *Across the Wide Missouri* (Boston: Houghton Mifflin, 1947); Henry Nash Smith, *Virgin Lands* (Cambridge: Harvard University Press, 1950); William Goetzmann, *New Lands, New Men* (New York: Viking, 1986); William Goetzmann, *Exploration and Empire* (Austin: Texas State Historical Association, 2000); D. W. Meinig, *The Shaping of America,* vols. 2 and 3 (New Haven: Yale University Press, 1993, 1998).

TWO

Big Bone Lick

Epigraph: William Hunter, "Observations on the Bones Commonly Supposed to Be Elephant's Bones, Which Have Been Found near the River Ohio, in America," *Phil. Trans.,* vol. 58 (1769): 34–45.

1. Robert Kincaid, *The Wilderness Road* (New York: Bobbs-Merrill, 1947); Roberta Ingles Steele and Andrew Steele Ingles, eds. *Escape from Indian Captivity: The Story of Mary Draper Ingles and Son Thomas Ingles* (Radford, Va.: privately printed, 1982); James Alexander Thom, *Follow the River* (New York: Ballantine, 1981).

2. Louis Figurier, *Le Terre Avant le Deluge* (Paris, 1863), 300.

3. Interestingly, with the modern ban of the sale of elephant ivory, aimed at cutting down the poaching of rapidly dwindling natural populations, fossil ivory from Siberia is once again starting to come on the market.

4. Franklin to Robert Hunter Morris, June 13, 1756, *Franklin Papers,* vol. 6: 455–56.

5. William Franklin to Benjamin Franklin, December 17, 1765, *Franklin Papers,* vol. 12: 403–6; Samuel Wharton to Benjamin Franklin, September 30, 1767, *Franklin Papers,* vol. 14: 257–60.

6. It was Collinson who largely introduced Franklin's experiments about electricity to the Royal Society in London.

7. Alan W. Armstrong, in Nancy E. Hoffmann and John C. Van Horne, eds., *America's Curious Botanist: A Tercentennial Reappraisal of John Bartram, 1699-1777,* Memoirs of the American Philosophical Society, vol. 249 (Philadelphia: American Philosophical Society, 2004), 23–51.

8. William Darlington, *Memorials of John Bartram and Humphrey Marshall* (Philadelphia: Lindsay and Blakiston, 1849), 238.

9. This second letter (in ibid., 239) makes it unclear whether Collinson had received actual specimens from Greenwood and Croghan.

10. Wright to Bartram, August 22, 1762, quoted in George Gaylord Simpson, "The Beginnings of Vertebrate Paleontology in North America," PAPS, vol. 86 (1942): 140.

11. Bartram to Collinson, December 3, 1762, in Darlington, *Memorials,* 243.

12. "The Journal of Col. Croghan," *Monthly American Journal of Geology and Natural Science,* vol. 1 (1831): 257-72.

13. Max Savelle, *George Morgan, Colony Builder* (New York: Columbia University Press, 1932), 28.

14. Collinson, "A List of the Teeth and Bones Sent over by George Croghan, Esquire, February 7, 1767, from Philadelphia," *Phil. Trans.,* vol. 57 (1768): 467. Franklin stated elsewhere that he had received four teeth and that "one of [the tusks] is 6 Feet long & in the thickest part near 6 inches in diameter."

THREE
Franklin, Jefferson, and the Incognitum

1. Benjamin Franklin to George Croghan, August 5, 1767, *Franklin Papers,* vol. 14, ed. Leonard W. Labaree (New Haven: Yale University Press), 221-22.

2. Quoted in George Gaylord Simpson, "The Beginnings of Vertebrate Paleontology in North America," PAPS, vol. 86 (1942): 134.

3. *Philosophical Transactions of the Royal Society,* vol. 29 (1714): 62-71.

4. Peter Collinson, "An Account of Some Very Large Fossil Teeth, Found in North America," *Phil. Trans.,* vol. 57 (1768): 464-67; "Sequel to the Foregoing Account of the Large Fossil Teeth," *Phil. Trans.,* vol. 57 (1768): 468-69.

5. William Hunter, "Observations on the Bones Commonly Supposed to Be Elephant's Bones, Which Have Been Found near the River Ohio, in America," *Phil. Trans.,* vol. 58 (1769): 34-45.

6. To l'Abbe Chappe d'Auteroche, January 31, 1768, *Franklin Papers,* vol. 15, ed. William Wilcox (1972), 33-34.

7. Thomas Jefferson, *Notes on the State of Virginia; Written in the Year 1781, Somewhat Corrected and Enlarged in the Winter of 1782, for the Use of the Foreigner of Distinction, in Answer to Certain Queries . . .* (Paris: privately printed, 1782, American edition 1784).

8. Five to six times the volume of an elephant would mean it was 1.75 times the size in its linear dimensions.

9. "Description of BONES etc. Found near the RIVER OHIO," *Columbian Magazine,* vol. 1, part 3 (1786): 106.

10. "Memoir on the Extraneous Fossils Designed to Show that They Are the Remains of More than One Species of Non-descript Animal," TAPS, vol. 4 (1799): 510–18.

11. Claudine Cohen, *The Fate of the Mammoth* (Chicago: University of Chicago Press, 2002).

FOUR
Jefferson's "Great-Claw" and a World About to Change

1. "A Memoir on the Discovery of Certain Bones of a Quadruped of the Clawed Kind in the Western Parts of Virginia," TAPS, vol. 4 (1799): 246–60.

2. The idea that it was "possibly of the lion kind" was suggested to him by his friend John Stuart when he sent him the bones; Julian P. Boyd, "The Megalonyx, the Megatherium, and Thomas Jefferson's Lapse of Memory," PAPS, vol. 102 (1958): 420–35.

3. "A Description of the Bones Deposited by the President in the Museum of the Society, and Represented in the Annexed Plates," TAPS, vol. 4 (1799): 526–31.

4. George Gaylord Simpson, one of the most brilliant vertebrate paleontologists of the twentieth century, said that Wistar's paper was "a model of cautious, accurate scientific description and inference, an achievement almost incredible in view of the paleontological naiveté of his associates and of the lack of comparative materials. The objective part of the paper is so clear, complete, and correct that it has never been significantly bettered for the same or similar subjects." Simpson, "The Beginnings of Vertebrate Paleontology in North America," PAPS, vol. 86 (1942): 153.

5. The specimen was actually from Lujan, Argentina. The best description of its discovery and description is in George Gaylord Simpson, *Discoverers of the Lost World* (New Haven: Yale University Press, 1984), 3–12.

6. Boyd, "The Megalonyx." The mounted skeleton of *Megatherium* in a life pose was the first such reconstruction ever attempted, and the illustration of the complete fossil was also a first. See also Martin J. S. Rudwick, *Bursting the Limits of Time* (Chicago: University of Chicago Press, 2005), 357.

7. Barbe-Marbois later was Napoleon's minister of the treasury, and it was he who negotiated with Jefferson the Louisiana Purchase of 1803.

8. Thomas Jefferson, *Notes on the State of Virginia; Written in the Year 1781, Somewhat Corrected and Enlarged in the Winter of 1782, for the Use of the Foreigner of Distinction, in Answer to Certain Queries* . . . (Paris: privately printed, 1782, American edition 1784).

9. Georges-Louis Leclerc de Buffon, *Époques de la Nature,* supplement 5 to *Histoire Naturelle* (Paris, 1778), 432. It was in this work that Buffon fully espoused the idea of an extremely ancient, changing world (and hinting that his seven major phases—Époques—of earth history might correspond to the seven "days" of creation in Genesis).

10. William Robertson, *The History of America; the Twelfth Edition, in Which Is Included the Posthumous Volume, Containing the History of Virginia, to the Year 1668; and of New England, to the Year 1652*, 4 vols. (London, 1812).

11. Howard C. Rice Jr., "Jefferson's Gift of Fossils to the Museum of Natural History in Paris," PAPS, vol. 95 (1951): 597–627. See also Caspar Wistar, "An Account of Two Heads, Found in the Morass, Called the Big Bone Lick, and Presented to the Society, by Mr Jefferson," TAPS, new series, vol. 1 (1809): 375–80.

FIVE

The First American Dinosaurs

1. William Parker Foulke, "Statement Respecting the Fossil Bones, Shells, and Wood Presented by Him to the Academy this Evening," PANS, vol. 10 (1858): 213.

2. Glenn Matlack, personal communication.

3. Charles Willson Peale, Diary 15, *Peale Papers*, APS; also *The Selected Papers of Charles Willson Peale and His Family*, ed. Lillian B. Miller, vol. 2, part 1 (New Haven: Yale University Press, 1988), 173.

4. Leonard Warren, *Joseph Leidy: The Last Man Who Knew Everything* (New Haven: Yale University Press, 1998), 56.

5. *Peale Family Papers, Memoranda of the Peale Museum*, Historical Society of Pennsylvania.

6. E. S. Lull, "Triassic Life of the Connecticut Valley," *Connecticut State Geological and Natural History Survey*, Bulletin 24 (1915).

SIX

Fossils and Show Business

Epigraph: Rev. Nicholas Collin, "An Essay on Those Inquiries in Natural Philosophy, Which Are at Present Most Beneficial to the United States of North America," TAPS, vol. 3 (1793): xv–xxvii.

1. TAPS, vol. 4 (1799): xxxvii–xxxix.

2. Toby A. Appel, "Science, Popular Culture, and Profit: Peale's Philadelphia Museum," *Journal of the Society for the Bibliography of Natural History*, vol. 9 (1980): 619–34.

3. Whitfield Bell, "A Box of Old Bones: A Note on the Identity of the Mastodon, 1766–1806," TAPS, vol. 93 (1849): 169–76.

4. *Exhumation of the Mastodon* (Collection of the Maryland Historical Society).

5. Charles Willson Peale, "Skeleton of the Mammoth Is Now to Be Seen at the Museum," *The Selected Papers of Charles Willson Peale and His Family*, ed. Lillian B. Miller, vol. 2, part 1 (New Haven: Yale University Press, 1988), 378.

6. *Aurora*, February 18, 1802.

7. Paul Semonin, *American Monster* (New York: New York University Press, 2000).

8. Charles Willson Peale to Sir Joseph Banks, June 22, 1802, *Peale Papers*, ed. Miller, vol. 2, part 1, 436.

9. *The American Citizen and General Advertiser*, May 18, 1802, quoted in Semonin, *American Monster*.

10. John Godman, *American Natural History* (Philadelphia: Carey and Lea, 1826-28), vol. 2, 67.

11. Anonymous, *Climenole*, no. 8, Philadelphia, March 24, 1804.

12. John Adams to F. A. van der Kemp, September 1, 1800, *Adams Papers*, Massachussetts Historical Society.

13. *The Port Folio* (Philadelphia, 1802), vol. 2: 71-73.

14. *The Embargo, or Sketches of the Times* (Boston: printed for the author, 1808).

15. Lillian Miller, "Charles Willson Peale as History Painter," *American Art Journal* 13 (1981): 47-68; Katherine C. Woltz (work in progress).

16. George Gaylord Simpson and H. Tobien, "The Rediscovery of Peale's Mastodon," PAPS, vol. 98 (1954): 279-80.

17. Madison to Barton, letter to M. Lacépède of Paris, *Philadelphia Medical and Physical Journal*, vol. 2 (1805): 58-60; Benjamin Smith Barton, "Facts, Observations, and Conjectures, Relative to the Elephantine Bones (of Different Species) that Are Found in Various Parts of North-America; In a letter from the Editor to Mr G. Cuvier, of Paris," *Philadelphia Medical and Physical Journal*, vol. 3, first supplement (1808): 22-35.

18. Barton, "Facts, Observations," 26.

19. David A. Wells, *First Principles of Geology* (New York, 1861), 316. This vindicated Collinson's information, dismissed by Bartram, that some at least of the skeletons at Big Bone Lick had been preserved upright. For modern data see A. Dreimans, "The Extinction of Mastodons in Eastern North America: Testing a New Climatic-Environmental Hypothesis," *Ohio Journal of Science*, vol. 68 (1968): 257-72.

SEVEN

Fossils and Extinction

Epigraph: John Adams to F. A. van der Kemp, January 26, 1802, *Adams Papers*, Massachussetts Historical Society.

1. Robert Plot, *The Natural History of Oxford-shire* (Oxford, 1677).

2. Georges Cuvier, *Recherches sur les Ossemens Fossiles de Quadrupèdes* [Researches on the Fossil Bones of Quadrupeds] (Paris, 1812). The "Preliminary Discourse" was later published in English as *Essay on the Theory of the Earth*, with translation and notes by Robert Jameson; quotations here from the 1818 New York edition.

3. Charles A. Miller, *Jefferson and Nature* (Baltimore: Johns Hopkins University Press, 1988).

4. Keith Thomson, *Fossils: A Very Short Introduction* (Oxford: Oxford University Press, 2005).

5. Keith Thomson, *Before Darwin: Reconciling Science and Religion* (New Haven: Yale University Press, 2005).

6. Thomas Burnet, *Sacred Theory of the Earth* [orig. *Telluris Theoria Sacra*, 1681], trans. Basil Willey (Carbondale: Southern Illinois University Press, 1965).

7. James Hutton, "Abstract of a Dissertation" (Royal Society of Edinburgh, April 1785), expanded as "The Theory of the Earth," *Transactions of the Royal Society of Edinburgh*, vol. 1 (1788): 209–304.

8. Benjamin Franklin, "Conjectures Concerning the Formation of the Earth," TAPS, vol. 3 (1793): 1–5.

9. Alan Cutler, *The Seashell on the Mountaintop* (New York: Dutton, 2003).

10. Martin J. S. Rudwick, *Bursting the Limits of Time* (Chicago: University of Chicago Press, 2005).

11. Charles Lyell, "Presidential Address," *Quarterly Journal of the Geological Society of London*, vol. 7 (1851): lxxiii.

12. Thomson, *Before Darwin*.

EIGHT

Mary Anning's World

1. Eric Buffetaut, *A Short History of Vertebrate Palaeontology* (London: Croom Helm, 1987).

2. Philippe Taquet and Kevin Padian, "The Earliest Restoration of a Pterosaur and the Philosophical Origins of Cuvier's *Ossemens Fossiles*," *Comptes Rendues*, vol. 3 (2004): 157–75.

3. Martin J. S. Rudwick, *Georges Cuvier, Fossil Bones, and Geological Catastrophes* (Chicago: University of Chicago Press, 1997).

4. Gideon Mantell, *Fossils of the South Down or Illustrations of the Geology of Sussex* (London, 1822).

5. Georges Cuvier, *Recherches sur les Ossemens Fossiles: Où l'on Rétablit les Charactères de Plusieurs Animaux dont les Révolutions du Globe ont Détruit les Espèces* [Researches on Fossil Bones: Describing the Character of Many Animals Destroyed by Revolutions of the Globe], 3rd ed. (Paris, 1825).

6. Edward Lhwyd, *Lithophylacii Botannici Ichnographia* (Oxford, 1699). Lhwyd described some fossil teeth from Oxfordshire, one of which was probably *Megalosaurus* and another was probably from the sauropod now known as *Cetiosaurus*.

7. P. E. Olsen, A. R. McCune, and K. S. Thomson, "Correlation of the Early Mesozoic Newark Supergroup by Vertebrates, Principally Fishes," *American Journal of Science*, vol. 282 (1982): 1–44.

8. Some authors, for example Patsy Gerstner in "The 'Philadelphia School' of Paleontology: 1820–1845" (unpublished dissertation, Case Western Reserve University, 1967), have concluded that a strength of early American paleontology was that it was conducted by physicians rigorously trained in anatomy. I would argue that this is either false or insufficiently true. They lacked the forensic, analytical approach to anatomy that had distinguished Wistar and the geological base of Cuvier.

9. Ibid.

10. John D. Godman, "Description of a New Genus and New Species of Extinct Mammifierous Quadruped," TAPS, vol. 3 (1830): 478–85.

11. Patsy Gerstner, "Vertebrate Paleontology, a Nineteenth-Century Transatlantic Science," *Journal of the History of Biology*, vol. 3 (1970): 137–48.

12. Carl Zimmer, "The Equation of a Whale," *Discover Magazine*, vol. 19 (April 1998).

13. Richard Owen, "Report on the Missourium Now Exhibiting at the Eygptian Hall, with an Inquiry into the Claims of the Tetracaulodon in Generic Distinction," *Proceedings of the Geological Society of London*, vol. 3 (1843): 689–95.

14. Edward Hitchcock, "Ornithichnology: Description of the Foot Marks of Birds (Ornithichnites) on New Red Sandstone in Massachusetts," *American Journal of Science*, vol. 29 (1836): 307–40.

15. Edward Hitchcock, *Ichnology of New England* (Boston, 1858); *Supplement* (1865).

16. Isaac Lea, "On a Fossil Foot-Mark in the Red Sandstone of Pottsville, Schuylkill County, Penna.," TAPS, vol. 10 (1852): 302–17.

NINE

An American Natural Science

1. Featherstonhaugh's journal lasted only one year.

2. Benjamin Smith Barton, *A Discourse on Some of the Principal Desiderata in Natural History* (Philadelphia, 1807).

3. Barton, "Letter to M. Lacépède, of Paris, on the Natural History of North America; Philadelphia, October 31, 1802," *Philosophical Magazine*, vol. 22 (1805): 97–102, 204–11. Incidentally, Barton thought *Megalonyx*, with its large claws, was a polar bear.

4. Barton, *Desiderata*.

5. Benjamin Smith Barton, "Memorandum Concerning the Earthquakes of North America," *Philadelphia Medical and Physical Journal*, vol. 1 (1785): 60. The first to make the observation was David Rittenhouse, "Some Observations on the Structure of the Earth in Pennsylvania and the Adjoining Countries," *Columbian Magazine*, vol. 1 (1786): 49.

6. Thomas Nuttall, "Observations on the Geological Structure of the Valley of the Mississippi," JANS, vol. 2 (1821): 14–64.

7. Alexander Wilson, *American Ornithology; or, The Natural History of the Birds of the United States: Illustrated with Plates, Engraved and Colored from Original Drawings Taken from Nature,* vol. 3 (Philadelphia, 1811), xiii.

8. Letter to Samuel George Morton, April 3, 1830, quoted in Nathan Reingold, *Science in Nineteenth-Century America* (Chicago: University of Chicago Press, 1964), 34-38.

9. Map (no title), Bartram Papers, APS.

10. C. F. Volney, *A View of the Soil and Climate of the United States of America,* translated by C. B. Brown (Philadelphia, 1804).

11. William Maclure, *Observations on the Geology of the United States of America* (Philadelphia: Conrad, 1809). Also, revised edition, *Observations on the Geology of the United States of America; with Some Remarks on the Effect Produced on the Nature and Fertility of Soils, by the Decomposition of the Different Classes of Rocks; and an Application to the Fertility of Every State in the Union, in Reference to the Accompanying Geological Map* (Philadelphia, 1817, and TAPS, new series, vol. 1 [1818]: 1-91).

12. Maclure, *Observations* (1809 ed.), 10-21.

13. Simon Baatz, "Philadelphia Patronage: The Institutional Structure of Natural History in the New Republic, 1800-1833," *Journal of the New Republic,* vol. 8 (1988): 111-38; Patricia Tyson Stroud, "The Founding of the Academy of Natural Sciences of Philadelphia in 1812 and Its Journal in 1817," *Archives of Natural History,* vol. 2 (1995): 221-33.

14. John C. Greene, "The Development of Mineralogy in Philadelphia, 1780-1820," PAPS, vol. 113 (1969): 283-95.

15. Katherine Woltz, personal communication.

16. Benjamin Silliman, "Review of an Elementary Treatise on Mineralogy and Geology . . . by Parker Cleaveland," AJS, vol. 1 (1818): 35-56.

TEN

An American Geology

1. Robert Jameson, *Manual of Mineralogy: Containing an Account of Simple Minerals, and also a Description and Arrangement of Mountain Rocks* (Boston, 1816).

2. Patricia Tyson Stroud, *Thomas Say, New World Naturalist* (Philadelphia: University of Pennsylvania Press, 1992).

3. Thomas Say, "Observations on Some Species of Zoophytes, Shells, etc., Principally Fossil," AJS, vol. 1 (1818): 381-87.

4. Samuel G. Morton, "Geological Observations on the Secondary, Tertiary, and Alluvial Formations of the Atlantic Coast of the United States of America; Arranged from the Notes of Lardner Vanuxem," JANS, vol. 6 (1828): 59-71.

5. George P. Merrill, *The First Hundred Years of American Geology* (New Haven: Yale University Press, 1924), 94, 101; Walter B. Hendrickson, "Nineteenth-

Century State Geological Surveys: Early Government Support of Science," *Isis*, vol. 52 (1961): 357-71.

6. G. W. Featherstonhaugh, *Geological Report of an Examination Made in 1834 of the Elevated Country Between the Missouri and Red Rivers* (Washington, D.C., 1835).

7. David Dale Owen, *Report of a Geological Reconnaissance of the State of Indiana, Made in the Year 1837, in Conformity to an Order of the Legislature*, part 1 (Indianapolis, 1859).

8. David Dale Owen, *Report of a Geological Survey of Wisconsin, Iowa, and Minnesota; and Incidentally of a Portion of Nebraska Territory; Made Under Instructions from the United States Treasury Department* (Philadelphia, 1852).

9. Torrey to Hall, New York, July 21, 1841, quoted in Nathan Reingold, *Science in Nineteenth-Century America* (Chicago: University of Chicago Press, 1964), 166-67.

10. Hall to Silliman, March 28, 1842, quoted in Reingold, *Science*, 169. Hall wrote a similar letter to John Torrey a few days later; Merrill, *First Hundred Years*, 668.

11. Charles Lyell, *Travels in North America, in the Years 1841-2; with Geological Observations on the United States, Canada, and Nova Scotia* (New York: Wiley and Putnam, 1845).

ELEVEN
Bad Lands

1. William B. Scott, *The Leidy Commemorative Meeting* (Philadelphia: Academy of Natural Sciences, 1923), 39.

2. Leonard Warren, *Joseph Leidy: The Last Man Who Knew Everything* (New Haven: Yale University Press, 1998).

3. An earlier collection from the river bluffs at Natchez had been given to the American Philosophical Society in 1836 by William Henry Huntington; APS Minutes, April 1, 1836. In 1852 Leidy described the first North America lion from a mandible in that collection; TAPS, vol. 10, new series (1852): 319-21.

4. George Gaylord Simpson, *Discoverers of the Lost World* (New Haven: Yale University Press, 1984), 21-39.

5. PANS, vol. 3 (1846): 106-7; PANS, vol. 4 (1847): 262-66.

6. AJS (2), vol. 2 (1846): 288-89; amplified in a second paper in 1847, AJS (2), vol. 3 (1847): 248-50.

7. PANS, vol. 3 (1848): 322.

8. The larger group Titanotheria has representatives in America, Asia, and Africa.

9. They were legally married in 1859 in Peoria, Illinois; Edward Harris, *Up the Missouri with Audubon: The Journal of Edward Harris*, ed. John Francis McDermott (Norman: University of Oklahoma Press, 1947).

10. PANS, vol. 3 (1848): 47.

11. PANS, vol. 3 (1847): 315; PANS, vol. 4 (1848): 52.

12. Edward Hitchcock, "A New Theory of the Earth," *North American Review*, vol. 28 (1829): 265.

13. Joseph Leidy, "A Flora and Fauna Within Living Animals," *Smithsonian Contributions to Knowledge*, vol. 5 (1853): 5–67.

14. Dennis Murphy, personal communication, 2005.

15. Owen, *Report of a Geological Survey of Wisconsin*, 189.

16. Ibid., 192–201.

17. Frank Lloyd Wright and Lewis Mumford, *Thirty Years of Correspondence* (Princeton: Princeton Architectural Press, 1935).

18. Owen, *Report of a Geological Survey of Wisconsin*.

19. Scott, *Leidy Commemorative Meeting*, 41.

20. Alexander Culbertson had succeeded Kenneth McKenzie as superintendent of Fort Union in 1840.

21. Thaddeus A. Culbertson, "Journal of an Expedition to the Mauvaises Terres and the Upper Missouri in 1850," ed. John Francis McDermott, *Smithsonian Institution Bureau of American Ethnology, Bulletin*, vol. 147 (1952): 1–164.

22. Joseph Leidy, "The Ancient Fauna of Nebraska: or, A Description of Remains of Extinct Mammalia and Chelonia, from the Mauvaises Terres of Nebraska," *Smithsonian Contributions to Knowledge*, vol. 6 (1853): 1–126.

23. Joseph Leidy, "Notes of Some Remains of Fishes Discovered by Dr John E. Evans," PANS, vol. 8 (1856): 256–57.

24. JANS, vol. 7 (1869): 1–472; "Report of the Vertebrate Fossils of the Tertiary Formations of the West," in *Preliminary Report of the United States Geological Survey of Wyoming and Portions of Contiguous Territories* (Washington, D.C., 1871), 340–70.

25. Louis Agassiz, *Geological Sketches* (Boston, 1866).

26. Joseph Leidy, "Notice of Remains of Extinct Reptiles and Fishes, Discovered by Dr F. V. Hayden, in the Bad Lands of the Judith River, Nebraska Territory," PANS, vol. 8 (1856): 72–73.

27. Frederick H. Burkhardt, ed., *The Correspondence of Charles Darwin*, vol. 7 (Cambridge: Cambridge University Press, 1992).

TWELVE

Dr. Leidy's Dinosaur

Epigraph: William Parker Foulke, "Statement Respecting the Fossil Bones, Shells, and Wood Presented by Him to the Academy this Evening," PANS, vol. 10 (1858): 215–18.

1. Ibid.

2. Ibid.

3. Earl Spamer, "The Great Extinct Lizard: *Hadrosaurus foulkii,* 'First Dinosaur' of Film and Stage," *The Mosasaur,* vol. 7 (2003): 109–25; Richard C. Ry-

der, "Hawkins' Hadrosaurs: The Stereographic Record," *The Mosasaur*, vol. 3 (1986): 169–80; John Phillips, *Notices of Rocks and Fossils in the University Museum, Oxford* (Oxford, 1863).

THIRTEEN
Ferdinand Vandiveer Hayden

Epigraph: F. V. Hayden, *Preliminary Report of the United States Geological Survey of Wyoming and Portions of Contiguous Territories* (Washington, D.C., 1871).

1. Mike Foster, *Strange Genius: The Life of Ferdinand Vandiveer Hayden* (Niwot, Colo.: Roberts Reinhart, 1994).

2. Ibid., 38.

3. *Proceedings of the American Association for the Advancement of Science* (1854): 290.

4. Hayden to Leidy, Albany, January 11, 1854, ANS.

5. Hayden to Leidy, Danville, January 10, 1856, ANS.

6. F. V. Hayden, *Sun Pictures of Rocky Mountain Scenery, with a Description of the Geographical and Geological Features, and Some Account of the Resources of the Great West; Containing Thirty Photographic Views Along the Line of the Pacific Rail Road, from Omaha to Sacramento* (New York: J. Bien, 1870).

7. *Transactions of the Albany Institute*, vol. 14 (1859): 1–16; PANS, vol. 11 (1859): 8–30.

8. Foster, *Strange Genius*, 60–61.

9. Joseph Leidy, "The Cretaceous Reptiles of the United States," *Smithsonian Contributions to Knowledge*, vol. 14 (1865): 1–135.

10. *Geological Magazine of London*, vol. 5 (1865): 432–35.

11. *Science, Correspondence*, vol. 75 (1932): 584.

12. Cope to his father, Haddonfield, December 30, 1868, quoted in Henry Fairfield Osborn, *Cope: Master Naturalist* (Princeton: Princeton University Press, 1931). The paper Cope had in mind was probably his encyclopedic review, finally published after many revisions as *A Synopsis of the Extinct Batrachia and Reptilia and Aves of North America*, TAPS, vol. 14 (1871): 1–252.

13. Joseph Leidy, "The Extinct Mammalian Fauna of Dakota and Nebraska, Including an Account of Some Allied Forms from Other Localities, Together with a Synopsis of the Mammalian Remains of North America, Illustrated with 30 plates," JANS (2), vol. 7 (1869): 1–472.

14. Karl Waage, "Deciphering the Basic Sedimentary Structure of the Cretaceous System in the Western Interior," *Geological Association of Canada*, Special Paper 13 (1975): 55–91.

15. F. V. Hayden, *Preliminary Report of the United States Geological Survey of Wyoming* (Washington, D.C., 1871), 142.

16. F. V. Hayden, *Final Report of the United States Geological Survey of Nebraska and Portions of the Adjacent Territories, Made Under the Direction of the*

Commissioner of the General Land Office (Washington, D.C., 1872); F. V. Hayden, *Preliminary Field Report of the United States Geological Survey of Colorado and New Mexico* (Washington, D.C., 1869).

17. Joseph Leidy, *Contributions to the Vertebrate Fauna of the Western Territories; Report of the United States Geological Survey of the Territories, 1873*, part 1 (Washington, D.C., 1873): 15–16.

18. Quoted in Robert Ellison, *Fort Bridger, Wyoming: A Brief History* (Cheyenne: Historical and Landmark Commission of Wyoming, 1931), 34.

19. "Diary of Judge William Carter," *Wyoming Historical Society, Annals of Wyoming*, April 1939, 79–110.

20. Ellison, *Fort Bridger*, 44.

21. Leidy originally used the European spelling rather than the simplified American *Paleosyops* that became common later.

22. PAPS, vol. 11 (1870): 316.

23. F. V. Hayden, *Fourth Annual Report of the U.S. Geological Survey of the Territories, Embracing Nebraska* (Washington, 1867, printed 1873), 52.

24. Louis Agassiz, *Geological Sketches* (Boston: Ticknor and Fields, 1866), 1–28.

25. Hayden to Leidy, Colorado City, August 10, 1869, ANS.

FOURTEEN

Kansas and a New Regime

Epigraph: R. S. Elliott, "Experiments in Cultivation on the Plains Along the Line of the Kansas Pacific Railway," in F. V. Hayden, *Preliminary Report of the United States Geological Survey of Montana and Portions of Adjacent Territories* (Washington, D.C., 1872).

1. During the Civil War, the federal government closed the mail service from St. Joseph, Missouri, to Salt Lake City and beyond to Sacramento, because parts of the route lay in Confederate territory.

2. Oscar Osburn Winther, *The Transportation Frontier* (Albuquerque: University of New Mexico Press, 1964).

3. Kenneth J. Almy, ed., "Thof's Dragon and the Letters of Capt. Theophilus H. Turner, M.D., U.S. Army," *Kansas History*, vol. 10 (1987): 170–200.

4. LeConte was a member of a famous scientific family. A geologist and entomologist, he later became chief clerk of the U.S. Mint in Philadelphia.

5. Michael J. Everhart, *Oceans of Kansas: A Natural History of the Western Interior Sea* (Bloomington: Indiana University Press, 2005).

6. Samuel Williston, *The University Geological Survey in Kansas*, vol. 4: Paleontology, part 1 (Topeka, 1898).

7. Ibid.

8. Ibid., 28.

9. Edward Drinker Cope, "On the Fossil Reptiles and Fishes of the Cretaceous Rocks of Kansas," in F. V. Hayden, *Preliminary Report of the United States*

Geological Survey of Wyoming and Portions of Contiguous Territories (Washington, D.C., 1873), 385-431.

10. Edward Drinker Cope, "On the Geology and Paleontology of the Cretaceous Strata of Kansas," in F. V. Hayden, *Preliminary Report of the United States Geological Survey of Montana and Portions of Adjacent Territories* (Washington, D.C., 1872), 318-19.

FIFTEEN

Entry of the Gladiators

1. There are two biographies of Cope: Henry Fairfield Osborn, *Cope: Master Naturalist* (Princeton: Princeton University Press, 1931), and Jane Davidson, *The Bone Sharp: The Life of Edward Drinker Cope* (Philadelphia: Academy of Natural Sciences, 1997).

2. Osborn, *Cope*, 39.

3. "Amphibamus Grandiceps, a New Batrachian from the Coal Measures," PAPS, vol. 17 (1865): 134-37.

4. "Remarks on the Remains of a Gigantic Extinct Dinosaur, from the Cretaceous Green Sand of New Jersey," PAPS, vol. 18 (1866): 275-79.

5. The only full-length biography of Marsh is Charles Schuchert and Clara Mae LeVene, *O. C. Marsh, Pioneer in Paleontology* (New Haven: Yale University Press, 1940).

6. Ibid., 25.

7. Louis Agassiz, "Highly Interesting Discovery of New Sauroid Remains," AJS (2), vol. 33 (1862): 138.

8. O. C. Marsh, "Description of the Remains of a New Enaliosaurian (Eosaurus Acadianus), from the Coal Formation of Nova Scotia," AJS (2), vol. 34 (1862): 1-16.

9. Marsh notebooks, quoted in Shuchert and LeVene, *O. C. Marsh*, 61.

10. David B. Weishampel and Luther Young, *Dinosaurs of the East Coast* (Baltimore: Johns Hopkins University Press, 1998).

11. "On Some Cretaceous Reptilia," PANS, vol. 20 (1868): 233-42.

12. Minutes, March 24, 1868, PANS, vol. 20 (1868): 92-93. Cope published a virtually identical account in AJS (2), vol. 46 (1868): 263-64.

13. PANS, vol. 22 (1870): 9-10; AJS (2), vol. 49 (1870): 392.

14. AJS (2), vol. 50: 140-41; TAPS, vol. 14 (1870): 114.

15. David Rains Wallace, *The Bonehunters' Revenge: Dinosaurs, Greed, and the Greatest Scientific Feud of the Gilded Age* (Boston: Houghton Mifflin, 1999); Mark Jaffe, *The Gilded Dinosaur: The Fossil War Between E. D. Cope and O. C. Marsh and the Rise of American Science* (New York: Crown, 2000).

16. The party also included Samuel Hooper of Boston and J. P. Usher, secretary of the interior under Lincoln; Elizabeth Cary Agassiz, *Louis Agassiz*, vol. 2 (Boston, 1886), 661.

17. Turner, August 23, 1868, Fort Wallace, in "Thof's Dragon and the Letters of Capt. Theophilus H. Turner, M.D., U.S. Army," ed. Kenneth J. Almy, *Kansas History*, vol. 10 (1987): 193.

18. Turner, September 13, 1868, Fort Wallace, in ibid.

19. Hayden to Leidy, September 18, 1868, ANS. *Mylodon* is a Pleistocene relative of the sloths and armadillos.

20. Agassiz to Marsh, February 12, 1873, Yale.

21. Agassiz to Marsh, February 2, 1872, Yale.

SIXTEEN
Riding the Rails

1. It should be noted that the transcontinental lines were by no means the first railroads in the West. Many of the very first lines were short affairs running the length of a town's main street and operating by means of horse-drawn "cars." The first railroad in California operated out of Sacramento in 1852. When the Kansas Pacific Railroad finally crossed the Missouri and reached Topeka in 1865, it joined up with a spiderweb of smaller, preexisting lines. By 1870 the Kansas Pacific had reached Denver, and two years later the state had at least eight companies operating sections of track: the Union Pacific (two branches); Kansas Pacific; Atchison, Topeka, and Santa Fe; Missouri, Kansas, and Texas; Kansas City, Emporia, and Southern; Kansas City, Burlington, and Santa Fe; Leavenworth, Lawrence, and Galveston; and Memphis, Kansas, and Colorado. But most of these operated in the eastern half and southeastern corner of the state carrying local commerce and agricultural freight. Roughly west of Junction City, only the three principal national mail lines operated.

2. F. V. Hayden, *Second Annual Report of the United States Geological Survey of the Territories for the Years 1867, 1868, and 1869* (Washington, D.C., 1873), 76.

3. F. V. Hayden, *Preliminary Report of the United States Geological Survey of Wyoming, 1870* (Washington, D.C., 1871), 143.

4. Culbertson estimated the total Indian population of the Upper Missouri and its tributaries at 54,550.

5. Hayden to Leidy, Fort Pierre, February 9, 1855, ANS.

6. Michael F. Kohl and John S. McIntosh, eds., *Discovering Dinosaurs in the Old West: The Field Journals of Arthur Lakes* (Washington, D.C.: Smithsonian Institution Press, 1997), 92-94.

SEVENTEEN
The First Yale College Expedition

Epigraph: Charles Betts, "The Yale College Expedition of 1870," *Harper's New Monthly Magazine*, vol. 43 (October 1871): 663-71.

1. Edward Drinker Cope, "Report, 1871," in F. V. Hayden, *Preliminary Report of the United States Geological Survey of Wyoming and Portions of Contiguous Territories* (Washington, D.C., 1873), 386.

2. Mudge to Marsh, Manhattan, December 28, 1866, Yale.

3. *Geological Magazine* (4), vol. 1 (1894): 337-39.

4. O. C. Marsh, Autobiographical fragment, Yale.

5. George Bird Grinnell, unpublished memoir, Yale.

6. R. B. Marcy, *The Prairie Traveler: A Handbook for Overland Expeditions with Illustrations, and Itineraries of the Principal Routes Between the Mississippi and the Pacific, and a Map* (New York: Harper and Bros, 1859; second edition, London, 1863). Marcy was much influenced by the experiences of the British Army in India and the English explorer Sir Richard Burton. He had served in the Mexican War as well as the Civil War, and was a vastly experienced plains traveler and explorer. In 1852 he led a seventy-man expedition to Texas and Oklahoma to survey and find the source of the Red River.

7. Betts, "Yale College Expedition."

8. George Bird Grinnell, "An Old-Time Bone Hunt," *Natural History*, vol. 23 (1923): 330.

9. Meek and Hayden, PANS, vol. 13 (1861): 14-47.

10. Grinnell, "Old-Time Bone Hunt," 331.

11. Ibid., 329-30.

12. Betts, "Yale College Expedition."

13. "Professor Marsh's Rocky Mountain Expedition: Discovery of the Mauvaises Terres Formation in Colorado," AJS (2), vol. 50 (1870): 292.

14. Grinnell, "Old-Time Bone Hunt," 334.

15. Hayden to Leidy, Big Sandy Creek, September 7, 1870, ANS.

16. Hayden to Leidy, Fort Bridger, October 1, 1870, ANS.

17. Hayden to Leidy, Cheyenne, November 13, 1870, ANS.

18. Judge Carter to Marsh, Fort Bridger, August 21, 1870, Yale.

19. Betts, "Yale College Expedition."

20. Ibid.

21. Hayden to Leidy, November 13, 1870, ANS.

22. Henry Farnham, unpublished memoir, Yale.

23. Letter from son of James Wadsworth to Ernest Howe, would-be Marsh biographer, August 30, 1931, Yale.

24. Ibid.

25. Henry Farnham memoir, Yale.

EIGHTEEN

The Competition Begins

1. O. C. Marsh, Autobiographical fragment, Yale.

2. An excellent firsthand account of this trip exists in the diaries of student participant George Lobdell: see Mary Faith Pankin, "George G. Lobdell, Jr., and

the Yale Scientific Expedition of 1871 at Fort Bridger," *Annals of Wyoming,* vol. 70 (1998): 25-44; and "The Yale Scientific Expedition of 1871: A Student's-Eye-View," *Oregon Historical Quarterly,* Winter 1998-99, 374-436.

3. Charles Schuchert and Clara Mae LeVene, *O. C. Marsh, Pioneer in Paleontology* (New Haven: Yale University Press, 1940), 124.

4. Marsh, Autobiographical fragment.

5. Schuchert and LeVene, *O. C. Marsh,* 124.

6. B. D. Smith wrote to Marsh from "Smith's Fork, Wyoming." There are two "Smith's" or "Smith" Forks in western Wyoming. One is a tributary of the Bear River, and the second runs parallel to Black's Fork, east of Fort Bridger, and is presumably the one from which Smith wrote. Both streams are said to be named by, or for, the famous mountain man Jedediah Smith, who hunted in the Green River region in the spring of 1824 and 1825. B. D. Smith was not a relative.

7. Thomas Condon to Marsh, November 21, 1870, Yale.

8. Condon to Marsh, Eugene, November 8, 1890, Yale.

9. AJS (3), vol. 3 (1872): 56-57.

10. Cope to Vaux, Wyandotte Cavern, August 27, 1871, Vaux Archive, APS.

11. Cope to Vaux, Fort Wallace, September 25, 1871, APS.

12. Cope, Topeka, September 6, 1871, quoted in Henry Fairfield Osborn, *Cope: Master Naturalist* (Princeton: Princeton University Press, 1931), 160.

13. Cope, Topeka, September 7, 1871, Osborn, *Cope,* 161.

14. "Note of Some Cretaceous Vertebrata in the State Agricultural College of Kansas," PAPS, vol. 12 (1871): 168-70.

15. Cope, September 7, 1871, Osborn, *Cope.*

16. Cope to J. P. Lesley, published in PAPS, vol. 12 (1871): 174-76.

17. "On Two New Ornithosaurians from Kansas," PAPS, vol. 12 (1872): 420-22. Indeed, that eventually turned out to be the case for both of Cope's "species."

NINETEEN

Buffalo Land

1. Charles H. Sternberg, *The Life of a Fossil Hunter* (New York: Henry Holt, 1909).

2. William E. Webb, *Buffalo Land: An Authentic Narrative of the Adventures and Misadventures of a Late Scientific and Sporting Party upon the Great Plains of the West; With Full Descriptions of the Country Traversed, the Indian as He Is, the Habits of the Buffalo, Wolf, and Wild Horse, etc., etc.* (Cincinnati: Hannaford, 1872).

3. *Kansas Magazine,* vol. 2 (1872): 100.

4. Edward Drinker Cope, "On the Geology and Vertebrate Paleontology of the Cretaceous Strata of Kansas," in F. V. Hayden, *Preliminary Report of the United States Geological Survey of Montana and Portions of Adjacent Territories* (Washington, D.C., 1872), 318-49. Jane Davidson, "Edward Drinker Cope, Professor Paleozoic, and Buffalo Land," *Transactions of the Kansas Academy of*

Science, vol. 106 (2003): 177–91, shows that the extract quoted was taken from an early draft (not the final published version) of Cope's paper.

5. There are other books where one can spend happy hours trying to identify the coded portraits: two classic examples being David Lodge's *Small World* (New York: Macmillan, 1984), about the world of English literature, and Gordon MacCreagh's *White Waters and Black* (New York: Century, 1926), about an expedition to the Amazon.

6. Davidson, "Edward Drinker Cope"; Robert Taft, *Artists and Illustrator of the Old West, 1850–1900* (New York: Scribner, 1953). Mark Jaffe in *The Gilded Dinosaur* accepts *Buffalo Land* as a true story and argues for Cope as Professor Paleozoic. Jane Davidson, *The Bone Sharp: The Life of Edward Drinker Cope* (Philadelphia: Academy of Natural Sciences, 1997), while more circumspect, also points to Cope as the professor.

7. Compare Charles Betts, "The Yale College Expedition of 1870," *Harper's New Monthly Magazine,* vol. 43 (October 1871): 663–71; and Theodore R. Davis, "Buffalo Hunting," *Harper's Weekly,* December 14, 1867, 792.

8. David Rains Wallace, *The Bone Hunter's Revenge* (Boston: Houghton Mifflin, 1999).

9. Webb, *Buffalo Land,* 37.

10. Webb to Cope, January 31, 1870, ANS.

11. Webb, *Buffalo Land,* 106–7.

12. Charles Lyell, "On the Evidence of Fossil Footprints of a Quadruped," AJS (2), vol. 2 (1846): 25–27.

13. Minutes, PAPS, vol. 11 (1870): 311.

TWENTY

1872: The Year of Conflict

1. Joseph Leidy, "Contribution to the Extinct Vertebrate Fauna of the Western Territories," *Report of the United States Survey of the Territories,* vol. 1 (1873): 14–358.

2. PANS, vol. 14 (1872): 167–69. Also AJS (3), vol. 4 (1872): 239–40.

3. On the dates of Professor Cope's recent publications: AMNAT, vol. 7 (1873): 303–6. Also AJS (3), vol. 5 (1873): 235–36. This paper was also sent to the Academy of Natural Sciences.

4. Leidy to Hayden, September 2, 1872, ANS.

5. Hayden to Leidy, Yellow Stone Valley, October 2, 1872, ANS.

6. Hayden to Leidy, Denver, July 8, 1872, ANS.

7. "On the Existence of Dinosauria in the Transition Beds of Wyoming," PAPS, vol. 12 (1872): 48–51.

8. Cope to Brig. Genl. E. O. C. Ord, June 22, 1872, in Henry Fairfield Osborn, *Cope: Master Naturalist* (Princeton: Princeton University Press, 1931).

9. James Van Allen Carter to Marsh, Fort Bridger, July 6, 1872, Yale.

10. Cope to his father, Church Buttes, July 28, 1872, in Osborn, *Cope.*

11. James Van Allen Carter to Leidy, Fort Bridger, June 11, 1873, Leidy Papers, ANS.

12. B. D. Smith to Marsh, undated (circa July 1872), Yale.

13. B. D. Smith to Marsh, July 21, 1872, Yale.

14. Chew to Marsh, Fort Bridger, July 16, 1872, Yale.

15. Cope letter, Camp near Church Buttes, July 28, 1872, in Osborn, *Cope*.

16. Cope to his father, Camp on Fontanelle Creek, September 8, 1872, in Osborn, *Cope*.

17. Cope to his father, Camp on Green River, mouth of Fontanelle Creek, September 8, 1872, in Osborn, *Cope*.

18. Cope to his sister, Ham's Fork, September 8, 1872, in Osborn, *Cope*.

19. Cope summarized the results of his year's field work in a one-hundred-page report for Hayden: "On the Extinct Vertebrata of the Eocene of Wyoming, Observed by the Expedition of 1872, with Notes on the Geology," *Sixth Annual Report of the USGS of the Territories* (Washington, D.C., 1873), 545-649.

20. Annie to her parents, September 21, 1872, in Osborn, *Cope*.

21. Cope to his father, October 12, 1872, in Osborn, *Cope*.

22. Marsh correspondence, Yale.

23. "Preliminary Description of New Tertiary Mammals, Part I," AJS (3), vol. 4 (1872): 122-28; Parts II, III, IV, ibid., 202-24.

24. AJS (3), vol. 4 (1872): 322-23.

25. "Notice of a New and Remarkable Fossil Bird," AJS (3), vol. 4 (1872): 344; "On a New Subclass of Fossil Birds (Odontornithes)," AJS (3), vol. 5 (1873): 161-62.

26. Carter to Marsh, Fort Bridger, June 2, 1872, Yale.

TWENTY-ONE

The Case of the Great Horned Mammals

Epigraphs: O. C. Marsh, "The Fossil Mammals of the Order Dinocerata," AMNAT, vol. 7 (1873): 147-53; Edward Drinker Cope, letter of September 8, 1872, in Henry Fairfield Osborn, *Cope: Master Naturalist* (Princeton: Princeton University Press, 1931).

1. AJS (3), vol. 4 (1872): 202-24.

2. PAPS, vol. 12 (1872): 515.

3. PAPS, vol. 12 (1872): 590.

4. Cope to his father, September 8, 1872, in Osborn, *Cope*.

5. B. D. Smith to Marsh, August 28, 1872, Yale.

6. PAPS, vol. 12 (1872): 576-77, 578-79.

7. AJS, vol. 5 (1873): 117, 293, 310; AMNAT, vol. 7 (1873): 52, 146, 217.

8. PAPS, vol. 12 (1872): 577.

9. PAPS, vol. 13 (1873): 39-74.

10. PAPS, vol. 13 (1873): 255-56.

11. AMNAT, vol. 7 (1873): 304.

12. Cope also started issuing papers in a series of his own devising that he called *Paleontological Bulletins*. His first Bridger paper was number 1 in this series, and also had a nominal date of July 29, 1872.

13. O. C. Marsh, *Dinocerata: A Monograph of an Extinct Order of Gigantic Mammals* (Washington, D.C., 1886).

TWENTY-TWO
Going Separate Ways

1. Hayden to Leidy, October 2, 1872, ANS.

2. Hayden to Leidy, April 23, 1873, ANS.

3. Leidy to James V. Allen Carter, Philadelphia, April 25, 1873, ANS.

4. Dolley MS, 1935, ANS.

5. Ibid.

6. Clifford to Marsh, May 3, 1873, Yale.

7. Clifford to Marsh, Red Cloud Agency, October 24, 1875, Yale.

8. Smith to Marsh, Henry's Fork, January 2, 1874, Yale.

9. Carter to Marsh, Fort Bridger, July 10, 1877, Yale.

10. Sam Smith to Marsh, Green River, December 1, 1873, Yale.

11. John M. Peterson, "Science in Kansas: The Early Years, 1804-1875," *Kansas History*, vol. 10 (1987): 201-40.

12. Mudge to Marsh, Manhattan, February 4, 1874, Yale.

13. Cope to his father, Greeley, Colorado, July 6, 1873, in Henry Fairfield Osborn, *Cope: Master Naturalist* (Princeton: Princeton University Press, 1931), 195.

14. Cope Field Diaries, American Museum of Natural History.

15. Smith to Marsh, Henry's Fork, January 2, 1874, Yale.

16. Carter to Marsh, Fort Bridger, October 13, 1873, Yale.

17. A memo from Wheeler, preserved in Cope's field notebook, shows that each civilian man was to be assigned rations for the season of "75 pounds of fresh beef ['or mutton,' added in pencil], 25 of ham, 25 of bacon, 150 of flour, 20 of beans, 2 1/2 of rice, 12 of coffees, 1 1/2 of tea, 20 of sugar, 2 of candles, 2 of soap, 3 1/2 of salt, 1/2 gallon of vinegar, 2 ounces of pepper, 2 quarts of pickles, 10 boxes of yeast powder and 25 pounds of potatoes (when they can be purchased)."

18. Cope to his wife, Summit of Sangre de Christo Pass, July 29, 1874, in Osborn, *Cope*.

19. Cope to his wife, San Ildefenso, August 15, 1874, in Osborn, *Cope*.

20. Cope to his father, Santa Fe, August 17, 1874, in Osborn, *Cope*.

21. Cope to his wife, Canon d.l.Haguas, September 6, 1874, in Osborn, *Cope*.

22. Cope to his wife, Tierra Amarilla, September 14, 1874, in Osborn, *Cope*.

23. Cope to his wife, camp near Nacimiento, New Mexico, October 11, 1874, in Osborn, *Cope*.

24. G. G. Simpson, "Hayden, Cope, and the Eocene of New Mexico," *PANS*, vol. 103 (1951): 1-22.

TWENTY-THREE
Two into Four Won't Go

1. Hayden to Leidy, Fort Bridger, September 30, 1870, ANS.
2. Hayden to Marsh, May 14, 1867, Yale.
3. Hayden to Marsh, February 22, 1867, Yale.
4. Hayden to Marsh, Washington, February 8, 1869, Yale.
5. William B. Scott, *The Leidy Commemorative Meeting* (Philadelphia: Academy of Natural Sciences, 1923), 42.
6. Hayden to Leidy, October 13, 1870, ANS.
7. Hayden to Leidy, Cheyenne, Wyoming, November 13, 1872, ANS.
8. Carter to Marsh, July 6, 1872, Yale.
9. Hayden to Leidy, Yellow Stone Valley, October 2, 1872, ANS.
10. Hayden to Marsh, Washington, December 8, 1873, Yale.
11. Leidy to Hayden, Philadelphia, September 2, 1872, postscript, ANS.
12. Hayden to Leidy, Corinne, Utah, October 16, 1872, ANS.
13. Joseph Leidy, *Contributions to the Vertebrate Fauna of the Western Territories; Report of the United States Geological Survey of the Territories, 1873,* part 1 (Washington, D.C., 1873).

TWENTY-FOUR
To the Black Hills

1. The perimeter of the area designated for the Sioux ran from the east bank of the Missouri River at 46 degrees north, south to the Nebraska line, then west to the 104th parallel west, north to the 46th parallel, and back to the starting point— or roughly the entire western half of what is now South Dakota.
2. In his 1857 expedition Lieutenant Warren had reported that the same geological formations were present in the Black Hills as were already producing gold in economic quantities in the Wind River and Bighorn Mountains of Wyoming.
3. General Sheridan to Marsh, Chicago, June 3, 1873, Yale.
4. T. H. Stanton to Marsh, Cheyenne, June 3, 1874, Yale.
5. Sheridan to Marsh, Chicago, August 15, 1874, Yale.
6. Stanton to Marsh, Fort Laramie, October 3, 1874, Yale.
7. Stanton to Marsh, Cheyenne, October 1, 1874, Yale.
8. Ord to Marsh, St. Louis, October 6, 1874, Yale.
9. Ord to Marsh, Omaha, December 13, 1874, Yale.
10. Stanton to Marsh, Cheyenne, August 20, 1875, Yale.

TWENTY-FIVE
To the Judith River

1. Edward Drinker Cope, *The Origin of the Fittest* (New York: Appleton, 1887).
2. Clifford to Marsh, Red Cloud Agency, Nebraska, May 2, 1876, Yale.

3. Gary E. Moulton, ed., *The Journals of the Lewis and Clark Expedition* (Omaha: University of Nebraska Press, 1983-2001).

4. Ferdinand Vandiveer Hayden, "Geological Sketch of the Estuary and Fresh Water Deposit of the Bad Lands of the Judith, with Some Remarks upon the Surrounding Formations," TAPS, vol. 11 (1859): 123-38.

5. Cope to his wife, Fort Claggett, September 2, 1876, in Henry Fairfield Osborn, *Cope: Master Naturalist* (Princeton: Princeton University Press, 1931).

6. Moulton, ed., *Journals of Lewis and Clark*, May 31, 1805.

7. Cope to his wife, October 3 and October 8, 1876, in Osborn, *Cope*, 228.

TWENTY-SIX
The Rise of Dinosaurs

1. Lakes to Marsh, Golden City, Colorado, April 2, 1877, Yale.

2. Arthur Lakes, journal entry for March 26, 1877, in Michael F. Kohl and John S. McIntosh, eds., *Discovering Dinosaurs in the Old West: The Field Journals of Arthur Lakes* (Washington, D.C.: Smithsonian Institution Press, 1997). (Cited hereinafter as Lakes Journal.)

3. Lakes to Marsh, June 13, 1876, Yale.

4. Lakes Journal, 11.

5. Lakes Journal, 24-25. Mudge was exactly sixty.

6. AJS (3), vol. 14 (1877): 514-16.

7. Lakes Journal, 15.

8. PAPS, vol. 15 (1875): 82-84; AMNAT, vol. 11 (1877): 629.

9. Mudge to Marsh, Cañon City, August 12, 1877, Yale.

10. Mudge to Marsh, Cañon City, August 15, 1877, Yale.

11. "Harlow" and "Edwards" to Marsh, July 19, 1877, Yale; quoted in Charles Schuchert and Clara Mae LeVene, *O. C. Marsh, Pioneer in Paleontology* (New Haven: Yale University Press, 1940), 196. This letter no longer exists in the Marsh archive at Yale.

12. There is an unsubstantiated report by Samuel Williston that Marsh was shown a dinosaur bone during his 1868 stop at Como Station; W. D. Matthew, ed., *Dinosaurs, with Special Reference to the American Museum of Natural History, Handbook, no. 5* (New York: American Museum of Natural History, 1915), 124-31.

13. Lakes Journal, 95.

14. Carlin to Marsh, Washington, December 24, 1877, Yale.

15. Carlin to Marsh, Washington, December 29, 1877, Yale.

16. The suggestion has been made that the article was actually planted by Carlin; John H. Ostrom and John S. McIntosh, *Marsh's Dinosaurs: The Collections from Como Bluff* (New Haven: Yale University Press, 1966), 16.

17. Carlin to Marsh, Como, Wyoming, March 23, 1878, Yale.

18. Carlin to Marsh, Como, Wyoming, May 23, 1878, Yale.

19. Carlin to Marsh, Como, Wyoming, September 19, 1878, Yale.

20. Lakes Journal, 130-31.

21. Lakes Journal, August 11, 1879, quoted in Ostrom and McIntosh, *Marsh's Dinosaurs*, 29.

22. Cope to his daughter, Como, Wyoming, August 8, 1879, in Osborn, *Cope.*

23. Lakes Journal, 147.

24. Reed to Marsh, September 1, 1881, Yale.

TWENTY-SEVEN
The Good, the Bad, and the Ugly

1. Alfred Sherwood Romer, "Cope Versus Marsh," *Systematic Zoology,* vol. 13 (1964): 201-7.

2. Ronald Rainger, "The Rise and Decline of a Science: Vertebrate Paleontology at Philadelphia's Academy of Natural Sciences, 1820-1900," PAPS, vol. 136 (1992): 1-32.

3. Ibid.

4. T. H. Huxley, *Lectures on Evolution* (New York: Humboldt Library of Popular Science Literature, 1882).

5. Peter Bowler, "Edward Drinker Cope and the Changing Structure of Evolutionary Theory," *Isis,* vol. 68 (1977): 249-65.

6. Edward Drinker Cope, field journals, American Museum of Natural History.

7. Baldwin–Marsh correspondence, Yale.

8. Smith to Marsh, April 5, 1883, Yale.

9. Normally only incoming letters for Marsh had been saved, but for really important replies he kept copies of his drafts in his files.

10. Hayden to Marsh, April 2, 1874, Yale.

11. Letter quoted in Elizabeth Noble Shor, *Fossils and Flies: The Life of a Compleat Scientist, Samuel Wendell Williston* (Norman: University of Oklahoma Press, 1971), 118-19.

TWENTY-EIGHT
Going Public

1. David A. Wells, *First Principles of Geology* (New York, 1861), 262.

2. Baird to Marsh, Washington, August 2, 1875, Yale.

3. Baird to Marsh, Washington, November 22, 1875, Yale.

4. Marsh to Baird, draft, December 20, 1875, Yale.

5. Quoted in Charles Schuchert and Clara Mae LeVene, *O. C. Marsh, Pioneer in Paleontology* (New Haven: Yale University Press, 1940), 385.

6. Ronald Rainger, *An Agenda for Antiquity: Henry Fairfield Osborn and Vertebrate Paleontology at the American Museum of Natural History, 1890-1935* (Tuscaloosa: University of Alabama Press, 1991).

7. *16th Annual Report U.S. Geological Survey,* Part I (Washington, D.C., 1889), 133-244.

8. The *Dinoceras* reconstruction was put on display in 1888; *Proceedings of the Boston Society of Natural History*, vol. 32 (1888): 342-43.

9. Martin Rudwick, *Scenes from Deep Time* (Chicago: University of Chicago Press, 1992).

10. See, for example, *Life Through the Ages: A Commemorative Edition*, ed. Stephen Jay Gould and Phillip J. Currie (Indianapolis: Indiana University Press, 2001).

11. Camille Flammarion, *Le Monde Avant la Creation de l'Homme* (Paris, 1886).

12. *New York Journal*, December 11, 1898.

TWENTY-NINE

1890: The End of the Beginning

1. Edward Drinker Cope, *The Vertebrata of the Tertiary Formations of the West* (Washington, D.C., 1883).

2. William Berryman Scott, *Some Memories of a Palaeontologist* (Princeton: Princeton University Press, 1939), 48.

3. Ibid., 62.

4. James Van Allen Carter to Marsh, Fort Bridger, June 27, 1877, Yale.

5. Carter to Marsh, Fort Bridger, June 16, 1877, Yale.

6. Carter to Marsh, Fort Bridger, July 10, 1877, Yale.

7. Carter to Marsh, Fort Bridger, August 6, 1877, Yale.

8. William Berryman Scott, *A History of Land Mammals in the Western Hemisphere* (New York: McMillan, 1913).

9. Ronald Rainger, *An Agenda for Antiquity: Henry Fairfield Osborn and Vertebrate Paleontology at the American Museum of Natural History, 1890-1935* (Tuscaloosa: University of Alabama Press, 1991).

10. John S. McIntosh, "The Second Jurassic Dinosaur Rush," *Earth Sciences History*, vol. 9 (1990): 22-27.

11. Tom Rea, *Bone Wars: The Excavation and Celebrity of Andrew Carnegie's Dinosaur* (Pittsburgh: University of Pittsburgh Press, 2001).

12. Elizabeth Noble Shor, *Fossils and Flies: The Life of a Compleat Scientist, Samuel Wendell Williston* (Norman: University of Oklahoma Press, 1971).

13. J. B. Hatcher, "Some Localities for Laramie Mammals and Horned Dinosaurs," AMNAT, vol. 30 (1896): 112-20.

14. Jules David Prown, ed., *Discovered Lands, Invented Pasts: Transforming Visions of the American West* (New Haven: Yale University Press, 1992); Andrew Wilton, ed., *American Sublime: Landscape Painting in the United States, 1820-1880* (London: Tate Publishing, 2002).

15. Rebecca Bedell, *The Anatomy of Nature: Geology and American Landscape Painting, 1825-1875* (Princeton: Princeton University Press, 2001).

16. Lakes Journal, 108.

17. Capt. Clarence E. Dutton, "The Physical Geology of the Grand Canyon District," *Annual Report of the United States Geological Survey to the Secretary of the Interior*, 1880-81, 85.

18. George Bird Grinnell, "An Old-Time Bone Hunt," *Natural History*, vol. 23 (1923): 329–36.

19. Charles Betts, "The Yale College Expedition of 1870," *Harper's New Monthly Magazine*, vol. 43 (October 1871): 663–71.

20. Roderick Frazier Nash, *Wilderness and the American Mind* (New Haven: Yale University Press, 1967).

21. Scott, *Some Memories of a Palaeontologist*.

22. Robert V. Bruce, *The Launching of Modern American Science, 1846–1876* (Ithaca, N.Y.: Cornell University Press, 1987), 98.

23. W. J. T. Mitchell, *The Last Dinosaur Book* (Chicago: University of Chicago Press, 1998).

Index

Note: Page numbers followed by *i* or *m* indicate illustrations or maps, respectively.

Academy of Natural Sciences (Philadelphia), 112, 123, 124–25, 132, 196–97, 210, 215, 238, 239, 297, 310; Cope and, 134, 151, 155, 156, 158–59, 248–49, 254, 298; curators at, 156, 248–49, 298; fossil collections of, 34, 40, 44–45, 96, 107, 144, 255; founding members of, 94–96; journal of natural science established by, 96; Leidy and, 106; sponsorship of collecting expeditions, 128, 129, 136, 255
Account of the Skeleton of the Mammoth (Peale), 48–49
acquired characteristics, 271, 299, 342–43
Adams, John, 51, 57
Agassiz, Alexander, 306
Agassiz, Louis, xi, 121, 128–29, 133, 143, 149, 153, 160, 165–67, 169, 200, 205–7, 294, 306
Agathaumas, 275, 279
Agathaumas sylvestris, 218
Age of Enlightenment, 60, 67, 68
Age of Mammals, The (Zallinger), 315
Age of Reptiles, The (Zallinger), 315
Ainsworth (topographer), 251, 253
Alberti, Friedrich August von, 336
Alien and Sedition Acts, 93
Allosaurus, 286, 289–90
alluvial deposits, 82, 92, 93, 99, 335, 336
amateur naturalists, 86
Amblypoda, 230
American Association for the Advancement of Science, 129, 179

American Century (magazine), 316
American Fur Company, 110, 113, 114, 170
American incognitum, 17, 24–33, 37, 48, 58. *See also* mastodon
American Journal of Science, 160, 163, 185, 195, 215, 234, 236, 297
American Journal of Science and the Arts, 96–97, 99, 105, 108
American Museum (journal), 41
American Museum of Natural History (New York), 53, 297, 312, 315, 322, 323
American Naturalist (journal), 236, 240, 297
American Philosophical Society, 32, 34, 40–41, 47–48, 87, 94, 198, 199, 209, 233, 235–39, 238, 283, 295, 297, 298, 354n3
Amherst, Jeffrey, 11
ammonites, 59, 76
Amphibamus, 157, 296
Anatomical Museum, University of Pennsylvania, 106
Anchisaurus, 45
Ancient Fauna of Nebraska, The (Leidy), 119
Anderson, Joseph, 247
Andrews, Roy Chapman, 332
animals, American versus European, 30–31, 38–39, 49, 51
Annan, Robert, 47
Anning, Mary, 76–79, 314
Anning, Richard, 77

371

Antarctic, 332
Antrodemus, 290
Apache Indians, 148
Apatosaurus, 284, 290, 292
Apatosaurus grandis, 287
Appalachian frontier, 10–12
Arapaho Indians, 148, 269
Archaeopteryx, 73, 73*i,* 141, 156, 226, 310–11
Arctic, 332
Aristotle, 64, 67
Arkansas River, 170
army. *See* U.S. Army
Artiodactyla, 237
artists: and nature, 326; paleontological, 312–13, 314–15
Art Students League, 314
Atchison, Topeka, and Santa Fe Railroad, 6, 359n1
Atherton (professor), 268
Atlantosaurus, 285, 314
Audubon, John James, 117, 148

badlands, 108–9, 170, 184, 221, 272, 274–75
Bad Lands, 108–10, 113–20, 118*m,* 126–30, 185, 263, 267, 321
Bad Lands National Park, 115
Baird, Spencer Fullerton, 106, 117, 128, 137, 149, 217, 255, 257, 260–61, 311–12
Baldwin, David, 285–86, 301–2
Ballou, William Hosea, 308
Baltimore and Ohio Railroad, 171
Banks, Joseph, 49
Bannock Indians, 139
Barbe-Marbois, François, Marquis de, 39, 348n7
Barnum, P. T., 43, 208
Barosaurus, 290
Barton, Benjamin Smith, 40, 53–54, 87–90, 94, 97, 102
Bartram, John, 19–21, 87, 92, 350n19
Bartram, William, 87
Basilosaurus cetoides, 83, 85
bathmism/bathymogenesis, 342
Bathmodon, 233, 235, 241
Bavaria, 73

Beadle, Erastus, 201
Beagle, H.M.S., 103
Bear Wolf, 275
Beaver Head, 275
Beckwith, Henry, 280–81
Behemoth, 80
Belgian Royal Museum of Natural History (Brussels), 315
Belgium, 315
Bement, Clarence, 242–43
Benton, Wyoming, 179
Betts, Charles (Charlie), 177, 182, 184, 188–89, 329
Bierstadt, Albert, 326
Big Bone Lick, 13, 15–33, 40, 47, 107, 350n19
biological paleontology, 71
birds, 123, 226, 284
Blackfeet Indians, 273
Black Hills, 244, 263–69, 272–73
Blumenbach, Johann Friedrich, 27
Bodmer, Karl, 153, 326
Boll, Jacob, xi
Bone Hunters' Revenge, The (Wallace), 208
Boquet (colonel), 20
Bozeman Trail, 171, 263
Brackett, K. C., 320
Braddock, George, 11
Bridger, Jim, 109, 139–40
Bridger Formation, 139, 141, 180, 211, 257–58, 261
British Association for the Advancement of Science, 312
British Geological Survey, 314
British Museum (London), 84, 310, 315
Brongniart, Alexandre, 75
Brontops, 313
Brontosaurus, 37, 284, 292, 313, 316
Brontotherium gigas, 227, 230
Brown, Barnum, 323
Brush, George Jarvis, 159
Bryant, William Cullen, 52
Buckland, William, 75, 79–80, 123, 125, 284, 294, 316
buffalo, 10, 13, 32, 174, 197, 203, 208
Buffalo, New York, 11

Buffalo Land (Webb), 202*i*, 202–10

Buffon, Georges-Louis Leclerc, comte de, 31, 38–40, 51, 58, 60, 68, 348n9

Buntline, Ned (pseudonym of Edward Judson), 201, 203

Bureau of Indian Affairs, 263–64, 268

Burnet, Thomas, 62–63

Burr, Aaron, 82

Burton, Richard, 140

Butler (captain), 198

California Trail, 171

Camarasaurus, 292

Camarasaurus supremus, 285

Cambridge University, 90

camels, 110

Camptosaurus, 292

Canada, 11, 272

Cañon City, Colorado, 285–87

Carlin, William Edward, 288–93, 301

Carnegie, Andrew, 323

Carnegie Museum of Natural History (Pittsburgh), 322–23, 325

Carpenter (lieutenant), 301

Carroll, J. M., 250

Carson, Christopher "Kit," 140

Carter, James Van Allen, 140–41, 194, 211, 213, 215–16, 219–20, 223, 226–27, 235, 244–47, 249–50, 257, 259, 301, 320–21

Carter, William A., 140–41, 186–87, 194, 211, 219–20, 223, 235, 245, 247, 321

catastrophism, 64, 68

Catesby, Mark, 25, 92

Catlin, George, 3

Centennial Exposition (Philadelphia, 1876), 125, 311–12

Central Pacific Railroad, 172

Central Park, New York City, 124–25

Ceratopsia, 275

Cetiosaurus, 284

Chapman, Henry, 243

Chappe d'Auteroche, Jean-Baptiste, 29–30

Cherokee Trail, 149, 171

Chew, John, 194, 219–21, 223–25, 227, 245–47, 271, 293, 329

Cheyenne Indians, 148, 150, 183, 198–99, 269

Chicago fire, 243

Chickasaw Indians, 15

China, 332–33

Civil War, 134, 156, 160

Claosaurus, 279

Clark, Malcolm, 174

Clark, William, 40, 275–76

Clarke, F. W., 313

Cleaveland, Parker, 98–99

Clidastes, 162

Clifford, Hank, 244–46, 266, 272–73, 329, 332

Coal Measures, 157

Cody, William F. "Buffalo Bill," 183, 189, 203, 207, 266

Coelurus, 292

collectors. *See* fossil collectors

Collin, Nicholas, 46, 47, 87, 96

Collinson, Peter, 19–23, 27–28, 30, 346n6, 350n19

Colorado, 137, 279–87, 293, 320

Columbian Exposition (Chicago, 1893), 313

Columbian Magazine, 32

Columbia University, 322

Comanche Indians, 148

Como Bluffs, Wyoming, 172–74, 287–92, 294

Condon, Thomas, 194–95, 302

Conkling, Roscoe, 165, 207

Connecticut, 45, 100

Connecticut River, 85

Conybeare, William, 78, 80–81, 336

Cooper, James Fenimore, 201

Cope, Alfred, 155–56

Cope, Annie Pim, 157, 217, 224, 248–49, 251–52, 273, 276, 297

Cope, Edward Drinker: and Academy of Natural Sciences, 134, 151, 155, 156, 158–59, 248–49, 254, 298; and *Buffalo Land,* 206–7, 209–10; collections of, 322; and dinosaurs, 125, 151, 279, 282–87, 290–93, 323; early career of, 155–59, 161; and *Elasmosaurus* affair, 162–64; and evolutionary theory, 271, 299, 341–43; Hayden and, 144, 216–19, 233, 250, 257; later work of, 318–19; Leidy and,

Cope, Edward Drinker (*continued*)
156–58, 163–64, 259, 262; Marsh's feud
with, 162–65, 168–69, 190, 195, 219–41,
256, 260–61, 271, 282, 295, 301, 302–4,
306–9; Mudge and, 154, 197–98, 232;
The Origin of the Fittest, 299, 341; *Paleontological Bulletins,* 364n12; personality of, 158, 247, 297; *The Primary
Factors of Organic Evolution,* 299, 341–
42; publications by, 364n12; and reconstructions of fossil animals, 311, 314;
and review of Leidy's book, 134–35; and
significance of, 295–300, 331; and systematic paleontology, 300; *The Vertebrata of the Tertiary Formations of the
West,* 318; and the West, 7, 141, 153–54,
173, 177–78, 189, 195–201, 203, 206,
209, 213, 215–33, 241–42, 245–46, 248–
53, 256, 259, 261–62, 271–76, 294, 318,
329–30; will of, 248–49; and younger
paleontologists, 319–20
Cope, Julia, 157, 217, 224, 249, 251
Cope, Thomas Pim, 155
Cope's Bible, 318
Cope's Law, 300, 343
Cope's Rule, 300, 343
Corson, Joseph K., 140–41, 211, 213, 215,
216, 219, 223, 224, 243, 245, 257
Coryphodon, 241
Cow Island, 274–76
Crazy Horse, 264, 269
Creation, 58, 62, 66. *See also* life, origins
of
Cretaceous period, 69, 95*m,* 114, 130, 131,
136, 143, 152–53, 157, 206, 218, 252
*Cretaceous Reptiles of the United States,
The* (Leidy), 134–35, 259
Croghan, George, 18–25, 27, 47, 244
Crow Indians, 273
Crystal Palace exhibitions (England), 124,
310, 311, 312
Culbertson, Alexander, 110, 117, 119, 244
Culbertson, Joseph, 109–10
Culbertson, Samuel, 109
Culbertson, Thaddeus, 117, 119, 174, 213
Cunningham (fossil collector), 165

Custer, George Armstrong, 7, 148, 264–
65, 269, 272–73
Cuvier, Georges, Baron, 4, 27, 33, 37, 47,
50, 53, 58–60, 72–75, 78–80, 91, 103,
232

Dakota Sioux Indians, 263
Dana, James Dwight, 91, 159
Darwin, Charles, 71, 81, 90, 103, 107, 113,
121, 125, 156, 161, 271, 298, 341–43
Darwin, Erasmus, 71, 112
Darwinism, 7
Daubenton, Louis-Jean-Marie, 16–17, 27,
28, 30, 36
Dear, J. W., 266
De Beauvois, Ambrose-Marie-François-
Joseph Palisot, 42, 45
Deinodon, 120
deism, 59, 96, 339
De Kay, James Ellsworth, 90, 96
De la Beche, Henry, 314, 336
Delano, Columbus, 268
Delaware Indians, 13, 19
Dennie, Joseph, 51
De Pauw, Cornelius, 40
Descartes, René, 60
Desnoyers, Jules, 336
Dickeson, M. W., 107
dictionary, American, 87
Dinoceras, 235, 236–37, 240–41
Dinocerata, 185, 228, 230, 235–37, 240–41,
313
Dinocerata (Marsh), 306
dinosaurs: in America, 41–45, 60, 85, 120,
122–25, 130, 173, 218, 275, 279–94;
bipedal, 123, 315; birds and, 123, 284; in
Canada, 272; classification of, 284; di-
versity of, 284; eggs of, 332; in England,
79; excavation of, 122–23; exhibition of,
124–25, 310–13, 322; naming of, 80;
popularity of, 310–11, 315–16, 322–23,
331; spectacular discoveries of, 279–94,
315
Dinosaurs of North America, The (U.S.
Geological Survey), 313
Diplodocus, 284, 286–87, 290

Diplodocus carnegii, 323, 325
Diracodon, 292
Discosaurus, 163
Discourse on Some of the Principal Desiderata in Natural History, A (Barton), 88
disease, causes of, 112
Dolley, Charles, 243
Dollo, Louis, 315
Douglass, Earl, 325
Draper, Bettie, 12
Dryosaurus, 290
Dryptosaurus, 232
dry screening, 325
Dudley, Joseph, 26
Duria Antiquor (de la Beche), 314
Dutton, Clarence, 327–28
Dystropheus, 279

earth: age and history of, 29, 57, 61, 64–66, 68–70, 73–75, 81, 130–31, 136, 335–37, 348n9; surface of, 62–64, 337; theories about, 60, 62–66, 339
Eaton, Amos, 101–2, 105
Edward VII, king of England, 323
Edwards, Milne, 113
Egan (captain), 268
Elasmosaurus, 151, 162–64, 163*f*, 177, 200, 259
elephants, 10, 15–16, 24, 27–28, 33, 49, 185, 193, 230, 237
Elliott, R. S., 147
Embargo (Bryant), 52
England, 11
Enlightenment, Age of, 60, 67, 68
Enquiry into the Original State and Features of the Earth, An (Whitehurst), 31
Eobasileus, 235–39, 241
Eocene period, 137–38, 252–53, 297
Eohippus, 185, 299
Eosaurus acadianus, 160, 296
Époques de la Nature (Buffon), 40, 348n9
erosion, 63
Essay on the Theory of the Earth (Cuvier), 75, 91, 103
eugenics, 300, 322

Evans, John, 114–16, 118, 120, 128, 165, 169
evolution: concept of, 70–71; Cope and, 271, 299, 341–43; as "development," 204; fossil evidence for, 70–71, 121; Great Chain of Being and, 68; of horses, 298–99; Leidy and, 338–40; and origin of life, 112; theory of, 121, 298–99, 322
Ewing, Samuel, 51–52
exhibitions: controversy over, 311–14; of dinosaurs, 124–25, 311–15, 322; of mastodons, 48–50, 53, 310. *See also* mounted skeletons
Exhumation of the Mastodon (Peale), 52–53
extinction: concept of, 57–58; evolution and, 71; fossil evidence for, 59, 81; and Great Chain of Being, 67; Leidy on, 339–40; of mastodon, 29, 32–33, 58; religion and, 58; scientific support for, 58–59, 88–89
Extinct Mammalian Fauna of Dakota and Nebraska, The (Leidy), 120
"the eye," 77

fakes, 84–85, 208–9
fauna, American versus European, 30–31, 38–39, 49, 51
Featherstonhaugh, G. W., 86–87, 100–101, 143
Federalists, 53, 96
Felker (professor), 153
Ferdinand II de Medici, 65
Field Museum (Chicago), 125, 315
field paleontology, 292–93, 297
fish, 70, 120, 137, 141, 199
Flammarion, Camille, 315–16
footprints, debate over, 209
forensic approach, 36
Fort Benton, 129, 170, 273–74
Fort Bridger, 6, 139–40, 142*m*, 149, 170–73, 180, 185–88, 193, 211, 213, 214*m*, 216, 218–19, 224, 227, 244, 246, 249, 320, 321
Fort Claggett, 275
Fort Custer, 321

Fort D. A. Russell, 184
Fort Duquesne, 11, 20
Fort Hays, 165, 171–72, 203
Fort Kearney, 171
Fort Laramie, 6, 109, 170, 266
Fort Laramie Treaty (1868), 263
Fort Leavenworth, 133
Fort McPherson, 177, 182–83, 244, 246, 264
Fort Pierre, 6, 109, 114, 117, 129, 133, 170
Fort Pitt, 20, 21
Fort Riley, 6, 147
forts, 170–71
Fort Union, 119, 133, 170
Fort Wallace, 6, 148, 150, 151, 154, 165, 172, 177, 189, 191, 196–98
fossil collectors: competition among, 169; Cope and, 271–72, 292; daily life of, 189; dangers for, 133–34, 175; image of, 329; management of, 292–93; Marsh and, 194, 219–21, 245–48, 292; significance of, 317. *See also* paleontologists
fossils: in America, 81–82; from Big Bone Lick, 13, 15–33, 40, 47, 350n19; biological evidence from, 70–71, 80–81; commerce in, 76–78, 293–94; detection of, 77, 184, 215; discovery of sites for, 173; early discoveries of, 5, 13, 25–26, 41–44, 60; evolutionary evidence in, 70–71, 121; extinction and, 59, 81; fame gained from, 78–79, 131, 161; geologic evidence from, 66, 68, 73–75, 92, 99; and Great Chain of Being, 67; as heart of scholarship, xiii–xiv; illustration of complete, 348n6; increase in discoveries of, 147; Indian beliefs about, 184; in Kansas, 150–54; in Massachusetts, 85; Matlack-Wistar specimen, 41–45; from New Jersey, 41–45; as political issue, 50–52, 295, 307; popularity of, 293–94, 310–11, 331; railroad men and, 173; recovery techniques for, 325; theories about, 61–62; weight of, 124; in the West, 4–7, 60, 100, 105–6, 113, 137–41, 168–70, 172, 184, 200
Foulke, William Parker, 44, 122–23

France: American antipathy for, 93; and French-Indian wars, 11; science in, 38–40
Franklin, Benjamin, 18–20, 22–27, 29–30, 37, 41, 64, 87, 92, 175, 346n6
free will, 342
French-Indian wars, 11–12, 19
French Revolution, 91
French Royal Academy of Sciences, 15
frontier, 317–18
fur trade, 109, 170

Galileo Galilei, 65
Gall, 269
Geikie, Archibald, 256
General Land Office, 137, 305
Genesis, book of, 57, 61, 63, 348n9
geological column, 69, 93, 335–37
geology: controversies in, 337; discipline of, 96–104; of Europe, 143; fossil evidence about, 65–66, 73–75, 99; popular interest in, 102; surveys of, 100, 114, 132; time-scale column established in, 335–36; of United States, 89–90, 92–94, 99–104, 130–31, 143, 166; of the West, 180, 182. *See also* earth
Geology and Mineralogy Museum (Darmstadt), 53
germ theory, 112
Gilliams, Jacob, 94
Gilmore, Charles, 325
Gist, Christopher, 17–18
God, 14, 21, 58, 62, 66–68, 80–81, 112, 326, 330, 336, 339
Goddard, Paul Beck, 44–45
Godin, Canon, 72
gold: in Black Hills, 263, 265, 267, 365n2; in California, 126
Goldfuss, Albert, 152–53
Goodman, John, 50, 84, 136
Grant, Ulysses S., 268–69
Great Chain of Being, 67–68, 89
great-claw, 34–37, 39–40, 59, 348n2
Great Exposition (London, 1851), 124
Green River, 137–39, 166, 172, 186–87, 193, 261–62
Greenwood, Joseph, 19–20, 23

Greg (general), 252
Grey, Zane, 179
Grinnell, George Bird, 181, 183, 185, 265
Guettard, Jean-Etienne, 15

Haddonfield, New Jersey, 42, 44, 122–23
Hadrosaurus, 42, 44, 45, 122–25, 135, 151,
   164, 279, 284, 310–12, 315
Hall, James, 101, 103–5, 126–29, 133, 135,
   208
Halloy, Jean-Baptiste d'Omalius d', 336
Hamilton, Alexander, 82
Harger, Oscar, 245, 324
Harlan, Richard, 83–85, 136, 152–53
Harmon, Adam, 14
Hartwell, Martin, 199
Harvard University, Museum of Compar-
   ative Zoology, xi–xiii, 153, 165
Hatcher, John Bell, 275, 324–25
Hawkins, Benjamin Waterhouse, 124–25,
   135, 164, 310–12, 314
Hayden, Ferdinand Vandiveer, 120, 149,
   165, 168–69, 173–76, 200, 206–7, 248–
   49, 272, 274; and Agassiz, 166; as col-
   lector, 126–31, 136–37, 254; Cope and,
   144, 216–19, 233, 250, 257; and di-
   nosaurs, 279, 281; early years of, 127;
   Leidy and, 127, 129–35, 138, 141, 143–
   44, 166, 211, 242, 257–62; Marsh and,
   141, 144, 185–87, 189, 194, 255–61, 304–
   6; and photography, 131–32; significance of, 331; and surveys, 131–33, 137–
   39, 141, 143–44, 304–6
Hays, Isaac, 83–84, 136
Helix leidyana, 185
Hemings, Sally, 52
Henry, Joseph, 106, 137, 149, 217, 257, 270
herpetology, 151
Hesperornis regalis, 192*i*, 193, 195, 226–27,
   313
Hickock, "Wild" Bill, 203
hierarchy of life, 67
Hilliard, A. W., 173
hippopotamus, 17, 30
Histoire Naturelle (Buffon), 38
*Historical Disquisition on the Mammoth,
   An* (Peale), 49–50

History of America (Robertson), 40
*History of Land Mammals in the Western
   Hemisphere, A* (Scott), 321
Hitchcock, Edward, 85, 91, 161
Hoffmann, J. L., 72–73
Hooke, Robert, 63
Hope, Bob, 205
Hope, Thomas Charles, 90–91
Hopkins, John, 122
horned mammals, 144, 185–86, 193, 215,
   222, 224, 227, 229–41, 245
horses, 107, 180, 185, 193, 262, 298–99
Howe (senator), 268
Hubbell brothers, 291–92
Hudson River, 26, 82
humans, evolutionary place of, 70
Hunter, John, 57
Hunter, William, 10, 28–29, 32
Huntington, William Henry, 354n3
Hutchins, Thomas, 22
Hutton, James, 63–64, 66, 91, 103, 337
Huxley, Thomas Henry, 113, 123, 134–35,
   284, 296, 298–99, 315, 321–22, 330
Hydrarchos harlani, 84
Hylaeosaurus, 80

ichnology, 85
Ichnology of New England (Hitchcock), 85
Ichthyornis, 226, 313
ichthyosaurs, 77, 80, 152, 154, 310–11
Iguanodon, 80, 123, 279, 284, 310, 315–16
Illinois, 19
illustrations, of fossil animals, 313–15
incognitum, American, 17, 24–33, 37, 48,
   58. *See also* mastodon
*Index to the Geology of the Northern United
   States* (Eaton), 101
Indiana, 196
Indian Removal Act (1830), 175
Indian Ring scandal, 266
Indians: beliefs of, concerning fossils, 184;
   in Black Hills, 263–69, 272–73; in *Buf-
   falo Land,* 205; and conflicts with other
   Indians, 174; eastern settlers' relations
   with, 4; Hayden and, 132, 133; in
   Kansas, 148, 150, 166, 198–99; in Ohio,
   18–19; peace arrangements with, 139,

Indians (*continued*)
150, 175, 263, 268; railroads admired by, 243; reservations for, 139–40, 175, 263–64, 268–69, 365n1; treaties with, 18–19; in Upper Missouri territory, 359n4; western explorers' and settlers' relations with, 6, 174–75; women married to white men among, 244. *See also names of specific nations*
Ingles, George, 10, 12
Ingles, Mary Draper, 10, 12–15, 148
Ingles, Thomas, 10, 12
Ingles, William, 10, 12
Iowa, 100
Iroquois, 19
Isaac, J. C., 273, 276
ivory, 346n3

jackets, for specimens, 297
Jackson, Andrew, 175
Jackson, William Henry, 132
James, Frank, 176
Jameson, Robert, 91, 98
Jefferson, Thomas, 4, 13, 30, 34–40, 46–53, 58–59, 67, 86–89, 106, 199, 333, 348n2, 348n7
John Day deposits, Oregon, 194–95
Johnson, Andrew, 263
Johnson, William, 19
Journal of the Academy of Natural Sciences, 96
Judith River, 130, 136, 170, 272, 274–76
Judson, Edward (pseudonym, Ned Buntline), 201, 203
Jurassic period, 294, 314

Kansas, 73, 120, 147–54, 166, 177, 200–201, 247
Kansas City, Burlington, and Santa Fe Railroad, 359n1
Kansas City, Emporia, and Southern Railroad, 359n1
Kansas Pacific Railroad, 6, 147, 149, 171, 198, 200, 203, 359n1
Kansas River, 149, 204
Kansas State University, 324
Kaup, J. J., 53

Kennedy, Edward, 292
Kickapoo Indians, 22
Kimball, Moses, 43
kinetogenesis, 341
King, Clarence, 144, 305–6
Kiowa Indians, 148
Kirtland, Jared, 127
Klarge, J., 320
Knight, Charles, 314, 316
Koch, Albert, 84, 136
Koch, Robert, 112

Labrosaurus, 290
Lacépède, Bernard-Etienne-Germain de la Ville-sur-Illon, comte de, 88
Laelaps, 125, 134, 157, 162, 164, 232, 279, 314
Lake Como, Wyoming, 180
Lakes, Arthur, 279–83, 286, 288–89, 291–93, 326–27
Lakota Sioux Indians, 244, 263–64, 269, 298
Lamarck, Jean-Baptiste de, 71, 271, 299, 342–43
Lamothe (major), 187
Lancaster Indian Treaty (1744), 175
Laosaurus, 290
Laramie City Daily Sentinel (newspaper), 290
Leavenworth, Lawrence, and Galveston Railroad, 359n1
LeConte, John Lawrence, 151, 206–7, 357n4
legends, about prehistoric animals, 14, 20–21, 48
Leidy, Allwina, 106
Leidy, Anna Harden, 106
Leidy, Joseph, 199, 301, 354n3; *The Ancient Fauna of Nebraska*, 119; Cope and, 156–58, 163–64, 259, 262; *The Cretaceous Reptiles of the United States*, 134–35, 259; and dinosaurs, 42, 44–45, 122–23, 130, 279, 284, 315; and *Elasmosaurus* affair, 163–64; and evolutionary theory, 338–40; *The Extinct Mammalian Fauna of Dakota and Nebraska*, 120; first fossil treatise written

by, 6; Hayden and, 127, 129–35, 138, 141, 143–44, 166, 211, 242, 257–62; and horses, 107–8; Marsh and, 262; personality of, 106–7, 161; retirement of, 254, 256, 262; review of book by, 134–36; scholarly character of, 110–12; significance of, 105–6, 298; and the West, 109–10, 113, 116–17, 119–21, 139–41, 149, 168–69, 180, 186–87, 194, 203, 211–13, 215–18, 220–21, 223, 226–32, 235–36, 241–44, 257, 259, 261, 329; and younger paleontologists, 320

Lenard, Henry, 12

Lesley, J. P., 233

Lesquereux, Leo, 132–33, 137, 141, 149, 206–7

LeVene, Clara Mae, 285

leviathan, 84

Lewis, Meriwether, 40, 274

Lewis and Clark expedition, 5, 51, 60, 83, 152–54, 274

Lhwyd, Edward, 59

life, origins of, 111–12, 338–40. *See also* Creation

Life of a Fossil Hunter, The (Sternberg), 272

lions, 34–36, 348n2, 354n3

Little Bighorn, Battle of the (1876), 7, 244, 269, 273

London, 27

Longueuil, Charles Le Moyne, Baron de, 15–16

Los Angeles County Museum of Natural History, 315

Louisiana Purchase, 3, 51, 60, 348n7

Louis-Philippe, king of France, 53

Loxolophodon, 233–37, 234*i,* 241

Lucas, Oramel, 284–86

Lyell, Charles, 64–65, 70, 103–5, 113, 143, 209, 336

Lyme Regis, Dorset, England, 76–78, 80

Maclure, William, 92, 93–94, 96, 98, 101, 114, 157, 335

McKenzie, Kenneth, 117, 244

McKenzie, Owen, 117

Madison, James (bishop), 53–54

Maine, 100

"mammal quarry," 294

mammals, fossils of North America, 109–10, 114, 116–17, 119–20, 294, 311. *See also* horned mammals

mammoth, 15–16, 25, 28–29, 31–33, 49

Manifest Destiny, 4

Mann, Camillus, 94

Manospondylous gigas, 323

Mantell, Gideon, 75, 78–80, 123, 284, 315

Manual of Mineralogy (Jameson), 98

Marcy, R. B., 360n6

Marsh, Othniel Charles, 209; Agassiz and, 160, 166–67; and *Buffalo Land,* 207–8; Cope's feud with, 162–65, 168–69, 190, 195, 219–41, 256, 260–61, 271, 282, 295, 301, 302–4, 306–9; death of, 248; and dinosaurs, 279–93; *The Dinosaurs of North America,* 313; early career of, 159–61; and employee relations, 194, 285, 290–91, 293, 301–2, 308, 324; Hayden and, 141, 144, 185–87, 189, 194, 255–61, 304–6; and horse evolution, 298–99; and Indians' treatment, 267–69, 298; later work of, 318; Leidy and, 262; Mudge and, 177–78, 186, 193, 226, 232, 247–48, 282–83, 286, 289, 324; and National Academy of Sciences, 270, 297, 298, 304–6; and Peabody Museum, 255, 265, 269, 313, 315; personality of, 161, 296–97; *Preliminary Descriptions of New Tertiary Mammals,* 232; production of *Dinocerata* and *Odontornithes* volumes, 306, 313, 324; and reconstructions of fossil animals, 311–14; significance of, 295–300, 331; and the West, 176, 177–95, 197–201, 206, 211, 213, 215–32, 241, 244–50, 256–62, 264–69, 271–73, 275, 294, 329–30; and younger paleontologists, 319–20

Maryland, 100

Massachusetts, 85, 100

Masten, John, 47–48

mastodon: carnivore/herbivore debate about, 17, 24, 28–30, 32–33, 37, 49, 53–54; complete fossil of, 54; elephants related to, 10, 15–16, 24, 27–28, 33; exhib-

mastodon (*continued*)
  tion of, 48–50, 53, 310; extinction of,
  29, 32–33, 58; fossils of, 15–33, 39–40,
  47–48, 54, 82, 84; interpretations of,
  24–25, 27–33; legends about, 14, 20–21,
  48; molar of, 16*i;* naming of, 27; as po-
  litical issue, 50–52; reconstruction of,
  50*i;* symbolic value of, 10, 49, 51
Mather, Cotton, 26
Matlack, Timothy, 41–43
Matlack, William, 42
Matlack-Wistar specimen, 41–45
Mauvaises Terres. *See* Bad Lands
Maximilian zu Wied-Neuwied, Prince,
  152–53
medical education, 127
Medicine Lodge Treaties (1867), 148
Meek, Fielding Bradford, 128–29, 132–33,
  137, 141, 152, 169, 200, 206, 218
Megaceratops, 237, 241
Megacerops, 241
Megalonyx, 37, 107
Megalosaurus, 80, 123, 125, 279, 284, 294,
  316
Megatherium, 37, 40, 348n6
Memphis, Kansas, and Colorado Rail-
  road, 359n1
Merrill (professor), 153
Merycoidodon, 110
Mesozoic period, 92, 294, 314, 315
Metropolitan Museum of Art (New York),
  314
Miami Indians, 18
Michaelis, Christian Frederick, 47
Middle Eocene, 186
Miller, Jacob, 326
Millspaw, Peter, 48
mind, evolution of, 342
mineral collecting, 294
mineralogy, 89, 96–98
Miniconjou Indians, 267
Minnesota, 100
Minor (fossil collector), 165
Mississippi, 107
Missouri, 94, 117
Missouri, Kansas, and Texas Railroad,
  359n1

Missouri Pacific, Denver, and Rio Grande
  Railroad, 6
Missouri River, 5, 6, 8*m,* 92, 94, 109, 119,
  129–30, 149, 170, 273–74, 275–76
Missourium theristodon, 84
Mitchill, Samuel Latham, 90, 100–102
Moa, 85
Monde Avant la Creation de l'Homme, Le
  (Flammarion), 315–16
Monoclonius, 275
Montana, 120, 130, 272–74
Monthly American Journal, 86
Monthly Magazine, 37
Moody, Pliny, 85
Moran, Thomas, 326
Morgan, George, 22, 47
Morgan, John, 47
Mormon Trail, 171
Mormon War, 139
Morrison, Colorado, 280–84, 289, 293
Morton, Samuel George, 90, 99, 105, 157
mosasaurs, 72–73, 83, 152–53, 157, 200,
  272, 281
Mountain Crow Indians, 275
mountains: artists and, 326; formation of,
  63–64
mountain men, 5, 109, 139, 329, 361n6
mounted skeletons, 48, 124–25, 312, 315,
  348n6. *See also* exhibitions
Mudge, Benjamin Franklin, 171, 200, 206,
  257, 271–72, 275, 293; background of,
  152; as collector, 152–54, 168; Cope
  and, 154, 197–98, 232; Marsh and, 177–
  78, 186, 193, 226, 232, 247–48, 282–83,
  286, 289, 324
Murchison, Roderick Impey, 336
Murphy, Dennis, 112
Museum of Comparative Zoology, Harvard University, xi–xiii, 153, 165
Musquattime Indians, 22

names: of geological strata, 335–37; of
  species, 223–24, 231–38, 285, 301
Nanosaurus, 286, 290
Napoleon Bonaparte, 72
Nassau, Prince of, 128–29
Natawista, 110

National Academy of Sciences (Washington), 217, 250, 270, 295, 297, 298, 304–6

National Land Company, 203

national parks, 132

natural history, 7, 88–89, 97, 127–28, 155

Natural History and Antiquities of Selborn (White), 86

Natural History of Carolina, Florida, and the Bahama Islands (Catesby), 92

Natural History Museum (London), 323

natural philosophy, 60

natural science, 7

natural selection, 71, 341

Natural Theology, 80, 127

nature: artists and, 314–15, 326; concept of, 59

Nature (journal), 333

Nebraska Territory, 108, 113, 137. *See also* Bad Lands

Neptunist model, 337

Netherlands, 72

Newberry, John Strong, 101, 105, 126–27, 133–35, 149, 255, 279

New Harmony, Indiana, 98, 114

New Jersey, 41–45, 100, 122, 157–58

New Mexico, 137, 250–53, 271, 285, 297

New Orleans, 11, 15, 22

New River, 10, 14

newspapers, 307–9, 316

New York Academy of Sciences, 96

New York Herald (newspaper), 308

New York Journal and Advertiser (newspaper), 316

New York Lyceum of Natural Sciences, 96

New York State, 26, 47–48, 54, 100–104

New York Tribune (newspaper), 267–68, 308

New York Weekly (newspaper), 203

New York World (newspaper), 316

Noah's Flood, 28–29, 32, 61–63

Nolan, Edward, 239

North, L. H., 265

North America, Late Cretaceous seaway over, 95*m*

North Carolina, 100

Northern Pacific Railroad, 6, 244, 269

Notes on the State of Virginia (Jefferson), 13, 30, 39, 51, 58

Noyes, John Humphrey, 300

Nuttall, Thomas, 91–92

Odontornithes (Marsh), 306, 313, 324

O'Fallon (Indian agent), 152–53

Ogallala Indians, 269

Ohio, 11, 18–19, 31, 100

Ohio Company, 18

Ohio River, 10, 12, 14–15

Oklahoma, 148

"Old Dutch Woman," 12–15

O'Loghland (professor), 119

Oneida Community, 300

Onondaga Giant, 208–9

On the Origin of Species (Darwin), 81, 121, 161, 343

Ord, E. C., 219, 264, 266–67

Oregon, 194

Oregon Trail, 8*m*, 149, 171, 183

Oreodon, 110, 185

Origin of the Fittest, The (Cope), 299, 341

ornithiscians, 284

Ornithochirus harpyia, 199

ornithology, 92

Osborn, Henry Fairfield, 312, 319–22

Owen, David Dale, 100, 114–17, 131, 213

Owen, Richard, 80, 84, 85, 113, 284, 310

Owen, Robert, 98, 100

Oxford University, 90

Oxford University Museum, 315

paintings, of prehistoric life, 314–15

Palaeoscincus, 120, 275

Palaeosyops, 141, 213, 230, 241

Palaeotherium, 75, 76*i*, 108, 108*i*, 232

Paleface (film), 205

Paleobiology (journal), 333

Paleontological Bulletins (Cope), 364n12

paleontologists: as adventurers, xiii–xiv; competition between, 103–4, 128, 160, 162–65, 167–69, 186–87, 189–90, 195, 213–41, 256–62, 285–86, 291, 293, 302–3; dangers for, 175–76; experience of, 201; fame sought by, 230–31; image of, 326–33; improper practices of, 302,

paleontologists (*continued*)
308; motivations of, 79; second generation of, 319–26; site selection by, 5; territorial rights of, 185–86, 232; in the
West, 4–7. *See also* fossil collectors
paleontology: American, 4, 34, 36–37,
81–83, 105, 352n8; biological, 71;
Cope's theoretical contributions to,
300; discipline of, 75; field, management of, 292–93, 297; glamour of, 309,
331–33; modern British, 75–76; novel
about, 202–10; popularity of, 293–94,
310–11, 315–16, 322–23, 331; questions
of, 25; scholarly practices in, 223–24,
231–40, 285, 301; scientific versus romantic aspects of, xiii, 326–33; secu-
rity concerns in, 293; strands of, 7;
systematic, 300
Paleozoic Museum, New York, 124–25,
164, 311
Pantodonta, 237
Parker, John D., 152, 177, 206–7
Parmentier, Nicholas, 94
Paso, El (steamboat), 119
Pasteur, Louis, 111–12, 112, 338
Patterson, Bryan, xiii
Pawnee Indians, 148, 184
Peabody, George, 159–61
Peabody Museum of Natural History
(Yale University), 255, 265, 269, 292,
313, 315
Peale, Charles Willson, 42–43, 45, 46–49,
51–53, 310
Peale, Rembrandt, 47, 48–50, 51, 310
Peale, Rubens, 47
Peale, Titian, 47
Peale's Museum (Philadelphia), 43, 44–45,
46, 47, 53, 89
Penn, Thomas, 19
Penn, William, 19
Pennant, Thomas, 32
Pennsylvania, 100
père des Boeufs (father of cattle), 13, 15
Perissodactyla, 237
Peru, 24
Peterson, O. A., 323–25
Petrified Fish Cut, 141

Philadelphia, sciences in, 89
Philadelphia academy. *See* Academy of
Natural Sciences
Philadelphia Centennial Exposition
(1876), 125, 311–12
Philadelphia Linnaean Society, 87
Philadelphia Museum (Peale's Museum),
43, 44–45, 46, 47, 53, 89
Phillips, William, 336
photography, 131–32
physiogenesis, 341–42
Piegan Indians, 275
Pine Ridge agency, 321
Pittsburgh, 11
plate tectonics, 64
Plato, 67, 89
Platte River, 149, 171, 183, 184, 248, 329
Playfair, John, 64
plesiosaurs, 78, 80, 151, 162–64
Plot, Robert, 58, 79
Poebrotherium, 110
politics: fossils and, 50–52, 295, 307; sur-
veys and, 305–7
Polycotylus latipinnus, 153
Pontiac Indians, 22
Pope, John, 196, 198
Porter, T. C., 243
Port Folio, The (magazine), 51
Portheus molossus, 199, 231–32
Powell, John Wesley, 144, 305–8
prairie, western, 3–4
Prairie Traveller, The (Marcy), 181,
360n6
*Preliminary Descriptions of New Tertiary
Mammals* (Marsh), 232
"Preliminary Discourse" (Cuvier), 75
press, 307–9, 316
Priestley, Joseph, 91
*Primary Factors of Organic Evolution,
The* (Cope), 299, 341–42
Princeton University, 125, 310, 314, 322;
geological expeditions from, 319–21,
329
Principles of Geology (Lyell), 103
Proboscidea, 237
*Proceedings of the Academy of Natural Sci-
ences*, 96, 111

Proceedings of the American Philosophical Society, 223, 239
Professor Paleozoic (literary character), 202*i*, 202–10
Prout, Hiram A., 105, 108–10, 232, 316
Pteranodon, 191–93, 199
pterodactyl, 73*i*, 73–74, 78, 310
pterosaurs, 310–11

Quakers, 156, 158

railroads, 6, 132–33, 149, 171–73, 182, 208, 359n1
Ramsay, Nathaniel, 47
rations, expedition, 364n17
Raynold, W. F., 133
Recherches sur les Ossemens Fossiles de Quadrupèdes (Cuvier), 59, 79–80
reconstructions. *See* exhibitions; mounted skeletons
Red Cloud, 263, 266–69, 298
Red Cloud Agency, 265–67
Redi, Francesco, 338
Reed, John, 329
Reed, Mayne, 201
Reed, William Harlow, 288–94, 323
religion: academic study and, in Europe, 90–91; evolution and, 339–40; extinction and, 58–60; fossils and, 7; Natural Theology, 80. *See also* Creation; God
Rensselaer Polytechnic Institute, 101
reptiles, 72
Republican River, 149, 153–54, 177, 178, 198, 204
reservations, Indian, 139–40, 175, 263–64, 268–69, 365n1
rhinoceros, 16
Rifle Rangers, The (Reed), 201
River Crow Indians, 275
Robertson, William, 40
rocks, study of, 65–66, 93, 336
Rocky Mountains: Early Tertiary depositional basins in, 138*m;* effect of, on paleontology, 170; formation of, 130, 136
Rodgers, John R. B., 41
Roemer, Carl Ferdinand von, 160
Romantic Movement, 326–33

Romer, Alfred Sherwood, xi–xiii
Roosevelt, Theodore, 332
Royal Scottish Museum (Edinburgh), 125, 310
Royal Society (London), 22, 24, 26–28, 346n6
Russell, Andrew Joseph, 132

saber-tooth tiger, 120
Sacred Theory of the Earth (Burnet), 63
salamanders, 180
Saline River, 177, 198, 205
Santa Fe Trail, 133, 149, 171
Sauranodon, 290
saurischians, 284
Saurocephalus, 83, 154, 199
Saurodon, 83
sauropods, 284
Say, Thomas, 94, 98–99, 105, 157
Scalp Hunters, The (Reed), 201
scholarship, priority and ownership in, 103–4
Schuchert, Charles, 285
science: American institution of, 91–94, 96–99; American versus European, 38, 86–91; career opportunities in, 156; fame and, 230–31; in nineteenth-century America, 326; originality in, 231; paleontology and, 329–33; political and religious reactions against, 90–91; professionalization of, 158–59, 298; revolution in nineteenth-century, 60–61. *See also* natural history
Science (journal), 333
Scientific and Descriptive Catalogue of Peale's Museum (Peale and de Beauvois), 45
Scott, William Berryman, 319–22, 325, 330
Scudder, S. C., 260
security, of field sites, 293
Sedgwick, Adam, 75, 336
Seybert, Adam, 90, 96
sharks, 120
Shawnee Indians, 12–14, 19, 21, 148
Shedd, W. G., 251, 253
Sheffield Scientific School, Yale University, 159

Shelburne, William Petty-FitzMaurice, Lord, 22–23
Sheridan, P. H., 181, 265
Sherman, William T., 264
Shinn, John, 94
Shoshone Indians, 139–40, 188, 245, 247
shrewlike creatures, 294
Siberia, 15–16, 25, 28–29, 31–33, 346n3
Silliman, Benjamin, 91, 96–98, 103
Silliman, Benjamin, Jr., 159
Simpson, George Gaylord, xiii, 253, 348n4
Sioux Indians, 109, 174, 183, 242, 244, 263–64, 266–67, 269, 273, 298, 365n1
Sitting Bull, 7, 266, 269, 273
Six Nations, 19, 175
skeletons. *See* mounted skeletons
skulls, 72–73
sloth, 37, 40
Smith, B. D., 194, 219–21, 223–25, 227, 235, 246, 293
Smith, Jedediah, 361n6
Smith, Robert, 18
Smith, Sam, 219–20, 245–47, 249–50, 271, 293, 301–2, 320–21, 329, 332
Smith, William, 66, 74, 93, 99, 101
Smith (general), 265
Smith's Fork, Wyoming, 361n6
Smithsonian Institution (Washington), 106, 117, 125, 137, 144, 149, 217, 255, 272, 307, 310–11, 314
Smoky Hill River, 133, 149, 153, 192, 199, 204
Smoky Hills, 152, 177, 189, 201
Smoky Hill Trail, 171
Solomon River, 133, 149, 153–54, 177, 178
South America, 332
South Carolina, 100
Southeast Asia, 332
Southern Pacific Railroad, 6
Spallanzani, Lazaro, 338
Speakman, John, 94
species: complexity of, 70; extinction of, 88–89; fossil evidence about, 70–71; naming of, 223–24, 231–38, 285, 301; new, 68, 70, 339; variation in, 299, 341–43
Speir, Frank, 319–21

spontaneous generation, 111, 338–39
Spotted Tail, 269
Spotted Tail Agency, 265–66
squaw men, 244
Stanley, David S., 242, 264
Stanton, T. H., 264–66, 268
State Agricultural College, Kansas, 247, 324
Stegosaurus, 284, 290, 292, 313
Steno (Niels Stenson), 65–66
Sternberg, Charles H., 169, 201, 271–73, 275–76, 293, 297
Sternberg, George M., 149, 152, 165, 169, 171, 177, 189, 245, 257, 271
Sternberg Museum of Natural History, Kansas, 199
Stirpiculture, 300
Stonesfield, Oxfordshire, England, 79–80, 294
Stono River, 25
Stuart, John, 348n2
"sublime," the, 326
surveys: commissioning of, 100; Hayden and, 132, 137, 304–6; paleontology and, 6, 114; political controversy over, 305–7. *See also* U.S. Geological Survey
Sutter, John, 126
Sutton, Willie, xiv
Swallow, George Clinton, 133
Sylvester, Lady, 78

Tecumseh, 148
teeth: bird, 226–27; dinosaur, 120, 130; horse, 107; mastodon, 13, 15, 16*i*, 17–18, 20, 22, 25–29, 47; recovery techniques for, 325
telegram, Cope's, 223, 233–34, 238
Tennessee, 100
Tenskwatawa, 148
Tertiary period, 69, 92, 99, 109, 114, 116, 130, 131, 136, 138*m*, 143, 315
Tetracaulodon, 84
Theory of the Earth (Whitehurst), 62
theropods, 284
thieves, 176, 243
time scale (geological column), 69, 93, 335–37

Tinoceras, 226, 227–28, 232, 235–37, 240–41

Tippecanoe, Battle of (1811), 148

Titanosaurus montanus, 283, 285, 286

Titanotherium, 109, 141, 185, 230, 232, 233, 235, 241, 354n3

Titanotherium anceps, 193, 226, 232

Torrey, John, 101–2

Trachodon, 120

tracks, study of, 85

tragedy, literary, 295

Transactions of the American Philosophical Society, 35, 163

Transition Series, 93, 102, 103, 105, 336

Treaty of Aix-la-Chapelle (1748), 11

Treaty of Paris (1763), 11

Treaty of Utrecht (1713), 11

Triceratops, 218, 275, 284, 313

trichinosis, 105

trilobites, 59

Troodon, 120

Troost, Gerard, 94

Turner, Frederick Jackson, 317–18

Turner, George, 32–33, 37

Turner, Theophilus, 150–51, 164, 165–66, 171, 201, 245

turtles, 109, 117, 119, 120, 213

tusks, 10, 13, 15, 22, 24, 27, 28, 48–50, 84, 346n3

Tweed, William Marcy "Boss," 125

Tyrannosaurus rex, 38, 284, 323

Uintamastrix taro, 215, 223, 232, 241

Uinta Mountains, 187–88

Uintatherium robustum, 215, 223, 228, 230, 232, 235–37, 240–41

Union Pacific Central Branch, 149

Union Pacific Eastern Division, 149

Union Pacific Railroad, 6, 132, 171–73, 179, 182, 189, 242, 272, 359n1

United States: animals in, superiority/inferiority of, 30–31, 38–39, 49, 51; fossils in, 4–7, 60, 81–82; geology of, 89–90, 92–94, 99–104, 130–31, 143, 166; inferiority of, in Europeans' eyes, 38–40; Jefferson's defense of, 30–31, 39; Late Cretaceous seaway over, 95*m*; national

pride in science of, 37–39, 49, 87; science in, 38, 86–94; westward expansion of, 3, 7, 12, 18–19, 60, 147–49, 168, 170–72, 201, 317–18

U.S. Army, 139, 148–49, 175, 264. *See also* U.S. Department of the Army

U.S. Army Corps of Engineers, 250

U.S. Census Bureau, 317

U.S. Coast and Geodetic Survey, 305

U.S. Department of the Army, 266, 305

U.S. Department of the Interior, 268–69, 305

U.S. Geological Exploration, 305

U.S. Geological Survey, 100, 132, 137, 217, 244, 313

U.S. Government Survey, 176

U.S. Treasury Department, 100

U.S. War Department, 181

University of Groningen, Netherlands, 47

University of Kansas, 324

University of Pennsylvania, 106, 112, 156

UPR Trail (Grey), 179

Ussher, James, 62

Utah, 147

Van der Kemp, Francis Adrian, 51

Van Patten Collection, 196–97

Van Rensselaer, Stephen, 101

Vanuxem, Lardner, 90, 99, 105, 157

Van Vliet, Stewart, 110, 119

Vaughn, Alfred, 129

Vaux, William Sansom, 196–98, 303

Vertebrata of the Tertiary Formations of the West, The (Cope), 318

vertebrate paleontology. *See* paleontology

Virginia, 10, 34, 100

Volney, Constantin-François de, Count, 92–93

Vulcanist model, 337

Wadsworth, James, 189

Wagner Free Institution of Science, 249

Wallace, David Rains, 208

Warder (professor), 153

Ward's Natural Science Establishment, 313

Warren, G. K., 132–33, 365n2

Washington, George, 11, 17, 94, 175

Washita, Battle of the (1868), 148
Webb, William E., 153, 198, 200, 202–10, 252
Webster, Noah, 87
Werner, Abraham Gottlob, 91, 335–37
West, the: early experience of, 3–4; expansion into, 3, 7, 12, 18–19, 60, 147–49, 168, 170–72, 201, 317–18; fossils in, 4–7, 60, 100, 105–6, 113, 137–41, 168–70, 172, 184, 200; geology of, 89–90, 93–94, 180, 182; narratives about, 211–13; paleontology in, 4–6; romance of, 326–29
western novels, 201, 329
Western Pacific Railroad, 6
wet screening, 325
whales, 83
Wheeler, George M., 250–53, 301, 305
White, Gilbert, 86
Whitehurst, John, 31–32, 62
Whitten (lieutenant), 198
Wilcox, Joseph, 242
will, evolution and, 299, 342
Williams, Ernest, xiii
Williston, Frank, 290, 292–93
Williston, Samuel, 285, 286, 288–90, 293, 308, 324
Wilson, Alexander, 92
Wilson, Joseph, 173
Wind River Reservation, 139–40
Wisconsin, 100, 131

Wistar, Caspar, 36–38, 40–42, 45, 82, 156, 348n4
Wolfe, James, 11
Woodbury, New Jersey, 41–45
Woodward, John, 26
World's Columbian Exposition (Chicago, 1893), 313
Wortman, Jacob L., 323
Wounded Knee massacre (1890), 269
Wright, Bryce M., 245
Wright, Frank Lloyd, 115
Wright, James, 21
Wyandotte Cave, 196
Wyoming, 120, 137, 138, 142m, 172–73, 179–80, 193, 214m, 221, 230, 232, 279, 287–94

Xiphactinius, 199, 232

Yale University, 45, 96, 159–61, 297, 322; geological expeditions from, 180–95, 201, 207, 211, 227, 244–45, 328–29
Yarrow, H. C., 251–53
Yellowstone (steamboat), 170
Yellowstone National Park, 132, 141
Yellowstone River, 5, 6, 129, 133, 186, 242, 264

Zallinger, Rudolph, 315
Zeigler, Harry, 191